Urban Sustainability

The Urban Sustainability Book Series is a valuable resource for sustainability and urban-related education and research. It offers an inter-disciplinary platform covering all four areas of practice, policy, education, research, and their nexus. The publications in this series are related to critical areas of sustainability, urban studies, planning, and urban geography.

This book series aims to put together cutting-edge research findings linked to the overarching field of urban sustainability. The scope and nature of the topic are broad and interdisciplinary and bring together various associated disciplines from sustainable development, environmental sciences, urbanism, etc. With many advanced research findings in the field, there is a need to put together various discussions and contributions on specific sustainability fields, covering a good range of topics on sustainable development, sustainable urbanism, and urban sustainability. Despite the broad range of issues, we note the importance of practical and policy-oriented directions, extending the literature and directions and pathways towards achieving urban sustainability.

The series will appeal to urbanists, geographers, planners, engineers, architects, governmental authorities, policymakers, researchers of all levels, and to all of those interested in a wide-ranging overview of urban sustainability and its associated fields. The series includes monographs and edited volumes, covering a range of topics under the urban sustainability topic, which can also be used for teaching materials.

Zeina ElZein · Abdelazim Negm
Editors

Recent Approaches of Sustainable Architecture in Arid and Semi-arid Cities

Best Practices for Materials, Water, Energy and Urban Greening in Challenging Climates

Springer

Editors
Zeina ElZein
Department of Architecture,
Faculty of Fine Arts
Helwan University
Cairo, Egypt

Abdelazim Negm
Water and Water Structures Engineering
Department, Faculty of Engineering
Zagazig University
Zagazig, Egypt

ISSN 2731-6483 ISSN 2731-6491 (electronic)
Urban Sustainability
ISBN 978-981-95-1644-5 ISBN 978-981-95-1645-2 (eBook)
https://doi.org/10.1007/978-981-95-1645-2

This Springer imprint is published by the registered company Springer Nature Singapore Pte Ltd.
The registered company address is: 152 Beach Road, #21-01/04 Gateway East, Singapore 189721, Singapore

Preface

Cities in arid and semi-arid regions have always faced tough environmental and social challenges, but these difficulties have become far more pressing in recent years. Climate change, combined with rapid population growth and expanding urban areas, has made it increasingly difficult for these cities to maintain reliable access to water, clean energy, livable public spaces, and sustainable infrastructure. In many cases, these issues are made worse by economic inequality and inefficient governance, placing additional strain on communities and their resources. These complex, interconnected challenges demand new ideas about how we plan, build, and manage urban spaces in some of the world's most fragile environments.

The volume *Recent Approaches of Sustainable Architecture in Arid and Semi-arid Cities: Best Practices for Materials, Water, Energy and Urban Greening in Challenging Climates* was developed in response to these urgent concerns. This book brings together a diverse group of researchers, architects, planners, and environmental specialists to offer a practical look at how sustainable design, innovative technology, and community-based approaches can help shape better, more resilient cities in dry climates. Across its chapters, the book highlights proven methods, applied research, and hands-on case studies meant to inspire decision-makers, urban developers, and local leaders to rethink how we manage resources and design cities for the future.

The opening chapter introduces the essential concept of sustainable architecture in dry regions, placing special emphasis on the growing role of Artificial Intelligence (AI) and emerging digital tools in improving urban resilience and infrastructure efficiency. By looking at case studies from Egypt, Malaysia, and China, it reveals both the opportunities and limitations of using AI to enhance city systems and address environmental challenges.

The second part turns to resilience in architecture and urban planning, tracing the history of this concept and highlighting how it has evolved to include not only the physical durability of buildings but also social resilience, cultural preservation, and disaster preparedness. This section underscores how good design can help cities and communities adapt to environmental and societal changes, preserve heritage, and support people through difficult times.

The third section focuses on building materials and the critical need to embrace circular economy principles in the construction industry. Contributors discuss the environmental, social, and financial benefits of using local materials, recycling plastic, adopting bio-based products, and implementing better waste management. Real-world projects in Egypt, Saudi Arabia, and Algeria show how sustainable material choices can improve a building's thermal performance, reduce emissions, and strengthen local economies.

Water scarcity is a dominant issue in dry regions, and the fourth part of the book addresses it head-on. Here, authors present creative and practical solutions for improving urban water management, including water-sensitive urban design (WSUD), greywater recycling, indigenous conservation practices like Qanats, and nature-based drainage systems. Beyond the technical solutions, this section emphasizes the importance of policy reform, local community involvement, and public awareness in achieving long-term, sustainable water security.

The fifth part highlights the importance of urban greening initiatives, offering solutions like haloculture (the use of saline water and soil for vegetation) and rooftop gardens to counter rising temperatures, poor air quality, and the lack of public green spaces. These chapters demonstrate how greenery can help regulate city temperatures, improve air quality, and make urban areas more livable. By using practical examples from Cairo and other dense cities, the authors outline how green infrastructure can be successfully integrated into even the most built-up neighborhoods.

Renewable energy's role in future cities is addressed in the sixth section, focusing on how decentralized, community-centered renewable energy projects can help ease dependence on fossil fuels and strengthen local economies. The Oasis Community Center in Giza is presented as an example of how hybrid energy systems, water-saving technologies, and community education can work together to improve sustainability and resilience in growing cities.

The book's seventh part turns to digital tools, exploring how blockchain, IoT, and cyber-physical systems (CPSs) can improve project transparency, data security, and operational efficiency in urban construction. While these technologies offer exciting possibilities, the authors stress the need for regulatory frameworks, digital literacy programs, and stronger collaboration between academic researchers, industry professionals, and local governments to ensure these tools are applied fairly and effectively.

The final chapter of the book pulls together the major lessons from each section, offering a list of 25 clear, practical recommendations for policymakers, designers, planners, and urban managers. These recommendations cover everything from promoting the circular economy and preserving traditional water systems to supporting renewable energy, scaling up urban greening, and fostering digital innovation in construction and urban services. Together, they provide a roadmap for arid and semi-arid cities to build a sustainable, resilient future.

The editors would like to express sincere thanks to all the authors who contributed their knowledge and experience to this book. Their work represents a valuable resource for those working to address some of the world's most difficult environmental and urban challenges. We are equally grateful to our publishing partners for their support and assistance in making this book a reality.

It is our hope that this volume will be useful not only to researchers and students but also to decision-makers, municipal leaders, and community advocates working to create livable, equitable, and climate-smart cities in some of the planet's most challenging landscapes.

Cairo, Egypt Zeina ElZein
Zagazig, Egypt Abdelazim Negm
May 2025

Contents

Introduction to Sustainable Architecture in Arid and Semi-arid Cities

Overview of Sustainable Architecture and Urban Practices in Cities and Communities

Zeina ElZein and Abdelazim Negm

Abstract The book explores emerging approaches to sustainable architecture and urban resilience in arid and semi-arid regions, where climate challenges such as extreme heat, water scarcity, and rapid urbanization demand innovative and context-sensitive solutions. It offers a multidisciplinary view of how built environments can become more adaptive, inclusive, and ecologically responsible. The chapters examine how artificial intelligence is shaping smart sustainability and climate-responsive urbanism. Following that, architectural resilience is discussed in relation to climate adaptation, displacement, and conflict. A deeper focus on arid regions highlights how social-ecological systems can inform design, policy, and community-based responses. Subsequent chapters explore the integration of urban agriculture in public spaces, resilient planning for food and water security, and nature-based solutions for sustainable development. Case studies shed light on water-sensitive design, climate-sensitive architecture, and sustainable greening in dry environments. Later contributions focus on materials and waste, addressing circular construction, plastic reuse, and sustainable building systems. The book also features frameworks for adapting to climate risks, with emphasis on indigenous knowledge, participatory planning, and digital tools like urban digital twins. The volume highlights best practices from across the Middle East, North Africa, and beyond. It underscores the need for climate-responsive design that merges innovation with tradition, equity with efficiency. Ultimately, the book offers valuable insights for architects, urbanists, researchers, and policymakers working to create sustainable futures in some of the world's most climate-vulnerable settings.

Keywords Sustainable architecture · Arid cities · Climate resilience · Resources management · Urban greening

Z. ElZein (✉)
Architecture Department, Faculty of Fine Arts, Helwan University, Cairo, Egypt
e-mail: Zeina.ElZein@f-arts.helwan.edu.eg; Zeinabaher@gmail.com

A. Negm
Water and Water Structures Engineering Department, Faculty of Engineering, Zagazig University, Zagazig, Egypt
e-mail: amnegm@zu.edu.eg; amnegm85@gmail.com

Z. ElZein and A. Negm (eds.), *Recent Approaches of Sustainable Architecture in Arid and Semi-arid Cities*, Urban Sustainability,
https://doi.org/10.1007/978-981-95-1645-2_1

1 Background

More than half of the world's population now lives in cities, and this number is expected to rise to around 6 billion by 2045. As urban areas continue to expand, the demand for essential resources such as water, energy, and food will grow significantly. To meet these challenges, cities must adopt innovative and sustainable development strategies that can support this rapid urban growth [1, 2].

Cities in arid and semi-arid areas are facing critical challenges. These include the impacts of rapid urbanization, socioeconomic inequality, resource scarcity, and climate vulnerability. High temperatures, little precipitation and vegetation, and increasing strain on water and energy systems are common in these settings. The vulnerabilities in these regions are made worse by the consequences of climate change, which include increased temperatures, extreme weather, water scarcity, and decreased agricultural productivity. In these types of climates, urban populations are disproportionately exposed to air pollution, heat stress, and subpar housing, with marginalized communities suffering the most negative environmental effects.

Across the world, many of today's buildings aren't well-suited to the climates they're built in. Instead of responding to local conditions, they follow outdated international design trends from the twentieth century, often using too much glass, relying on lightweight materials, and being poorly oriented. These choices make them prone to overheating, sometimes throughout the entire year. To compensate, they rely on energy-hungry cooling systems, drawing electricity from power grids that are becoming increasingly unreliable during extreme weather events. As the global climate continues to warm, these vulnerabilities are only becoming more pronounced.

Looking ahead, the buildings we design must be able to passively protect people from both heat and cold. Resilience in architecture is not just about materials or structure. It also has to reflect how people live, what they value, their habits, their cultures, and how all of these change over time. Resilient buildings need thoughtful, integrated design, often learned from older, historic buildings where communities adapted their built environments in harmony with the climate over generations [3].

Architecture in arid regions goes beyond the basic function of providing shelter; it must act as a resilient bridge between people and the environment. Scholars and practitioners have long emphasized that architecture reflects human behavior, societal values, and awareness of the environment [4]. Thoughtfully designed architecture goes beyond mere aesthetics,it influences the lives and well-being of those who live within it. Therefore, architecture in arid regions must be responsive to—and adaptable to—changing environmental, social, and economic conditions.

Human beings have an innate drive to build defense, shelter, and survival. Ranging from primitive huts to massive religious monuments, each period in history has conveyed its environmental and cultural intelligence through its architecture. Ancient civilizations such as Ancient Egypt developed sophisticated passive cooling, natural ventilation, and water management concepts. These strategies, born out of necessity, remain viable today and inform strong and sustainable urban planning.

By integrating the cultural, environmental, technological, and social facets of sustainability within the contemporary urban setting, this book capitalizes on this ongoing connection. Sustainable architecture is not only a design requirement but also a vital instrument for upholding environmental harmony and human dignity in arid and semi-arid regions where difficulties spur innovation. In these areas, architectural techniques need to take sociopolitical instability and environmental extremes into account. In the Middle East and North Africa, cities are dealing with deteriorating infrastructure, informal settlements are growing, and climatic hazards are a major concern. All stakeholders must be involved in finding comprehensive, locally based answers to these pressures.

In drylands, sustainable architecture necessitates combining several strategies. Participatory urban greening, adaptive reuse, circular material practices, energy and water efficiency, and context-sensitive design are all incorporated. Every intervention adds to the greater environmental and social framework, whether it's a rooftop garden, a shaded patio, or a wind-catching ventilation system. This book serves as an example of how these initiatives and interdisciplinary research can work together to promote real, significant change.

In architecture, resilience is more than just structural strength. It encompasses the built environment's capacity to promote wellbeing, maintain social continuity, and adjust to change. The public sphere is another area where resilience is demonstrated. Public areas in many dry cities are in danger due to misuse, neglect, or privatization. However, these areas are necessary for equitable resource allocation, environmental cooling, and civic life. Public spaces can develop into multipurpose hubs of resilience by implementing greening programs, creating inclusive leisure areas, and incorporating water-sensitive design. Examples of how urban agriculture, tree planting, and green corridors are transforming urban microclimates and fostering stronger ties between communities are presented in the book.

Water and energy, two of the most vital resources in arid cities, are often managed by siloed institutions and centralized systems. This book explores decentralized, integrated approaches that align with urban design principles. For example, greywater systems, rooftop solar panels, and passive design strategies can reduce reliance on external resources and enhance community autonomy. These methods not only advance environmental goals but also promote equity by improving access to essential services.

Contemporary technologies are equally crucial. Data-driven decision-making is made possible by digital technologies like virtual planning environments, AI-assisted resource monitoring, and GIS-based mapping. But rather than taking the place of local knowledge, these instruments must enhance them. This book examines the significance of striking a balance between local context and technical innovation.

Governance and policy are equally important. Even the most innovative ideas run the risk of being isolated or unsustainable in the absence of enabling governmental frameworks. This book highlights the need for cross-sectoral cooperation and adaptable regulatory frameworks for climate-resilient urbanism. It demands that sustainability be incorporated into urban planning from the beginning, providing guidelines for materials, emissions, land use, and public involvement.

This volume's emphasis on the integration of the technical and the cultural aspects is one of its main contributions. Design must respect and reflect local tradition in areas where architecture is strongly linked to identity, whether via form, material, or function. The past and future can be connected through the adaptive reuse of historic structures, the resuscitation of traditional forms, and the incorporation of regional handicraft. In heritage-rich towns like Cairo, Tunis, or Marrakesh, where the past and present mix in intricate ways, this strategy is especially pertinent.

In this book, architecture is examined as a storytelling medium. Every chapter adds to the overall story of adaptability, hope, and resiliency. The necessity of inclusive and sustainable urban development has been emphasized by international frameworks including the New Urban Agenda, the Sustainable Development Goals (particularly SDG 11 on sustainable cities), and UN-Habitat programs. While providing region-specific approaches that take into account the reality of local situations, this book is in line with these global ideals.

2 Objectives and Scope

This chapter introduces the book *Recent Approaches of Sustainable Architecture in Arid and Semi-arid Cities: Best Practices for Materials, Water, Energy and Urban Greening in Challenging Climates*. It sets the tone for what follows, offering readers a sense of the key questions, themes, and intentions that shaped this collection. At its core, the book brings together a range of voices including researchers, practitioners, and community advocates, who are all working to find better ways to build and live in dry, climate-stressed cities.

The motivation behind this work is the fact that cities in arid and semi-arid regions are under growing pressure from heat, drought, population growth, and environmental degradation. Traditional planning tools often fall short in these settings, and new approaches are urgently needed. This book responds to that need, sharing insights and examples from different parts of the world, with a strong focus on the arid and semi-arid cities.

Through diverse chapters, the book aims to:

- Share design ideas that respond to climate conditions, such as cooling strategies, reuse of old structures, and local building methods;
- Show how green spaces and natural systems can help make cities more livable and less vulnerable to climate extremes;
- Highlight how communities are finding creative ways to grow food, manage water, and adapt their neighborhoods;
- Explore how recycled materials and local resources can reduce waste and lower the environmental impact of construction;
- Look at how digital tools including AI, mapping, and simulations, can support better decision-making and planning.

Throughout, the book emphasizes that solutions must fit the local context. It promotes inclusion of cultural and social aspects, respecting place-based knowledge, and designing with care for both people and the planet.

The following section presents a summary of each of the book themes.

3 Chapters

3.1 Introduction to Sustainable Architecture in Arid and Semi-Arid Cities

In this part, the chapter *Artificial Intelligence for Achieving Urban Smart Sustainability and Mitigating Climate Change* takes a closer look at how artificial intelligence (AI) is beginning to shape the future of sustainable cities, using China, Malaysia, and Egypt as key case studies. As urban areas grow rapidly and environmental pressures mount, there's a clear need for new tools and strategies that can help cities manage resources more wisely and respond to climate risks more effectively. AI is increasingly part of that conversation—not just as a buzzword, but as a practical tool for real-time monitoring, data analysis, and improved planning [5, 6].

In China, national policy plays a major role in pushing forward AI integration. The country is leveraging AI in areas such as transportation, energy systems, and pollution tracking [7]. Malaysia, though working on a smaller scale, has shown promising developments in using AI and Internet of Things (IoT) technologies to build smart cities and reduce energy use. Both countries offer useful insights into how technology, when combined with strong public planning, can help reduce emissions and improve urban resilience [8, 9].

Egypt presents a different context—one marked by high temperatures, limited water, and growing urban needs. Here, AI is being introduced in smart building design and broader planning initiatives to cope with resource scarcity and climate vulnerability. Energy-efficient architecture and climate modeling tools are gaining traction, with the aim of mitigating environmental stress while promoting more adaptive urban environments [10, 11].

The chapter highlights both the promise and the challenges of using AI in urban sustainability. Concerns over AI's own carbon footprint, ethical data use, and the need for inclusive governance are all part of the conversation [12, 13]. It's clear that while AI offers powerful capabilities—whether for urban design, disaster preparedness, or water resource management—it must be applied with a strong ethical framework and community input (Park et al. 2023) [14, 15].

The chapter argues for a balanced approach: one that values innovation but stays rooted in the local realities of each city. AI is a tool, not a solution on its own, and its success depends on thoughtful implementation and broad collaboration.

3.2 Resilience in Architecture and Planning

The first chapter in this part delves into the evolving and deep human relationship between architecture and resilience, beginning with the idea that the instinct to build is embedded not only in humans but in many species. From early shelters made for protection to today's intricate urban environments, architecture has long served as both a necessity and a form of expression. Over time, it grew beyond its practical roots, becoming a mirror of cultural identity, societal priorities, and environmental conditions [16, 17]. As these conditions continue to shift due to conflict, climate change, or rapid urbanization, the role of architecture also evolves, asking architects and communities to rethink how and why we build.

The chapter takes a broad view, exploring how architecture intersects with disciplines like sociology, environmental science, politics, and anthropology. It argues that architecture doesn't stand alone; rather, it is woven into the fabric of daily life, influencing and being influenced by everything around it [18, 19]. This holistic understanding becomes crucial in times of crisis. As climate-related disasters and manmade conflicts escalate, architecture must become a proactive tool for survival, adaptation, and recovery. From ancient systems of passive cooling and natural lighting to modern approaches like greening, groundwater management, and adaptive reuse, the built environment continues to offer powerful responses to crisis [20, 21].

One of the chapter's core messages is that resilience is as much about community as it is about construction. Through stories from conflict-affected regions like Syria and Palestine, we see how people have turned destruction into acts of creation—rebuilding homes, schools, and shared spaces with the limited resources at hand [22, 23]. These examples speak to the enduring role of architecture as a source of dignity, resistance, and collective memory.

The chapter challenges the idea that resilient architecture has a single form. Instead, it shows that resilience is context-specific and shaped by culture, environment, and social structures. Indigenous knowledge, inherited practices, and locally sourced materials all play key roles. Ultimately, the chapter calls for a new architectural mindset, one rooted in empathy, adaptability, and respect for place, that equips us to meet the challenges of today while building thoughtfully for the future.

The second chapter explores Resilient Social-Ecological System in the Arab Region and the urgent need for a new framework that addresses the complex interrelationship between ecological resilience and social systems in arid and semi-arid regions of the Arab world. As climate pressures mount and population growth continues to rise, countries in the Middle East and North Africa are experiencing dramatic shifts in their urban forms and ecological landscapes. The foundations of this discussion are rooted in the Principles of Social-Ecological Urbanism (2013), which emphasize the need for intentional urban design and institutional frameworks that support both environmental and community well-being.

Arid and semi-arid lands in the Arab region are among the most climate-sensitive ecosystems globally, increasingly affected by rising temperatures, reduced precipitation, and frequent extreme weather events [24, 25]. These ecosystems are also under immense pressure from human activities such as rapid urbanization, overgrazing, deforestation, and land reclamation [26]. As a result, the Arab region now faces accelerating land degradation, biodiversity loss, and heightened risk of food and water insecurity [27]. According to Huang and Li [28], arid zones are warming significantly faster than other parts of the world, increasing drought frequency, reducing runoff, and threatening agricultural productivity.

This chapter reviews how certain cities in the region have begun to shift away from expansive, resource-intensive urban sprawl toward more sustainable and resilient planning. Drawing on interviews, archival research, and field observations, the study analyzes how policy frameworks, design practices, and community-led strategies are shaping a new model of urban resilience. Emphasis is placed on localized environmental practices and traditional knowledge, as well as how urban residents are adapting to climatic and socio-economic changes in ways that foster long-term sustainability.

The chapter further explores how drylands, despite their fragility, can become testing grounds for innovative land-use policies and adaptive governance structures. Through a deeper understanding of arid ecosystems and their vulnerabilities, the chapter proposes strategies for designing urban environments that support both ecological functions and community resilience. From biodiversity hotspots in the Mediterranean to overexploited soils in North Africa, the chapter highlights the urgent need to align environmental planning with local socio-political realities.

Ultimately, the chapter argues that building truly resilient social-ecological systems in the Arab world depends on bridging environmental science, inclusive urbanism, and regionally rooted design approaches. It provides critical insights and practical recommendations that can guide future development toward greater ecological sustainability and social equity.

The third chapter explores resilient spaces through urban agriculture practices and how urban agriculture (UA) can serve as a sustainable solution to the challenges faced by public spaces in rapidly growing cities. Urbanization, climate change, and political instability have created numerous pressures on cities, making resilience a key focus in urban planning. The chapter argues that integrating urban agriculture into public spaces can enhance a city's resilience by turning these areas into productive, adaptable, and socially inclusive environments.

Central to this discussion is the concept of urban resilience, which is outlined in the Rockefeller Foundation's City Resilience Index (CRI). This framework divides urban resilience into four dimensions: health and well-being, economy and society, infrastructure and environment, and leadership and strategy [29]. These dimensions illustrate the interconnectedness of various factors—social, economic, environmental, and governance—that contribute to building resilient cities. Urban agriculture plays a crucial role in strengthening these dimensions, particularly in vulnerable areas, by promoting food security, fostering community engagement, and supporting environmental sustainability.

The chapter also looks at urban agriculture's historical impact, with examples from World War II and the COVID-19 pandemic. During these crises, urban agriculture helped ensure food security, boosted morale, and supported the sustainability of cities [30, 31]. It discusses different forms of urban agriculture, such as private gardens, community farms, rooftop gardens, and edible walls, which all contribute to enhancing the social and ecological aspects of urban life [32–34].

The chapter also emphasizes the wide range of urban agriculture applications, from small-scale projects like kitchen gardens and edible schoolyards to larger urban farms and market farms. These initiatives not only provide food but also promote social interactions, environmental stewardship, and local economic development [35]. For example, urban agriculture in New York City emerged as a crucial survival strategy during times of crisis, and similar efforts have grown in response to modern challenges like economic recessions and high unemployment [30].

Ultimately, the chapter concludes that urban agriculture has the potential to transform public spaces into more resilient, adaptive environments. However, its successful implementation depends on collaboration between urban planners, governments, and local communities, as well as the development of policies that integrate agriculture into the urban landscape. By doing so, urban agriculture can become a vital tool for creating cities that are not only sustainable but also resilient in the face of future challenges.

3.3 Sustainable Building Materials and Circular Economy

The first chapter in the part is titled "*Sustainable Environmental Architecture: Approaches for Sustainable Building Materials and Waste Management in the Era of Climate Change*" and looks at the pressing challenges that arid and semi-arid cities face, particularly in relation to climate change and resource scarcity. In these regions, the need for sustainable environmental architecture has become increasingly clear. This approach helps address key concerns like sustainable building materials and efficient waste management strategies, both of which are essential for creating resilient and eco-friendly urban spaces.

Arid and semi-arid areas are particularly vulnerable to the effects of climate change. Issues like water scarcity, extreme heat, desertification, and rapid urban growth are all exacerbating the strain on already limited resources. In light of this, sustainable architecture, which integrates ecological and climate-sensitive principles into the design and construction of buildings, offers a crucial solution. These principles not only help buildings withstand the harsh climatic conditions of these regions but also reduce their overall environmental impact.

One of the primary focuses of the chapter is sustainable building materials. Choosing the right materials can make a huge difference in minimizing environmental harm and ensuring that buildings are both affordable and durable. In particular, natural materials such as bamboo, compressed earth blocks, and recycled plastics have proven effective in reducing both construction costs and environmental footprints [36]. These materials are not only sustainable but also align well with the climate challenges faced in arid regions, helping to create buildings that are more in harmony with their surroundings.

Alongside material choices, waste management is another critical aspect of creating sustainable communities. One of the biggest contributors to environmental degradation is plastic waste, which represents a significant portion of urban and rural waste. With over 5.4 million tons of plastic waste generated each year, the environmental and health risks are becoming increasingly alarming [37]. The chapter emphasizes the importance of reducing, reusing, and recycling plastic waste. It also focuses on low-tech, open-source recycling methods, which are especially useful in countries with less developed infrastructure. These methods provide a way to process plastic waste locally, reducing the environmental burden and turning waste into valuable resources for construction.

Countries like Kenya have led the way by implementing successful bans on plastic bags, while others are focusing on creating biodegradable alternatives [38]. The chapter highlights how such measures can reduce plastic pollution in urban areas. It also explores how plastic waste can be turned into building materials such as insulation, roofing, and wall panels, thanks to low-tech recycling methods [39, 40]. This transformation of waste into a useful resource plays a key role in creating more sustainable and circular urban economies.

Circular economy principles—such as reducing waste, reusing materials, and recycling—are crucial for effective waste management, particularly in regions that lack adequate waste management infrastructure. For example, decentralized waste-to-energy solutions, like those implemented in India, can help reduce waste in areas where proper waste management systems are lacking [41]. In addition, public education and awareness campaigns are essential for fostering behavioral change and promoting recycling at the community level [42].

Low-tech recycling methods, which are especially useful in arid and semi-arid regions, offer an innovative way to tackle plastic waste. These methods involve simple, manual processes like sorting, shredding, and washing plastic waste, which can then be used as raw material for construction projects. This approach not only reduces the amount of plastic waste going to landfills but also creates new economic opportunities and local employment. Additionally, using recycled plastic in construction offers benefits like improved insulation, durability, and water resistance, all of which are vital for buildings in these harsh climates.

This chapter underscores the importance of sustainable building practices and waste management strategies in arid and semi-arid cities. By adopting a circular economy approach and utilizing low-tech plastic recycling methods, these regions can reduce their environmental impact and build resilient, sustainable urban spaces. The chapter offers valuable insights and practical recommendations for tackling the growing challenge of plastic waste while contributing to the creation of a more sustainable future for communities facing climate change.

The second chapter in this part discusses comparative economic viability and barriers to sustainable building materials. The construction sector has long been one of the major contributors to environmental problems, and in recent years, there has been a notable shift towards more sustainable building practices. As part of this movement, there has been an increasing focus on the use of sustainable building materials. This chapter looks closely at how cost-effective these materials are, as well as the obstacles preventing their wider use. Through a thorough comparison of renewable, recycled materials with traditional construction materials, the analysis covers key factors such as raw material availability, manufacturing processes, long-term durability, and life cycle costs. It also delves into the economic, regulatory, and sociocultural challenges that impede the widespread adoption of sustainable materials, providing practical recommendations to overcome these hurdles and promote a more sustainable built environment.

The construction industry's environmental footprint has spurred the adoption of materials that have less impact across their entire life cycle, from raw material extraction to production, use, and disposal. These sustainable alternatives, which include recycled content, renewable resources, and low-impact materials, have emerged as promising solutions to environmental concerns [43]. However, it is not just environmental factors driving this transition—there is also growing recognition that the economic feasibility of these materials plays a critical role in their wider acceptance. In the long run, the use of sustainable materials creates both operational and environmental value. The challenge lies in the higher initial costs associated with these materials, which can pose a significant barrier. Therefore, to ensure that sustainable materials can truly take root in the construction industry, it's important to look beyond just initial costs and consider the entire life cycle of these materials [44].

The main goal of this chapter is to explore the economic viability of sustainable building materials through a comparison of their performance in various stages of the building life cycle. It also addresses the barriers—economic, regulatory, and sociocultural—that prevent these materials from becoming mainstream. A central part of the analysis is to fill knowledge gaps and provide insights for policymakers, industry professionals, and researchers to help overcome these obstacles. The discussion compares different aspects of renewable and recycled materials against traditional ones, factoring in raw material availability, manufacturing processes, durability, and life cycle costs. In the final sections, the chapter explores how these findings can lead to the development of policies and incentives that encourage the broader adoption of sustainable materials.

In the past, materials like concrete and steel were favored in construction because they were readily available and relatively inexpensive. However, as sustainability has become more central to construction practices, the industry is beginning to shift its focus. Materials such as cross-laminated timber (CLT), bamboo, and recycled options are seen as environmentally friendly alternatives, although they come with higher upfront costs [45]. The initial investment required for these materials can be as much as 20% higher than traditional options, mainly due to the cost of raw materials and specialized labor. However, these higher costs are often offset over the long term, as sustainable materials typically require less energy and maintenance. In fact, some studies suggest that sustainable materials, despite their higher initial costs, can ultimately result in lower life-cycle costs (LCC) than traditional materials like concrete, which tend to require more energy and maintenance over time. Sustainable materials can also enhance indoor environmental quality, improving occupant health and productivity, which helps justify the initial premium [46].

To fully understand the economic impact of construction materials, it is crucial to consider life-cycle cost analysis (LCCA). While traditional materials like concrete and steel are often selected based on their relatively low initial cost, these materials can lead to higher long-term expenses due to factors like poor energy efficiency and ongoing maintenance requirements [47]. In contrast, sustainable materials may have higher initial costs, but their energy efficiency and low maintenance needs result in significant savings over the building's life span. For example, materials such as CLT may have higher upfront costs, but they reduce energy consumption and maintenance needs in the long run.

Recent studies have highlighted the potential of sustainable materials to reduce life-cycle costs while addressing environmental concerns. Hasan et al. [48] noted that incorporating bio-based and green composite materials into construction projects can significantly lower embodied carbon and operational energy use. This approach aligns with the principles of life-cycle cost analysis, offering both economic and environmental benefits. Similarly, Meinardi et al. [49] found that using advanced luminescent solar concentrator glazing technologies can reduce energy consumption, particularly in climates where energy demands are high.

The third chapter discusses Sustainable Green Finishing Material and Their Impact on Internal Spaces. It explores the importance of environmentally sustainable interior design (ESID) is increasingly recognized, yet many architects and engineers remain unaware of the need to integrate sustainability into their material selection. Interior design has traditionally emphasized aesthetics over environmental and health concerns. Many finishing materials, such as paints, carpets, and PVC flooring, release harmful chemicals that contribute to indoor air pollution, which can be significantly worse than outdoor air quality. Indoor pollutants like carcinogens and respiratory irritants are common, highlighting the need for sustainable materials that reduce these risks. Furthermore, the production of many materials consumes excessive water and energy, exacerbating environmental degradation. Prioritizing renewable, locally sourced, and energy-efficient materials is vital for sustainability in interior design [50].

Despite the increasing availability of sustainable materials, their adoption remains limited. The selection process is often driven by aesthetic considerations rather than environmental or health impacts. This research emphasizes the importance of life cycle cost assessments and the efficiency of using local materials, considering both the environmental footprint and the long-term sustainability of the building. By promoting materials with lower environmental impact and focusing on their potential for recycling and reuse, designers can contribute to sustainable urban development.

Sustainable architecture focuses on minimizing environmental harm while optimizing resource use. Sustainable interior design extends this philosophy by selecting materials that have a lower environmental impact and contribute to occupant health and well-being. Global and regional rating systems like LEED, BREEAM, and GSAS evaluate building sustainability based on energy use, water consumption, and material choices. These systems emphasize the importance of sustainable materials in creating healthy, efficient spaces, and support the goals of reducing carbon emissions and promoting waste reduction. By adhering to these standards, architects and designers can help create buildings that are not only sustainable but also conducive to the health and safety of their inhabitants [51, 52].

3.4 Sustainable Water Management in Urban Settings

The first chapter in this part focuses on Sustainable Water Management in Buildings: Innovative Strategies for an Ecological Future. Water is a crucial resource for life and economic development [53], but it has become increasingly scarce worldwide due to climate change, population growth, pollution, and unequal distribution. This is particularly true in Algeria, where water scarcity is a significant issue, with the country classified as water-poor by international standards, particularly with a per capita availability of less than 600 m^3 annually. This scarcity has created tensions and challenges in both urban and rural settings, including in the city of Blida, which is located near the Mitidja plain at the foot of the Atlas Blidean mountains. This region is facing significant water management issues due to variable rainfall and increasing demand [54].

Blida's water consumption, particularly in the residential sector, is a concern. On average, each person uses about 150 L of water per day, primarily for activities such as toilet flushing, showering, and cleaning. While there are water resources available, including surface and underground sources like the Mitidja water table, they are being overused, leading to further depletion. The situation is compounded by the underutilization of surface water from local wadis and the limited infrastructure for water management in the city [55].

Given these challenges, the chapter discusses strategies to reduce the environmental impact of water consumption in Blida while maintaining the health and comfort of its inhabitants. Sustainable water management is seen as a necessary solution, one that integrates both technology and community participation. One such approach is the adoption of water-saving technologies like low-flow devices (e.g., faucets, shower heads, and toilets), which can significantly reduce water consumption without compromising daily activities. For instance, replacing traditional showerheads with low-flow models can save around 2,700 gallons of water per year for a typical family, while high-efficiency toilets can save up to 4,000 gallons annually compared to older models [56].

Additionally, rainwater harvesting systems are an innovative solution for reducing reliance on municipal water systems. By collecting rainwater for non-potable uses such as irrigation, households and buildings can ease pressure on freshwater resources. Green roofs, a key element of sustainable architecture, also help by absorbing rainwater and providing thermal insulation for buildings, further contributing to water conservation.

The chapter emphasizes that achieving sustainable water management requires not only technical solutions but also a change in how water is viewed by society. This includes valuing water as an economic resource and promoting policies that encourage efficient water use. Furthermore, it calls for collective action, where both individuals and institutions contribute to creating a more sustainable and equitable water management system [54, 55, 57].

The second chapter discusses the integration of indigenous knowledge for sustainable water management in buildings. The issue of water scarcity has become a critical global challenge, particularly in arid and semi-arid regions. These regions, which face severe water shortages, have developed traditional and creative ways to manage their water resources sustainably. This chapter explores the potential for integrating indigenous knowledge into modern building design to address water scarcity while enhancing sustainability and livability. Rather than advocating for the wholesale adoption of these indigenous methods, the chapter proposes a transdisciplinary approach, combining both traditional and modern knowledge to yield more effective solutions.

In examining the case of Egypt, the chapter highlights how indigenous water management practices have contributed to sustainable architecture but also underscores that these practices may not be directly applicable in modern construction. For instance, the use of local materials like Kashrif particles in the Siwa Oasis demonstrates the benefits of indigenous methods, but modern technologies must be integrated to improve resilience and adaptability [58]. This approach recognizes that while indigenous knowledge offers valuable insights, it must be adapted to modern needs and lifestyles.

A compelling example of how indigenous knowledge can be successfully integrated with modern technology is found in Seville, Spain. There, the ancient practice of qanat irrigation has been revived and adapted to cool modern buildings, reducing energy consumption and ambient temperatures. This application of traditional knowledge not only conserves water but also promotes sustainability and livability by addressing the challenges of urban heat islands, a common issue in arid regions [59]. The qanat system, originally developed in ancient Persia, demonstrates how indigenous techniques can be re-imagined to meet contemporary needs, providing a sustainable solution to water management in modern urban environments [60].

Water scarcity in the Middle East and North Africa (MENA) region is exacerbated by rapid urbanization, climate change, and overreliance on fossil fuels. These factors contribute to an increase in water demand and pollution, creating a vicious cycle of water depletion. Cities in the MENA region, particularly those heavily reliant on desalination technologies, face significant environmental and fiscal challenges. As the region's population continues to grow, innovative solutions that integrate indigenous knowledge with modern technologies are essential for achieving long-term sustainability [61].

Ultimately, the chapter advocates for a balanced, transdisciplinary approach that combines the wisdom of indigenous water management practices with modern scientific advancements. This synergy can create more sustainable, energy-efficient, and environmentally friendly water management solutions in buildings, particularly in regions facing severe water scarcity. By fostering collaboration between indigenous knowledge holders and modern technologists, it is possible to develop water management systems that are both culturally relevant and technologically viable, ultimately enhancing the quality of life for urban residents in arid and semi-arid regions [62, 63].

In conclusion, integrating indigenous knowledge into water management strategies is not just about preserving cultural heritage but also about ensuring the sustainable use of resources. This approach offers a promising pathway to address the complex challenges of water scarcity and environmental degradation while enhancing the livability of urban spaces.

3.5 Proposing Water Sensitive Urban Design Solutions in Residential Areas

The third chapter works on proposing Water Sensitive Urban Design (WSUD) solutions in residential areas. It delves into how WSUD can be applied to residential areas, focusing specifically on Dar Misr Residential Compound in Helwan, Cairo, Egypt. Water scarcity is a growing concern in Egypt, which has seen a dramatic decline in its per capita water availability over the years. In 1959, each person had access to 1,893 cubic meters of water annually, but by 2012, this had fallen to less than 700 cubic meters, and the situation continues to worsen. With rapid population growth, experts predict that by 2030, Egyptians will only have access to 534 cubic meters per person, a stark contrast to the international water poverty threshold, [1, 64]. This trend underscores the urgency of finding sustainable solutions to manage the country's water resources more efficiently.

The problem is compounded by unequal distribution, with major urban areas like Cairo and Alexandria bearing the brunt of increased water demand. Additionally, inefficient irrigation practices in agriculture waste large amounts of water, further depleting the available supply. This chapter proposes WSUD as a much-needed approach to improving water management in urban areas. By integrating rainwater harvesting, permeable pavements, green roofs, and constructed wetlands, WSUD aims to address both water scarcity and the negative impacts of urban development on the environment [64]. These techniques not only reduce water consumption but also improve urban resilience to climate change, while providing a more sustainable and livable environment for residents.

One of the key benefits of WSUD is its ability to foster better community engagement and improve public health. For example, using native plants in landscaping can dramatically reduce the need for irrigation, conserving precious freshwater resources [65]. By applying these principles to Dar Misr, this chapter demonstrates how urban areas can integrate sustainable water management practices to improve both the environment and the quality of life for residents. The incorporation of green infrastructure, such as permeable surfaces and rainwater collection systems, can reduce stormwater runoff, recharge groundwater, and create a more sustainable urban landscape.

The situation in Egypt is dire. The country's reliance on the Nile River for 95% of its water resources, coupled with increasing population pressure and upstream development projects, presents a major challenge. Efforts by the government, such as the construction of dams and promoting water-efficient agricultural practices, are crucial to mitigating water scarcity. However, the chapter emphasizes that WSUD is a crucial component in ensuring sustainable urban growth. It offers practical solutions for residential developments like Dar Misr to integrate sustainable water practices into their design.

This chapter illustrates the significant potential of WSUD in addressing Egypt's water crisis. By offering sustainable water management solutions that improve resilience, community well-being, and environmental health, WSUD presents a practical pathway for future urban developments in Egypt and other regions facing similar challenges.

3.6 Urban Greening in Cities

The first chapter focuses on Green Space Expansion Using Haloculture Technology to Combat Climate Change and Global Warming. It discusses the potential of haloculture technology as a solution to combat climate change and global warming in the Middle East and North Africa (MENA) region, where urban green spaces face numerous challenges due to rising temperatures, drought, and salinity. Haloculture, which makes use of saline water and soil, is introduced as a promising approach to expanding green spaces in such harsh environments. The chapter uses the PRISMA method to examine the impact of climate change on MENA cities, particularly the intensification of urban heat islands, and explores how haloculture can contribute to creating more sustainable urban areas [66, 67].

The region is grappling with issues like water scarcity, desertification, and salinization of soil and water, all of which worsen as a result of climate change. These environmental shifts disrupt agriculture, strain resources, and put public health at risk, threatening the stability of urban areas [33, 34]. The chapter also emphasizes how climate change is affecting MENA cities, highlighting the need for adaptive strategies to help them become more resilient. On top of physical challenges, climate change also has serious social and psychological impacts, leading to mental health problems like anxiety and trauma, especially in the wake of extreme weather events such as floods and heatwaves [68, 69]. Vulnerable groups, including low-income communities and minorities, are particularly at risk due to their limited access to resources and support [70].

Another significant issue in MENA cities is the urban heat island effect, where urban areas experience higher temperatures than surrounding rural regions, increasing energy consumption and negatively impacting health and productivity [71]. To mitigate this, several strategies are suggested, including the use of reflective materials and the expansion of green spaces. However, in dry and semi-dry regions, traditional methods of planting are hindered by limited water resources and soil salinization [72]. This is where haloculture shows promise, as it enables the cultivation of salt-tolerant plants without relying on freshwater, thus offering a sustainable way to green urban areas in challenging climates.

Ultimately, the chapter advocates for the adoption of haloculture in MENA cities as a practical solution for addressing climate change, enhancing urban resilience, and improving environmental quality. It also highlights the importance of educating communities and professionals through pedagogical content knowledge (PCK) to ensure effective management of green spaces and sustainable urban development.

The second chapter explores the impact of green roofs on buildings' thermal performance. It discusses how green roofs can improve the thermal performance of buildings, particularly in hot-arid climates, with a specific focus on Egypt, especially Cairo. Rapid urbanization and a lack of green spaces in Egypt have led to significant environmental issues, including the Urban Heat Island (UHI) effect, where cities experience higher temperatures than surrounding areas, leading to increased energy consumption. Cairo, with its densely packed population, suffers from one of the lowest green space ratios per person globally, worsening both environmental conditions and climate change impacts [73].

Green roofs are presented as a potential solution to these problems. These roofs, which involve planting vegetation on building rooftops, offer a range of environmental, economic, and social benefits. Green roofs can help combat UHI effects by lowering temperatures and improving energy efficiency, while also enhancing air quality. Additionally, they assist with stormwater management by absorbing rainfall and reducing runoff [74]. There are three types of green roofs: extensive, intensive, and semi-intensive, each varying in plant height, maintenance requirements, and the thickness of the layers [75].

Despite these advantages, green roofs are not widely used in Egypt. One reason for this is Cairo's limited green spaces, which have been shrinking rapidly. Between 2017 and 2020, the city lost a considerable amount of green space [73], and the available green space per person is far below international standards. Many residents only have access to less than 1 m^2 of green space per capita, while the World Health Organization recommends a minimum of 9 m^2 [76]. However, rooftops in Cairo are often neglected, used for storage or housing building infrastructure, yet these spaces present an opportunity to create much-needed green areas [77].

The chapter argues that green roofs could solve several urban challenges. They provide a nature-based solution by transforming unused rooftops into productive, eco-friendly spaces that help mitigate the effects of climate change, improve energy efficiency, and increase the city's green area. The research suggests that adopting green roofs in Cairo would not only improve building performance but also contribute to sustainability and the overall well-being of its residents.

3.7 Energy Transition in the Built Environment

The first chapter explores the application of renewable energy in the built environment. It addresses how renewable energy (RE) can be integrated into Egypt's built environment to support sustainable development, especially in community centers. With Egypt's rapidly growing urban population and rising energy demand, the shift to renewable energy is essential for tackling the challenges of climate change and energy security [78]. By 2050, approximately 75% of Egypt's population is expected to live in cities, which will significantly increase the demand for energy [79]. While urbanization brings economic opportunities, it also puts immense pressure on Egypt's energy infrastructure, which struggles with high transmission losses and unequal access to electricity between urban and rural areas [80].

Renewable energy, especially solar power, presents a promising solution to these issues. However, there are significant barriers to overcome, including Egypt's complicated regulatory system and the need for substantial investments. The country is expected to require around $70 billion to meet its renewable energy goals by 2030 [81]. Despite these challenges, Egypt is making progress in solar energy, positioning itself as a regional leader in this field [82]. The chapter emphasizes the importance of understanding the energy consumption patterns of different types of buildings—such as residential, commercial, industrial, and public—because each type has distinct energy needs. Tailored strategies to improve energy efficiency in these sectors are crucial for addressing the specific conditions in Egypt's urban and semi-arid regions.

The role of renewable energy in reducing greenhouse gas emissions, combating climate change, and enhancing energy security is highlighted International Energy Agency [83]. As global energy demand rises, transitioning to renewables not only reduces air pollution but also creates new job opportunities Intergovernmental Panel on Climate Change [84]. Innovations in renewable technologies, like perovskite solar cells, artificial photosynthesis, and floating solar farms, hold great promise for improving energy efficiency and expanding the use of solar energy [85].

The chapter concludes with a case study that demonstrates how renewable energy has been implemented in Egypt. This real-world example showcases both the successes and challenges of integrating renewable energy into Egypt's energy system, providing valuable insights into how the country can effectively use renewable resources to foster a sustainable built environment and contribute to long-term development goals.

The second chapter discusses the role of blockchain and cyber-physical systems (cpss) in execution of sustainable smart construction. It delves into how blockchain technology and cyber-physical systems (CPSs) can transform the construction industry, making it more sustainable, efficient, and transparent, particularly in arid and semi-arid regions. These regions face unique challenges due to harsh climates, such as extreme temperatures, low rainfall, and high winds. By adopting advanced technologies like blockchain and CPSs, these challenges can be mitigated, and a better quality of life can be achieved.

Blockchain operates as a decentralized and secure digital ledger that records transactions without relying on a central authority. This decentralized nature makes it particularly valuable in sectors like construction, where transparency and trust between stakeholders are crucial. The integration of blockchain allows for real-time monitoring and secure documentation of every step in the construction process, ensuring accountability, traceability, and minimizing the risk of fraud or miscommunication. Blockchain's ability to facilitate secure, tamper-proof exchanges makes it an ideal tool for the construction industry, where transparency and trust are often lacking. Blockchain helps create a more collaborative environment by fostering an open and transparent information exchange between all parties involved [86].

Cyber-Physical Systems (CPSs), on the other hand, combine physical processes, computing, and communication technologies to automate tasks and optimize workflows. CPSs improve construction efficiency by integrating IoT devices and sensors that provide real-time data on building performance, environmental conditions, and operational status. These systems make it possible to adjust construction activities on-the-fly based on this data, leading to more precise and sustainable building practices. The role of CPSs in construction is vital, as they automate the flow of information between stakeholders and integrate advanced systems for real-time monitoring, energy management, water conservation, and eco-friendly construction techniques [87, 88].

The adoption of Industry 4.0 in construction, which includes CPSs, blockchain, and digitalization (e.g., Building Information Modeling), is transforming the sector. However, the fragmented nature of construction projects—where different stakeholders handle separate tasks—has hindered widespread adoption. Blockchain helps address this by creating a decentralized, transparent framework for collaboration across all stages of construction [89].

In conclusion, the combination of blockchain and CPSs offers promising solutions for creating sustainable smart cities, especially in arid and semi-arid regions. These technologies enable more efficient resource management, reduce waste, and ensure that construction practices are more sustainable. Their integration into construction can lead to a future where buildings are smarter, greener, and more connected.

References

1. ElZein Z, Milad NA, Mohamed AS, Mahmoud N, Abdo N, Abdel-Ghafar HM (2024) Sustainable development on the basis of the WEF nexus in arid coastal areas for climate change mitigation: case study of Rabia community in Matrouh, Egypt. Energy Nexus 14:100299. https://doi.org/10.1016/j.nexus.2024.100299
2. Eizein Z, Ahmed Milad N, Sherif Mohamed A, Kotb NM, Abdo N, Negm AM (2024) Role of water, energy and food nexus in sustainable development of communities: case studies in the MENA region. In: Negm AM, ElZein Z (eds) Integration of core sustainable development goals in rural areas, earth and environmental sciences library. Springer Nature Switzerland, Cham, pp. 15–66. https://doi.org/10.1007/978-3-031-60149-1_2
3. Roaf S, Cheshmehzangi A (2024) Persian vernacular buildings—Lessons on resilient design. In: Cheshmehzangi A, Roaf S (eds) Persian vernacular architecture. Urban Sustainability. Springer, Singapore. https://doi.org/10.1007/978-981-96-1116-4_1
4. Shah V (2023) The role of architecture in humaity's story. [Online] Available at: https://thougheconomics.com/the-role-of-architecture-in-humanitys-story/
5. Al-Raeei M (2024) The smart future for sustainable development: artificial intelligence solutions for sustainable urbanization. Sustain Dev 32(1):1–10. https://doi.org/10.1002/sd.3131
6. Son TH, Weedon Z, Yigitcanlar T, Sanchez T, Corchado JM, Mehmood R (2023) Algorithmic urban planning for smart and sustainable development: systematic review of the literature. Sustain Cities Soc 94:104562. https://doi.org/10.1016/j.scs.2023.104562

7. Zheng X, Ma Z, Yuang Z (2024) Urban design and pollution using AI: implications for urban development in China. Heliyon 10:Article e37735. https://doi.org/10.1016/j.heliyon.2024.e37735
8. Parks D (2019) Energy efficiency left behind? Policy assemblages in Sweden's most climate-smart city. Eur Plann Stud 27(2):318–335
9. Wang D, Peng D, Huang D (2025) Application and prospects of large AI models in virtual power plants. Electr Power Syst Res 241:111403. https://doi.org/10.1016/j.epsr.2024.111403
10. Wu T, Zhang T (2022) Multi-objective optimization of energy, visual, and thermal performance for building envelopes in China's hot summer and cold winter climate zone. J Build Eng 52:105034. https://doi.org/10.1016/j.jobe.2022.105034
11. Adinkrah J, Kemausuor F, Tchao ET, Nunoo-Mensah H, Agbemenu AS, Adu-Poku A, Kponyo JJ (2025) Artificial intelligence-based strategies for sustainable energy planning and electricity demand estimation: A systematic review. Renew Sustain Energ Rev 210:115161. ISSN 1364-0321
12. Konya A, Nematizadeh P (2024) Recent applications of AI to environmental disciplines: a review. Sci Total Environ 906:167026
13. Ansari SPMA, Odongo JO, Nomani MZM, Salahuddin G, Hasan MF, Pallathadka LK (2022) Evaluating the role of environment management systems based on artificial intelligence. Mater Today Proc 56:2240–2244
14. Ramyar R, Ackerman A, Johnston DM (2021) Adapting cities for climate change through urban green infrastructure planning. Cities 117:103316. https://doi.org/10.1016/j.cities.2021.103316
15. Jain H, Dhupper R, Shrivastava A, Kumar D, Kumari M (2023) AI-enabled strategies for climate change adaptation: protecting communities, infrastructure, and businesses from the impacts of climate change. Comput Urban Sci 3(25). https://doi.org/10.1007/s43762-023-00100-2
16. PBS (2011) The Incredible Termite Mound. [Online] Available at: https://www.pbs.org/wnet/nature/the-animal-house-the-incredible-termite-mound/7222/
17. Jamei E, Vrcelj Z (2021) Biomimicry and the Built Environment, Learning from Nature's Solutions. [Online]. Available at: https://www.mdpi.com/2076-3417/11/16/7514
18. Cutieru A (2021) How Social Sciences Shape the Built Environment. [Online] Available at: https://www.archdaily.com/962538/how-social-sciences-shape-the-built-environment
19. McIntyre MH (2006) A literature review of the social, economic and environmental impact of architecture and design. s.n., s.l.
20. Sherif S (2022) Cairo's Green Sprawl: The Move of Urban Green Space towards exclusivity. [Online] Available at: https://aps.aucegypt.edu/en/articles/947/cairos-green-sprawl-the-move-of-urban-green-space-towards-exclusivity
21. Al-Ibrashy M, DINSOK (2022) Athar Lina toolkit for groundwater solutions in historic settings-Treatment and reuse of water. Megawra Papers, s.l.
22. Okasha A (2011) The Architecture of Resilience. Ahmed Bahaa al-Din assosiation, s.l
23. Dalal A (2022) From Shelters to Dewllings. s.n., s.l.
24. Aghsaei H, Mobarghaee Dinan N, Moridi A, et al (2020) Effects of dynamic land use/land cover change on water resources and sediment yield in the Anzali wetland catchment, Gilan, Iran. Sci Total Environ 712:136449. https://doi.org/10.1016/j.scitotenv.2019.136449. PMID: 32050376
25. Ban N, Schmidli J, Schär C (2014) Evaluation of the convection-resolving regional climate modeling approach in decade-long simulations. J Geophys Res Atmos 119:7889–7907. https://doi.org/10.1002/2014JD021478
26. Li Y, Hui G, Zhao Z, Hu Y (2019) The bivariate distribution characteristics of spatial structure in natural Korean pine broad-leaved forest. J Veg Sci 23(6):1180–1190. https://doi.org/10.1111/j.1654-11032012.01431
27. Dimobe K, Tondoh J, Weber JC, Bayala J, Ouédraogo K, Greenough KM (2018) Farmers' preferred tree species and their potential carbon stocks in southern burkina faso: implications for biocarbon initiatives. PLoS One 13(12):e0199488. https://doi.org/10.1371/journal.pone.0199488

28. Huang G-Q, Li S (2019) Review of the genus Falsotrachystola Breuning, 1950 (Coleoptera: Cer- ambycidae: Lamiinae: Morimopsini). Zootaxa 4555(1):45–55. https://doi.org/10.11646/zootaxa.4555.1.3
29. Silva J, Morera B (2012) City Resilience Index - The Rockefeller Foundation, Arup. https://www.rockefellerfoundation.org/wp-content/uploads/CRI-Revised-Booklet1.pdf
30. Olsson G, Kerselaers E, Kristensen LS, Primdahl J, Rogge E, Wästfelt A (2016) Peri-urban food production and its relation to urban resilience. Sustainability 8:1340. https://doi.org/10.3390/su8121340
31. Sharifi A, Yamagata Y (2014) Resilient urban planning: major principles and criteria. Energy Procedia 61:1491–1495. https://doi.org/10.1016/j.egypro.2014.12.154
32. Smit J, Nasr J, Ratta A (2001) Urban agriculture: food, jobs and sustainable cities. United Nations Development Programme, The Urban Agriculture Network, New York, USA. http://www.jacsmit.com/book/Chap02.pdf
33. FAO (2020) Emissions due to agriculture. Global, regional and country trends 2000–2018. FAOSTAT Analytical Brief Series No 18. Rome
34. FAO (2020) Urban food systems and COVID-19: the role of cities and local governments
35. Deelstra T, Girardet H (2000) Urban agriculture and sustainable cities. Growing Cities, Growing Food: Urban Agriculture on the Policy Agenda, 43–65. https://www.researchgate.net/publication/284992045_Urban_agriculture_and_sustainable_cities
36. McDonough W, Braungart M (2002) Cradle to cradle: rethinking the way we make things. North Point Press
37. UNEP (2023) Visual Feature | Beat Plastic Pollution. Retrieved from https://www.unep.org/interactives/beat-plastic-pollution/
38. Singh N, Walker TR (2024) Plastic recycling: a panacea or environmental pollution problem. NPJ Mater Sustain 2:17. https://doi.org/10.1038/s44296-024-00024-w
39. Plastic Odyssey (2024) The expedition of solutions against plastic pollution. PRESS KIT 2024. https://plasticodyssey.org/wp-content/uploads/2023/DP/EN-PRESS-KIT-2023-PLASTIC-ODYSSEY-CP.pdf
40. Arqlite (2022) 7 recycled plastic building materials revamping construction – Arqlite Smart Gravel - The light and sustainable alternative to gravel. Retrieved from https://smart.arqlite.com/recycled-plastic-building-materials/
41. Delterra (2023) Circular cities: the six dimensions of circular waste management. Delterra. https://delterra.org
42. Solar Impulse Foundation (n.d.) How to shape the future of waste? Retrieved August 29, 2024, from https://solarimpulse.com
43. Attia S, De Herde A, Hensen J (2017) A critical review of methodologies to assess the life cycle costing of buildings. Energy Build 140:35–48
44. Barbosa L, Bergerson J, Birkved M (2018) Life cycle cost analysis of sustainable building materials: a review. J Clean Prod 177:491–501
45. Annibaldi V, Cucchiella F, Rotilio M (2021) Economic and environmental assessment of thermal insulation: a case study in the Italian context. Case Stud Constr Mater. https://doi.org/10.1016/j.cscm.2021.e00682
46. Schneider-Marin P, Lang W (2020) Environmental costs of buildings: monetary valuation of ecological indicators for the building industry. Int J Life Cycle Assess 25(9):1637–1659. https://doi.org/10.1007/s11367-020-01784-y
47. Fantozzi F, Gargari C, Rovai M, Salvadori G (2019) Energy upgrading of residential building stock: use of life cycle cost analysis to assess interventions on social housing in Italy. Sustainability. https://doi.org/10.3390/SU11051452
48. Hasan MR, Khan MA, Wuest T (2024) Towards Industry 5.0: a systematic literature review on sustainable and green composite materials supply chains. J Cleaner Prod 421:129001. https://doi.org/10.1016/j.jclepro.2024.129001
49. Meinardi F, Bruni F, Castellan C, Meucci M, Umair AM, La Rosa M, Catani J, Brovelli S (2024) Certification-grade quantum dot luminescent solar concentrator glazing with optical communication capability for connected sustainable architecture. Sol Energy Mater Sol Cells 245:111832. https://doi.org/10.1016/j.solmat.2024.111832

50. Moxon S (2020) Sustainability in interior design, 2nd edn. Laurence King Publishing, London
51. Beckman SL, Barry M (2020) Design thinking and sustainable development goals. Emerald Publishing Limited, Bingley
52. Moxon S (2021) Sustainability in interior design. Laurence King Publishing Ltd, China
53. Elzein Z (2024) Sustainable built environments and the water, energy, and food nexus: toward an interdisciplinary approach to urban resilience. Chn J Urb Environ Stud 12:2450004. https://doi.org/10.1142/S2345748124500040
54. Fanack Water (2019) Water of the Middle East and North Africa. Algeria Water Report. Algeria's per capita water availability: Renewable water resources and water crisis. Published on July 30, 2019. Retrieved from https://water.fanack.com/algeria/waterresources
55. Euzen A, Jeandel C, Mosseri R (eds) (2015) Water uncovered (L'eau à découvert). CNRS Éditions. https://doi.org/10.4000/books.editionscnrs.9746
56. Green Building Consulting & Engineering (2023) Sustainable buildings: challenges and solutions for water management (Bâtiments durables: Enjeux et solutions pour la gestion de l'eau). https://fr.greendesignconsulting.com/single-post/b%C3%A2timents-durables-enjeux-et-solutions-pour-la-gestion-de-l-eau
57. Abdelli R (2008) Strategies for sustainable water management in Remchi (Des stratégies pour une gestion durable de l'eau à Remchi). Thesis for State Engineer in Urban Hydraulics, Abou Bekr Belkaid University, Tlemcen
58. He C, Liu Z, Wu J et al (2021) Future global urban water scarcity and potential solutions. Nat Commun 12:4667. https://doi.org/10.1038/s41467-021-25026-3
59. Garcia-Franco N et al (2018) Climate-smart soil management in semiarid regions. Soil Manag Climate Change, 349–368. https://doi.org/10.1016/b978-0-12-812128-3.00023-9
60. Pleitgen F et al (2021) The Middle East is running out of water. CNN. Available at: https://edition.cnn.com/2021/08/22/middleeast/middle-east-climate-water-shortage-iran-urmia-intl/index.html
61. Sahu P, Debsarma C (2023) Urban water scarcity: a global challenge and impending solutions. https://doi.org/10.1007/978-3-031-21618-3_7
62. Morante-Carballo F et al (2022) What do we know about water scarcity in semi-arid zones? A global analysis and research trends. Water 14(17):2685. https://doi.org/10.3390/w14172685
63. Jaad A, Abdelghany K (2021) The story of five mena cities: urban growth prediction modeling using remote sensing and video analytics. Cities 118:103393. https://doi.org/10.1016/j.cities.2021.103393
64. Abdou A, Gawad I, ElZein Z (2017) Sustainable water treatment methods to be used in urban communities. In: Negm A (eds) Unconventional water resources and agriculture in Egypt. The Handbook of Environmental Chemistry, vol 75. Springer, Cham
65. Kazemi F, Barton H (2021) Water sensitive urban design (WSUD) approach for mitigating groundwater depletion in urban geography. Water Supply 21(7):334–347. https://doi.org/10.2166/ws.2021.001
66. Parnian A, Parvizi H, Selmy S, Mushtaq Z (2024) Haloculture: a pathway to reduce climate change consequences for societies. In: Negm AM, ElZein Z (eds) Integration of Core sustainable development goals in rural areas. Earth and Environmental Sciences Library. Springer, Cham. https://doi.org/10.1007/978-3-031-60149-1_14
67. Mahbod M, Parnian A (2024) sustainable development in rural areas: climate change impacts on water, agriculture, energy, and psychological health consequences. In: Negm AM, ElZein Z (eds) Integration of core sustainable development goals in rural areas. Earth and Environmental Sciences Library. Springer, Cham. https://doi.org/10.1007/978-3-031-60149-1_3
68. Cianconi P, Betrò S, Janiri L (2020) The impact of climate change on mental health: a systematic descriptive review. Front Psych 11:74
69. Clayton S (2020) Climate anxiety: psychological responses to climate change. J Anxiety Disord 74:102263
70. Levy BS, Patz JA (2015) Climate change, human rights, and social justice. Ann Glob Health 81(3):310–322

71. Zou Z, Yan C, Yu L, Jiang X, Ding J, Qin L, Wang B, Qiu G (2021, October). Impacts of land use/land cover types on interactions between urban heat island effects and heat waves. Build Environ 204:108138. https://doi.org/10.1016/j.buildenv.2021.108138
72. Abulibdeh A (2021) Analysis of urban heat island characteristics and mitigation strategies for eight arid and semi-arid gulf region cities. Environ Earth Sci 80(7). https://doi.org/10.1007/s12665-021-09540-7
73. Aly D, Dimitrijevic B (2022) Public green space quantity and distribution in Cairo Egypt. J Eng App Sci 69:15. https://doi.org/10.1186/s44147-021-00067-z
74. Manso M, Teotonio I, Silva MC, Cruz CO (2021) Green roof and green wall benefits and costs: a review of the quantitative evidence. Renew Sustain Rev 135. https://doi.org/10.1016/j.rser.2020.110111
75. Kazemi M, Courard L (2021, November 1) Modeling hygrothermal conditions of unsaturated substrate and drainage layers for the thermal resistance assessment of green roof: effect of coarse recycled materials. Energy Build 250:111315. https://doi.org/10.1016/j.enbuild.2021.111315
76. WHO (2012) Health indicators of sustainable cities in the context of the Rio+20 UN Conference on Sustainable Development. WHO; Geneva, Switzerland
77. Abd El-Hafeez M, Elmokadem A, Megahed N, EL-Gheznawy D (2016) Green roof analysis and design software calculator in Egypt. Port Said Eng Res J 20(1):12–20. https://doi.org/10.21608/pserj.2016.33626
78. World Bank (2023) Urban Development. [online] World Bank. Available at: https://www.worldbank.org/en/topic/urbandevelopment/overview
79. Egyptian Ministry of Electricity and Renewable Energy (2023) Energy Consumption Forecast Report, Cairo. Available at: http://www.moee.gov.eg/english_new/home.aspx. Accessed 11 Aug 2024
80. UNDP (2024) Reimagining cooperation in a polarized world. [online] Available at: https://hdr.undp.org/system/files/documents/global-report-document/hdr2023-24reporten.pdf
81. IEA (2023) Egypt - Countries & Regions. [online] Available at: https://www.iea.org/countries/egypt. Accessed 10 Aug 2024
82. IRENA (2024) International Renewable Energy Agency (IRENA). [online] Irena.org. Available at: https://www.irena.org/. [Accessed: 10 Aug. 2024].
83. International Energy Agency (IEA) (2022) Renewables 2022: analysis and forecast for 2027. Available at: https://iea.blob.core.windows.net/assets/ada7af90-e280-46c4-a577-df2e4fb44254/Renewables2022.pdf. Accessed 20 Aug 2024
84. Intergovernmental Panel on Climate Change (IPCC) (2023) Climate Change 2021 – The Physical Science Basis: Working Group I Contribution to the Sixth Assessment Report of the Intergovernmental Panel on Climate Change. Cambridge University Press, Cambridge
85. NREL (2023) Emerging photovoltaic technologies. Available at: https://www.nrel.gov/pv/emerging-tech.html. Accessed 10 Aug 2024
86. Akinradewo O et al (2023) A roadmap for present focus and future trends of blockchain technology in the built environment. Afr J Sci Technol Innov Dev 15(2):153–165. Available at: https://doi.org/10.1080/20421338.2022.2046249
87. Jiang Z et al (2022) Applying blockchain-based method to smart contract classification for CPS applications. Digital Commun Netw 8(6):964–975. Available at: https://doi.org/10.1016/j.dcan.2022.08.011
88. Qureshi AH, Alaloul WS, Alzubi KM (2022) Internet of Things (IoT) for construction cyber-physical systems. In: Cyber-physical systems in the construction sector. CRC Press, pp 60–77
89. Ali A (2023) How Saudi is turning desert into Greenland, Earth. Available at: https://vocal.media/earth/how-saudi-is-turning-desert-into-greenland. Accessed 5 Aug 2024

The Role of Artificial Intelligence in Achieving Smart Sustainability and Mitigating Climate Change in Urban Contexts

Youqin Pan

Abstract Rapid urbanization and growing environmental challenges highlight the need for innovative, integrated solutions to combat climate change. In this context, the rise of smart cities and AI-driven smart sustainability represents a transformative approach to tackle climate change and promote sustainable urban development. The chapter conducts a comparative study of China, Malaysia, and Egypt, each with its unique circumstances in applying AI to combat climate change through smart sustainability. China and Malaysia are major contributors of greenhouse gas emissions in Asia, so it is meaningful to navigate the path of both countries to mitigate climate change through innovative solutions supported by Artificial Intelligence. China benefits from a unified national strategy for developing and applying AI in climate change mitigation, particularly in smart city development, energy optimization, and intelligent transportation systems. Malaysia is deploying smart energy systems powered by IoT and AI, while also building smart cities, reducing carbon footprints and enhancing sustainability. Unlike China and Malaysia, Egypt is in an arid region with scarce natural resources and heightened vulnerability to climate change. Efforts in Egypt emphasize developing smart cities, featuring many smart building projects that promote green construction and sustainability. This study also points out AI's limitations, such as AI-induced energy consumption, data limitations, and ethical considerations, which must be addressed to unlock AI's full potential in creating sustainable and climate-resilient cities.

Keywords Artificial intelligence · Decision-making · Smart city · Urban planning · Climate change · Smart sustainability

---

Y. Pan (✉)
Marketing and Decision Science, Salem State University, Salem, MA, USA
e-mail: ypan@salemstate.edu

Z. ElZein and A. Negm (eds.), *Recent Approaches of Sustainable Architecture in Arid and Semi-arid Cities*, Urban Sustainability,
https://doi.org/10.1007/978-981-95-1645-2_2

1 Introduction

Tackling climate change and promoting sustainability in urban planning are critical because cities are major sources of greenhouse gas emissions. For example, in large cities, with thousands of vehicles on the road daily, greenhouse gas emissions have intensified, and pollution has worsened, which calls for urgent sustainable solutions to combat climate change. Recent climate disasters across the world, such as the wildfires in California [1] and the heat waves in Mexico [2], serve as evidence of the ongoing climate crisis. The challenges faced by Los Angeles underline the urgent need for sustainable urban planning to build resilience to natural disasters and ensure safer and more sustainable cities for future generations.

Given the urgent need to combat climate change, AI-powered solutions have been proposed and leveraged to transform practices across various industries. AI is being leveraged in various fields to improve decision-making, optimize processes, and foster sustainable outcomes. Recent studies indicate that AI can offer help in various areas, including transportation [3, 4], agriculture [5] land use [6], business [7], and healthcare [8]. Applications of AI in urban context are especially diverse, spanning areas such as building environments [9], urban planning [10–12], public transportations systems [3, 4], urban green infrastructure [13], water resource management [14–16], smart energy systems [17, 18], disaster resilience [19], and waste management [20]. The above studies demonstrate that cities can significantly reduce their environmental impact by leveraging AI-powered solutions. For instance, the building sector accounts for 30% of worldwide energy use and contributes 27% of the energy sector's total CO_2 emissions [21]. Therefore, fine-tuning the design of buildings is essential since the design of a building's envelope significantly affects its energy demand [9]. Additionally, energy-efficient buildings that incorporate renewable sources like solar and wind help reduce energy use and CO_2 emissions.

Despite extensive research on the role of AI in urban environments [22], the connection between urban AI, sustainable development, and climate change remains under investigation. A few recent studies started exploring the link between AI and smart sustainable development and climate change. For instance, Al-Raeei [23] explores how artificial intelligence (AI) can advance sustainable urbanization, emphasizing its transformative potential in urban planning, resource and traffic management, disaster management, air quality monitoring, and more. AI provides data-driven insights that help policymakers and urban planners design more efficient, resilient, and sustainable cities. The author discusses the potential benefits and obstacles of integrating AI into urban settings. Son et al. [11] argue that it is crucial to combine human intelligence and artificial intelligence to deal with issues of urbanization in such a way as to achieve smart sustainability. Zheng et al. [12] examined how AI supports urban design and contributes to reducing pollution in various Chinese cities. They further examined how AI-powered urban design tools contribute to building more sustainable, efficient, and functional cities. Zhang et al. [24] introduced the "smart urban windcatcher," an AI-driven solution aimed at improving urban wind environments in real time. This system provides a scalable solution to urban heat and air stagnation challenges, and its implementation could significantly contribute to mitigating climate-related impacts in dense urban areas.

Despite AI's progress in urban settings, there are areas which need to be improved. For example, Son et al. [11] pointed out that the utilization of AI in urban planning is still understudied due to the divide between theory and real-world practice. Additionally, the fast-paced advancement and integration of AI and IoT have profoundly influenced areas such as smart sustainability, climate change, and urban development. However, few studies have systematically examined the intersections of these technologies with smart cities [25]. Recently, urban areas have become key platforms for AI experimentation and deployment, with a focus on automating the management of people, energy, mobility, and surveillance [26, 27]. There is insufficient research, however, on how AI contributes to smart sustainability and climate change mitigation in this context. Furthermore, AI's sustainability is a concern due to its significant energy use and CO_2 emissions, pointing to the necessity of eco-friendly AI solutions [28]. While AI has the potential to achieve smart sustainability and mitigate climate change, clear governance frameworks are needed to maintain transparency, safety, and ethical standards in decision making processes [29]. Thus, this study intends to fill the research gaps.

This study explored how AI-driven technologies can enhance energy efficiency, optimize resource utilization, and reduce greenhouse gas emissions (GHG) in cities. In particular, the study employs case study method to compare China, Malaysia, and Egypt, each facing environmental challenges and working to incorporate AI into their national strategies to combat climate change.

2 Artificial Intelligence in Urban Environments

2.1 Artificial Intelligence

In the 1950s, research on machine intelligence laid the foundation for artificial intelligence (AI). AI gained greater attention in the 2000s and 2010s, driven by big data, enhanced computing power, and advancements in deep learning. The launch of ChatGPT in November 2022 made it available to the public and sped up the adoption of AI in various industries. AI systems use neural networks and machine learning algorithms to perform tasks such as learning, problem-solving, and natural language understanding that typically require human intelligence. While most widely used AI systems function under human oversight, where humans maintain control over the analytical processes and outcomes, more AI systems are evolving from simple human proxies to fully autonomous operators [30]. Such autonomous systems can surpass human intelligence in various industries, providing a major competitive advantage [31]. For example, AI and machine learning (ML) enable cars to maneuver through roads and make autonomous driving decisions. Self-driving vehicles use computer vision to identify and interpret objects, while ML algorithms help them learn and adjust to new situations and road changes [32]. AI and ML algorithms also improve farming operations by forecasting weather, managing water use, and tracking crop health [33]. Similarly, the advancement of AI holds significant potential for combating climate change in urban settings. The next sections explore AI's role in lowering CO_2 emissions and combating climate change in urban contexts.

2.2 The Role of AI in Urban Planning

Tanic [34] conducted a seminal study examining the use of artificial intelligence to support and guide decision-making in urban planning. However, the use of AI in urban design and planning remained limited until recently. AI's practical applications have now become more evident, driven by the growing availability of advanced information and communication technology (ICT) systems and large-scale datasets [35]. AI can revolutionize cities and urban development processes in remarkable and transformative ways [36, 37].

Quan et al. [38] proposed a smart design framework, leveraging AI to help designers by generating design options, simulating performance through data analytics, and enabling more efficient and data-driven outcomes. AI also facilitates collaboration and communication between designers and other stakeholders, which fosters better mutual understanding and reduces design errors and costs. An increasingly diverse range of models and techniques from different AI subfields is being used to transform urban planning [10–12] and urban design [37, 39] in the context of sustainable smart cities [40]. Moreover, sustainable smart cities are increasingly leveraging various AI techniques including Machine Learning (ML), Computer Vision (CV), and Large Language Models (LLM) to solve complex challenges in urban planning and design through the Urban Digital Twin (UDT) framework. By harnessing these technologies, urban planners can manage vast datasets, identify significant trends, and uncover insights that drive actions and strengthen decision-making across diverse urban sectors. However, significant gaps persist in integrating these technologies into UDT frameworks, largely due to limited understanding of their role in analyzing urban flows and improving decisions that advance environmental goals in sustainable cities [41]. Recent research acknowledged AI as a critical technology for realizing smart city objectives [11, 25]. As cities are both major contributors to carbon emissions and particularly vulnerable to climate-related risks, urban planning plays a critical role in integrating AI-driven solutions to effectively address the challenges of climate change.

2.3 AI for Sustainability

The world is increasingly witnessing the impacts of climate change, manifested through extreme weather events such as intense heat and cold waves, heavy rainfall and flooding, more frequent and powerful hurricanes, tsunamis, wildfires, and prolonged droughts [1, 19], many countries are leveraging AI technologies to tackle environmental and urban challenges, recognizing AI as a powerful tool for sustainable development. For instance, AI-powered climate models predict extreme weather patterns and enhance disaster preparedness [19]. Smart energy grids utilize AI to optimize power distribution, minimize energy waste, and integrate renewable sources [17, 18, 42]. Artificial intelligence enhances the development of smart cities by optimizing traffic flow, water management, waste management, energy efficiency, and other urban systems, thereby contributing to more sustainable and resilient urban environments [19, 23, 37, 43, 44].

The role of AI is increasingly recognized as central to achieving a sustainable and climate-resilient future [45, 46]. AI supports the transition toward low-carbon energy systems through the integration of renewable energy sources and the enhancement of energy efficiency, both of which are critical in tackling climate change [47]. AI is also transforming the transport sector to support sustainable development by enhancing energy efficiency and reducing emissions through its applications in autonomous vehicles, public transit, and traffic management. Abduljabbar et al. [3] showed that various AI applications, ranging from optimizing traffic flow to enhancing intelligent urban mobility, help address challenges such as increasing travel demand, CO_2 emissions, safety concerns, and environmental degradation. The deployment of intelligent systems can lower carbon dioxide emissions by around 60%, making a significant contribution to creating a cleaner urban environment [48]. Corporate operations also benefit from the integration of artificial intelligence, as AI-driven decision-making and forecasting tools have substantially reduced carbon costs, thereby demonstrating the effectiveness of AI in advancing corporate sustainability strategies [7]. In the manufacturing sector, AI-enabled smart manufacturing processes can reduce energy consumption, waste, and carbon emissions by 30–50%, with comparable reductions attainable in building energy usage [48]. At both governmental and industrial levels, AI tools play a vital role in optimizing energy systems, managing waste, and ensuring the efficient use of resources. Vinuesa et al. [49] explored the role of AI in sustainable development and highlighted the fact that AI advancements enhance the understanding of climate change and improve modeling of its potential impacts. By utilizing AI-driven analytics to monitor and regulate public utilities and services, governments can promote sustainable urban operations, effectively reducing the carbon footprint of cities [46]. By leveraging AI, governments, businesses, and communities can transition toward a low-carbon, resource-efficient, and inclusive economy, ensuring technological progress benefits both people and the planet.

2.4 AI in Energy Management and Carbon Emissions

AI plays a transformative role in enhancing energy efficiency, facilitating the integration of renewable energy sources, and supporting climate change mitigation. By leveraging machine learning and predictive analytics, AI optimizes energy utilization in buildings, industries, and transport systems, reducing waste and lowering carbon emissions [18, 47, 50]. The building and transportation sectors consume a substantial amount of energy. AI-driven smart buildings and transport systems help cut energy use, thereby lowering carbon emissions. Furthermore, by optimizing the use of clean energy when available, AI contributes to reducing dependence on fossil fuels, resulting in a substantial decrease in carbon emissions [50].

AI is integral to smart grid technology in modern energy systems. Smart grids are essential for advancing sustainable energy systems. Wang et al. [18] provide a detailed study on the current applications and future directions of large-scale AI models in Virtual Power Plants (VPPs). Their study underscores the multifaceted contributions of these models in improving VPP operational efficiency and intelligence through high-precision load and generation forecasting. These AI models contribute to optimizing energy dispatch, minimizing energy waste, and reducing carbon emissions, thereby establishing a strong foundation for sustainable VPPs. AI also contributes to the discovery of new materials for clean energy technologies [51]. AI-driven algorithms strengthen the integration of renewable energy into current power grids [52], paving the way for a reliable and efficient low-carbon energy transition [18, 53]. Thus, AI-supported energy and its efficiency is one of the key areas that policymakers should consider in efforts to reduce CO_2 emissions and address the climate crisis. However, challenges such as cybersecurity risks and investment costs need to be addressed in the widespread adoption of smart grids.

2.5 Predictive Analytics for Climate Modeling and AI-Driven Carbon Footprint Reduction Strategies

Climate change presents complex challenges that require innovative and adaptive solutions. AI technologies, such as machine learning, neural networks, and deep learning, enable the analysis of large climate datasets to identify patterns and improve climate modeling and projections [19]. AI-driven climate modeling improves accuracy, speeds up climate data analysis, and offers critical insights into climate impacts, guiding policy development and mitigation efforts [42]. AI can also assist in identifying high-risk areas vulnerable to climate-related hazards, formulating adaptation plans for businesses and communities, predicting wildfires and floods, and detecting regions prone to landslides [54]. Furthermore, AI-driven urban digital twins can analyze climate patterns and optimize urban infrastructure for lower emissions [55].

AI also allows businesses to track their carbon footprint by analyzing data from various sources, including transportation, energy usage, and waste generation. With

AI-driven insights, businesses can efficiently identify areas where emissions can be reduced and set targets for minimizing their carbon footprint [55]. Leveraging predictions from the AI-enabled models, governments can take proactive measures to mitigate the impact of climate change, safeguarding lives and minimizing environmental damage. However, addressing ethical concerns and potential AI biases is vital in the development and application of these strategies.

2.6 Smart Cities and Climate Change

Urbanization has progressed rapidly. Emerging cities in Asia and Africa clearly illustrate this growth. These cities consume considerable amounts of resources, including energy and water. As major transportation hubs, cities emit significant amounts of carbon through extensive road networks, public transport systems and airports. Cities consume 78% of the world's energy and generate more than 60% of GHG emissions, and energy lies at the heart of the climate problem and holds the key to its solution [56]. A study published in *Frontiers in Sustainable Cities* by Wei et al. [57] revealed that only 25 mega-cities account for 52% of urban GHG emissions, highlighting the significant impact of urban areas on global emissions. Therefore, cities must promote sustainability by addressing environmental concerns, including carbon reduction, climate resilience, and ecosystem protection, where possible, early in urban growth and development [58].

Cities are central to efforts in mitigating and adapting to climate change [59]. "Cities will be key to survival on the planet." [60, p. 710]. Initially, Sustainable cities were designed to reduce emissions, improve air quality, and conserve resources. Recently, smart cities have emerged as a promising approach to tackling issues of sustainability, resource efficiency, and rapid urbanization [25]. Smart cities are encouraged to prioritize objectives that yield tangible, actionable outcomes in addressing climate change, while remaining aligned with their governmental responsibilities [61]. The Sustainable Development Goals (SDGs) established by the United Nations provide a structured framework to guide cities in addressing challenges related to climate change adaptation, pollution, ecosystem protection, and energy utilization [62, 63].

Smart cities are fundamentally linked to sustainability [64]. Smart technologies in urban environments can enhance building efficiency and reduce energy utilization [9, 48], improve waste management systems [65], and promote smart mobility and smart environments [66]. Moreover, building design and construction are key to advancing the sustainability of smart cities [59]. Selim et al. [67] emphasize the application of sustainable architecture in the development of Egypt's New Administrative Capital. It is evident that sustainability forms a foundational element of smart cities [68]. By utilizing energy-efficient materials and sustainable practices, buildings can lower their carbon footprint and enhance the city's overall sustainability.

As AI serves as a driving force in advancing smart cities, it rapidly transforms the technological and urban landscapes [69]. Yigitcanlar and Cugurullo [70] examined how AI is applied in urban settings, highlighting both its potential for sustainability and the challenges it presents for cities. Cities can leverage digital technologies to enhance infrastructure efficiency, minimize resource consumption, improve environmental quality, and reduce carbon emissions [71]. Recent studies have advanced the field. For example, Bibri [40] explores the transformative impact of AI on sustainable smart city planning and design through urban digital twins (UDTs). The study emphasizes GenAI's pivotal contribution to enhancing data-driven urban planning processes via UDTs. The findings demonstrate GenAI's considerable potential in advancing key smart city domains, including energy, transportation, mobility, and infrastructure management.

Cities can leverage AI-driven smart systems to dramatically respond to the needs of their communities, driving decisions and processes that are sustainable and adaptable to change. Based on 213 research articles from 2014 to 2022, Herath and Mittal [72] found that transportation and security-related AI applications dominate the smart city research landscape. In transportation, AI's applications in autonomous vehicles, public transit, and traffic management boost energy efficiency and reduce emissions. Moreover, AI-driven security applications help reduce cities' carbon footprints while ensuring safety and operational efficiency. By integrating AI-driven security systems with smart grids and IoT technologies, cities can create energy-efficient networks that optimize electricity usage across infrastructure, reducing the carbon footprint. Eldrandaly et al. [73] highlight how AI-optimized surveillance systems can minimize the number of cameras required for monitoring, thereby reducing overall costs and leading to lower electricity consumption. Since surveillance systems often run 24/7, reducing the number of active devices translates to a decrease in energy demand, which in turn lowers CO_2 emissions. AI-enhanced surveillance plays a crucial role in smart cities, supporting services related to safety, security, and environmental sustainability.

2.7 Opportunities and Challenges in the Transition to Sustainable Smart Cities

Transitioning to sustainable smart cities presents many opportunities across environmental, social, economic, and technological dimensions. In terms of environmental opportunities, the transition to sustainable smart cities emphasizes the enhancement of resource usage through advanced technologies such as AI and IoT. Smart energy systems, including smart grids and automated energy management platforms, enable cities to reduce electricity waste and improve overall energy efficiency. This is particularly important for countries like China, where such systems can significantly reduce energy use and help lower CO_2 emissions. Additionally, smart water management systems can detect leaks, monitor consumption patterns, and ensure the equitable distribution of clean water. For countries in arid and semi-arid regions facing severe water scarcity, these smart systems can help alleviate water stress and support the achievement of the United Nations' SDG 6 goals. Furthermore, green infrastructure such as green roofs and urban forests helps mitigate the urban heat island effect and improve air quality. It is essential for megacities around the world to adopt such practices to address their growing environmental challenges. When integrated into urban planning, these technologies reduce the environmental footprint of cities and promote long-term sustainability.

In terms of global collaboration, smart city development greatly benefits from international collaboration among governments, academic institutions, private sector players, and NGOs. For example, *e-ReWater*—an AI-powered platform supported by a $1 million Google grant—was developed to enhance wastewater reuse in the Middle East and North Africa [74]. Such partnerships enable cities to exchange best practices and adopt proven technologies. Global networks such as UN-Habitat and the World Smart Sustainable Cities Organization can serve as platforms for this collaboration, fostering innovation and accelerating the adoption of sustainable urban solutions.

China, Malaysia, and Egypt have each approached this transformation within their unique socioeconomic contexts and rapid urbanization. Building the required technological infrastructure, such as IoT devices, data centers, and advanced communication networks, involves significant upfront costs [75]. The success of leveraging AI in mitigating climate change also depends on effective governance, active community engagement, and interdisciplinary collaboration regarding technology and policy. For example, fragmented regulations and the lack of uniform smart city standards can slow progress. Therefore, coordination between local and national governments is required to streamline policies.

Increased AI and IoT integration raise concerns over data privacy and cybersecurity risks. AI empowers IoT devices to learn and adapt through intelligent analytics, enhancing their efficiency and functionality. In smart cities, AI-driven IoT systems analyze sensor data to optimize urban planning, traffic flow, and resource management [76]. Integrating AI with IoT for intelligent data analysis allows organizations to harness interconnected devices and advanced computing to enhance performance and optimize strategic decision-making [77]. However, as AI algorithms process sensitive data from interconnected devices, the risk of privacy breaches grows. Additionally, cyberattacks on critical infrastructure can disrupt smart city operations. Adopting privacy-preserving measures, including encryption, anonymization, and decentralized data processing, is therefore essential in addressing data privacy and security concerns [78].

The digital divide is also a critical challenge that must be addressed to achieve the goals of smart and sustainable cities, as urban areas increasingly depend on digital analytics, big data, and smart solutions to tackle key issues shaping the future of cities [79]. Anastasiu [80] highlights the fact that lack of digital literacy limits citizens' ability to engage in technology-driven urban experiences and exercise their digital rights within a smart city environment. Effective smart city initiatives depend on robust stakeholder and citizen engagement [81, 82]. The digital divide is a socioeconomic barrier that disproportionately affects vulnerable populations, limiting their access to AI-driven urban services and smart city technologies [75]. Therefore, bridging the digital divide is a key step toward the successful implementation of smart city initiatives.

Lastly, AI's rapid expansion in urban contexts presents potential environmental risks. The substantial energy consumption of advanced AI systems can paradoxically exacerbate the very issue they aim to address—an increase in carbon emissions. AI's environmental impact is closely linked to the cloud computing infrastructure, including data centers and networks, that powers it. Data centers supply the power, cooling, and infrastructure necessary for both AI training and inference. In 2018, data centers in China produced approximately 99 million metric tons of CO_2 emissions [83]. The escalating carbon emissions and energy consumption associated with data centers have become a significant challenge in the fight against climate change [84]. Thus, strict oversight is essential to mitigate unintended consequences. Regulatory authorities play a key role in developing frameworks that assess the environmental impact of AI technologies and ensure their alignment with the reduction of CO_2 emission and sustainable development. It is advisable for cities to set up specialized departments to oversee AI and sustainability initiatives. Similarly, cities should invest in developing or recruiting AI and sustainability experts to ensure the ethical and efficient application of AI in urban operations. Such a proactive approach will maximize the advantages of AI for climate action while minimizing climate risks, fostering responsible and forward-looking governance.

3 Methodology

The current study includes a comparative case study of AI-driven initiatives, such as smart city projects, and examines how these efforts contribute to reducing greenhouse gas (GHG) emissions in three countries. The case study primarily focuses on "why" and "how" questions to analyze complex links between events or behaviors. Applying a case study approach can enhance our understanding of smart city dynamics [85], including their goals, objectives, key stakeholders, and strategies. A comparative case study is used here to formulate or evaluate generalizations that apply across multiple cases. The case study analysis involves systematically comparing a small to medium number of cases using qualitative data and leverages theoretical insights from individual cases to derive broader generalizations about phenomena sharing certain characteristics [86].

Many countries are pursuing smart city initiatives to combat climate change, given the vital role of sustainable urban development in cutting carbon emissions [87]. Understanding the impact of urban differences on CO_2 emission reduction necessitates a detailed analysis of regional heterogeneity. In China, there are notable disparities across regions in terms of AI development priorities and carbon emission levels. Thus, conducting scientific assessments of the diverse impacts of AI technologies on carbon emissions across different regions is essential, particularly in the development of smart cities [88]. Given the above, this study utilizes a comparative case study approach, aiming to provide a multifaceted perspective on the role of AI in promoting smart sustainability and addressing climate change in China, Malaysia, and Egypt.

China serves as an ideal case for studying the impact of AI technologies on reducing air pollution and carbon emissions. As a developing country, it has experienced significant economic growth but has consequently faced severe air pollution. This has motivated China to make concerted efforts to drive technological innovation and promote sustainable development.

Malaysia was selected for this study due to its commitment to sustainable urban development through innovative, technology-driven solutions. The Malaysian government has introduced initiatives such as the Malaysia Smart City Framework (MSCF) and the National Artificial Intelligence Roadmap (2021–2025) to support this effort. Malaysia's development challenges, such as urbanization, energy use, and environmental sustainability, make it a strong candidate for studying AI's role in low-carbon transitions.

In contrast to China's role as a global manufacturing center and Malaysia's position as a leading producer of gas and oil, Egypt faces a distinct set of environmental challenges shaped by its unique economic and geographic conditions. Egypt's already arid climate makes it particularly vulnerable to climate risks. Moreover, there is insufficient research on how AI can be applied to address climate change in arid and semi-arid regions. Therefore, Egypt was selected in the current study to gain clearer insights into how AI can support climate adaptation in arid regions.

This study examines the cases of China, Malaysia, and Egypt, each differing in economic scale, resource availability, cultural backgrounds, and religious contexts, to provide a comparative perspective on how AI technologies can help address climate change and support low-carbon urban transformation across different economic and environmental contexts. Additionally, these cases offer valuable context-dependent knowledge and insights.

4 Case Studies and Best Practices

4.1 China

China has encountered severe environmental challenges, including air pollution and water scarcity, stemming from elevated CO_2 emissions associated with rapid economic development and urbanization. In 2023, China remained the world's largest carbon emitter, releasing 11.9 billion metric tons of carbon dioxide [89]. Figure 1 illustrates a steady increase in CO_2 emissions since 2010, following a sharp rise from 2000 to 2010. China highly depended on coal for power generation in the past, thus leading to high CO_2 emissions. To tackle this issue, China has committed to reaching carbon neutrality by 2060 [90], a goal that requires a major shift away from fossil fuels, especially in the energy and transportation sectors.

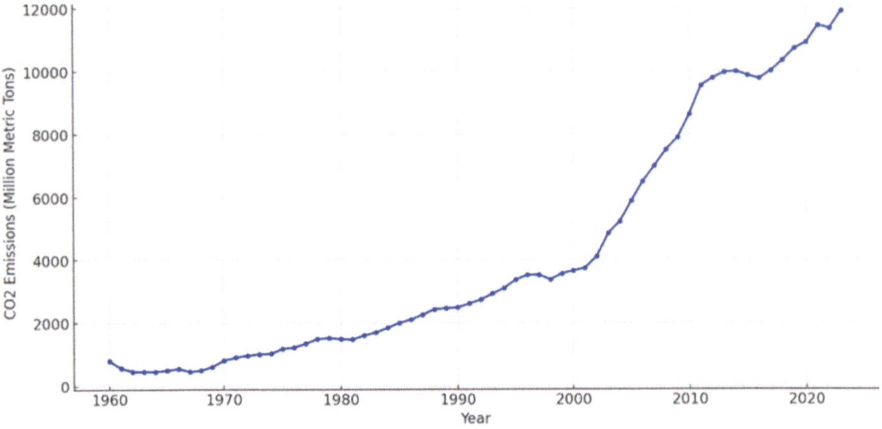

Fig. 1 CO_2 emissions from energy consumption in China from 1960–2023 [89]

In light of the above, China has embraced the smart city concept, utilizing advanced technologies to promote sustainability and enhance urban living standards [75]. The Chinese government is actively adopting AI technologies to reduce carbon emissions. At the national level, China has tried to integrate AI into its efforts in mitigating climate change. For example, the 2017 China AI development plan focus on AI applications to drive energy efficiency and tackle environmental challenges more effectively. In 2020, China established the National Carbon Neutrality goal to achieve carbon neutrality by 2060 [90]. To achieve its goals, China has made substantial investments in in AI-driven smart city initiatives designed to promote urban sustainability, reduce energy consumption and curb CO_2 emissions.

In response to the central government's directives, a variety of sustainable initiatives have been launched. Leveraging AI to monitor and predict air pollution levels enables cities and regions to reduce emissions and improve air quality through targeted interventions, as demonstrated by the Hangzhou-based City Brain AI initiative, which optimizes urban systems by improving traffic flow, cutting energy consumption, and managing waste more efficiently. These initiatives have led to less congestion, decreased emissions, and better air quality [91]. Recent studies also indicate that the mechanisms through which AI technologies impact air quality and CO_2 emissions vary significantly [92]. Additionally, Dong et al. [88] highlight that different regions of China experience rising carbon emissions for different reasons. Some may be due to high energy consumption from AI development, while others may require the replacement of traditional production methods with smarter technologies. Thus, understanding the root cause of rising CO_2 emissions is critical to finding the right approach to combat climate change.

4.1.1 AI-Powered Smart Cities for Climate Change Adaptation

Implementing smart cities is one possible way to further decrease CO_2 emissions and mitigate climate change, as smart cities are designed to focus on tangible actions to lower carbon emissions [61]. Wolniak and Stecula [93] provide an in-depth literature review on smart cities and evaluate how AI supports development in six key areas: smart governance, smart environment, smart economy, smart mobility, smart living, and smart people. Their study identifies objectives such as data use, citizen participation, automation, sustainability, security, new technology adoption, predictive insights, inclusivity, and economic benefits.

China's national strategy has included smart city development since 2012, leading to the creation of 290 pilot projects [94]. Major cities such as Shanghai and Hangzhou are spearheading the adoption of advanced technologies and innovative urban management strategies. Table 1 presents the major smart city development areas for Shanghai and Hangzhou. Although the focus of the smart city project aligns with the work of Wolniak and Stecula [93], Chinese smart cities have different priorities and face distinct challenges. Thus, implementing smart cities should consider the local urban context and the challenges they face [94]. Table 2 presents several key applications of smart city technologies, including IoT, cloud computing, big data, mobile internet, and AI. AI is vital to smart city development, as it works alongside other IT technologies to achieve smart sustainability and revolutionizes the way we address climate change. According to Guo et al. [95], smart city construction led to a substantial decrease in per capita CO_2 emissions, reducing them by around 18.42 log points in China. Similarly, Chu et al. [96] explored the relationship between smart city construction and ecological environmental quality, showing that China's smart city efforts between 2005 and 2017 reduced industrial waste gas and wastewater by 20.7% and 12.2%, respectively.

4.1.2 AI-Driven Smart Grid and Energy Management

China also prioritizes the energy sector, as energy consumption is a major contributor to CO_2 emissions. Cities have begun incorporating smart grids, green buildings, and smart lighting systems to reduce energy use and decrease carbon emissions. In China, the smart grid project began in Yangzhou City, Jiangsu Province. In 2010, Under the "Torch Program," the Ministry of Science and Technology named Yangzhou as the country's first "Smart Grid Industrial Base" [98]. Ahvenniemi et al. [59] highlight that emerging smart technologies are key to lowering CO_2 emissions and improving urban energy efficiency. Recent studies show that AI has made significant contributions to the energy sector. By optimizing the integration of solar, wind, and other renewables into the national grid, AI algorithms help reduce energy waste and support grid stability [99]. AI reshapes smart city energy systems, driving societal and economic progress [100]. For example, Wang et al. [18] examine how large-scale AI models support Virtual Power Plants (VPPs) by improving forecasting accuracy, optimizing

Table 1 Smart City Development Areas Identified in Smart City Plans

Smart City Plan	Major Smart City Development Areas
Shanghai Three-Year Action Plan for Smart City Development (2014–2016)	Public Services: Smart transportation, smart healthcare, smart education, smart elder care, smart culture, smart tourism, smart employment, smart weather, smart communities City Management: Management informatization, food safety informatization, environmental protection informatization, public security informatization, smart water, electricity, oil, and gas infrastructure Smart Economy: Online finance, smart logistics, smart commerce, smart manufacturing, smart enterprises, smart shopping districts, smart business parks
"Digital Hangzhou"—"New-Style Smart Hangzhou" Smart City Development Plan (2017)	Public Services: Smart education, smart healthcare, smart social security, smart community services, smart poverty alleviation, smart culture, smart tourism, smart agriculture, smart weather Precision Government: Smart transportation, smart policing, smart urban management, smart market supervision, smart surveillance, smart inspections, smart Party work, smart auditing, smart environmental protection, smart credit Economic Development: AI industry, big data industry, IoT industry, smart manufacturing, smart cross-border e-commerce

Adapted from U.S.-China Economic and Security Review Commission [97]

Table 2 Example Applications of Key Smart City Technologies

Technology	Example Smart City Applications
Internet of Things (IoT)	Networked video cameras, sensors, and actuators supporting street and traffic light automation; smart waste management; water, gas, and electrical infrastructure monitoring; environmental quality monitoring; smart electrical metering; smart manufacturing upgrades
Big Data	Social credit information platforms; traffic signal optimization; spatio-temporal GIS information platforms; police and fire incident analysis
Cloud Computing	Video surveillance data storage; cloud-based storage and processing of IoT sensor data; city- and district-level government data storage and computing centers; public cloud centers for business parks
Mobile Internet	E-government services; smart tourism services; community legal services; weather information; cashless payments
Artificial Intelligence (AI)	Smart surveillance cameras; streetlights; parking and public transportation systems; maintenance scheduling for smart infrastructure and smart manufacturing

Adapted from U.S.-China Economic and Security Review Commission [97]

energy dispatch, reducing waste, and lowering carbon emissions—thereby advancing sustainability goals.

4.1.3 AI-Driven Waste Management

AI is rapidly evolving and transforming industries such as waste management. Techniques like machine learning, neural networks, and optimization models are being used to enhance solid waste management processes [20]. Integrating AI and robotics into the design and operation of urban waste treatment plants is transforming solid waste management, enhancing operational efficiency and promoting sustainable practices [70, 101]. In May 2019, China's Ministry of Science and Technology designated Shanghai as a National Pilot Zone for the Innovative Development of New-Generation AI, known as the Shanghai AI Development Zone [65]. The Shanghai AI initiative focuses on urban transformation, targeting the city's economy, public services, and new management models. A clear example is the integration of AI in waste management, where autonomous vehicles, image recognition systems, and robotics manage waste collection, transport, recycling, and disposal with minimal human intervention [65]. This technology helps reduce landfill waste, improve recycling efficiency, and lower carbon emissions compared to traditional waste disposal methods. Furthermore, green space development is a key focus in China's urban planning to improve air quality and reduce urban heat islands. For instance, Shanghai has increased its green space by 50%, cutting cooling energy use [102]. By integrating AI into urban planning and infrastructure, Shanghai showcases how innovative solutions can drive sustainable development and support climate action efforts. Similarly, in Malaysia, waste-to-energy projects transform waste into electricity, decreasing landfill use and producing renewable energy. Under the Malaysia Smart City Framework, integrating digital technologies like big data analytics into urban planning can cut traditional energy use by as much as 15% [103].

4.1.4 AI-Driven Water Management

Water plays a critical role in achieving sustainable development and maintaining the balance of ecosystems and human life. [104]. However, a severe global water crisis continues, marked by water scarcity, water pollution, and inefficient management, calling for urgent and transformative change [105]. AI holds the potential to drive a shift toward a new water management paradigm. For instance, in the work by Sabri et al. [106], urban digital twins have been deployed to manage essential water infrastructure and lessen flood impacts in smart cities like Orange County, California, and Victoria, Australia. The research underscored the value of combining accurate location-based data with GIScience methods like GeoIoT and GeoAI, into smart infrastructure systems. These digital twins improve urban resilience in preparing for and reducing the impacts of climate-induced disasters by providing real-time monitoring, predictive modeling, and improved disaster response strategies. In China, climate change-induced shifts in precipitation patterns pose significant challenges for urban water management. Increased uncertainty in rainfall leads to more frequent sewer overflows, which discharge untreated wastewater, exacerbating water pollution, scarcity risks, and waterborne diseases. Digital twins offer a solution to reducing the vulnerability of China's urban water sector to climate change and enhancing resilience against its impacts.

4.1.5 Smart Transportation Systems

The transportation industry is a major driver of rising greenhouse gas emissions in China, with road transport being the largest source [107]. Reducing the number of private vehicles and promoting public transportation can help achieve carbon neutrality [108]. To achieve carbon neutrality, China implemented a major policy in 2009 to encourage the adoption of electric vehicles (EVs) in the public sector, resulting in an average reduction of 16.3% in urban transportation emissions. Moreover, this policy contributed to an average annual reduction of 28.87 kg per capita of urban transport emissions [81, 82]. Brzeziński [109] also argued that the adoption of electric and autonomous vehicles results in both lower CO_2 emissions and improved road safety. Similarly, vehicle-sharing and on-demand transport services reduce vehicle volume on roads, contributing to greenhouse gas emission reductions. China's success in developing electric and autonomous vehicles positions the country as a leader in the transition to sustainable transport, contributing meaningfully to global climate change mitigation.

4.2 Malaysia

Malaysia is a major oil and gas producer and a significant contributor to carbon emissions in Southeast Asia. Malaysia's CO_2 emissions have increased steadily, as shown in Fig. 2. In 2023, emissions from energy consumption amounted to 284.7 million metric tons, marking an increase compared to 2022. The CO_2 emissions from fossil fuel energy consumption reached their highest level in the same year [110]. Meanwhile, both urban and rural areas in Malaysia have experienced or are facing climate change, which has negatively impacted cities, as shown in Table 3.

4.2.1 Smart City Framework and Polices in Malaysia

Malaysia is integrating cutting-edge technologies to support urban decarbonization efforts. In 2021, the Malaysian government launched the Malaysia National Artificial Intelligence Roadmap 2021–2025, providing guidance for the development and implementation of AI technologies. This roadmap serves as a strategic framework for Malaysia's goal of becoming a global AI leader while ensuring its benefits are shared

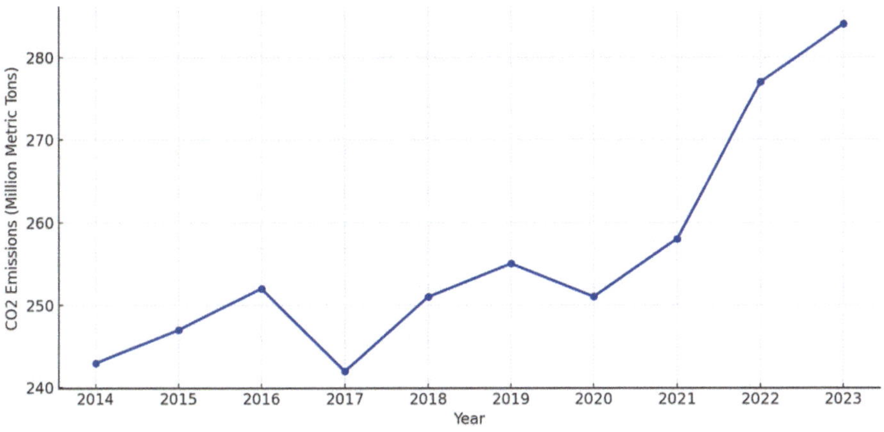

Fig. 2 CO_2 emissions from energy consumption in Malaysia from 2014–2023 [89]

Table 3 Summary of Climate Change in Malaysia

Category	Details
Temperature Increase	Between 1970 and 2013, Peninsular Malaysia, Sabah, and Sarawak experienced a surface mean temperature increase of 0.14 °C–0.25 °C per decade. Under a high greenhouse gas emissions scenario, average temperatures could rise by 3.11 °C by the 2090s
Flooding Vulnerability	Flooding is Malaysia's most damaging natural hazard. The frequency and severity of floods have increased and are projected to rise further with continued global warming
Sea Level Rise	Coastal areas may experience a 0.4 m–0.7 m sea level rise by 2100, with greater impacts in Sabah and Sarawak. This threatens agricultural production in coastal regions
Agriculture	Droughts and floods early in the rice-growing season could reduce yields by up to 60%. Drought conditions may also affect rubber, palm oil, and cocoa cultivation
Poverty & Inequality	Climate change is expected to exacerbate poverty and inequality, especially for low-income earners reliant on agriculture, fishing, and informal urban economies

Adapted from Ipsos [111]

across all sectors of society. Complementing this effort, the Malaysia Smart City Framework (MSCF) uses ICT to improve urban management and residents' quality of life. In line with Industry 5.0, it promotes human-machine collaboration focused on people's needs and experiences [75]. As cities become smarter, they unlock transformative opportunities for sustainability from streamlined urban management and reduced carbon emissions to more responsive public services. [112]. A smart city framework includes six key dimensions: Smart Economy, Smart People, Smart

Governance, Smart Mobility, Smart Environment, and Smart Living. These dimensions have been referenced in several studies [68, 109, 113]. They extend beyond theoretical concepts by illustrating how cities around the world are implementing practical strategies to address climate change through smart sustainability.

Cities are key drivers in combating and reversing the effects of climate change [59]. In Malaysia, Smart cities provide urban planners with effective tools to address challenges like energy use, pollution, waste, and crime. Table 4 summarizes smart city policies at different levels and the focus areas of smart city development in Malaysia. Lim et al. [113] examine how well practitioners in Malaysia understand and accept smart city policies, focusing on the domains of the Malaysian Smart City Framework (MSCF). Their findings indicate that although most domains were accepted, practitioners showed the most hesitation regarding the Smart Environment and Smart Digital Infrastructure. Additionally, Samsudin et al. [114] compared two smart city projects in Malaysian cities, Putrajaya and Iskandar Puteri, both of which have developed smart city blueprints—Putrajaya with 93 strategies (2018) and Iskandar Malaysia with 28 (2012). The authors point out that Putrajaya emphasizes micro-level implementation, while Iskandar Puteri takes a broader, regional (macro-level) approach. However, there are no clear methods to measure the success of their Smart City initiatives. A recent study by Bakhtiar and Samsudin [115] utilized the ESG model to evaluate the sustainability performance of smart cities in Malaysia, including Petaling Jaya, and emphasized the importance of local community awareness and engagement. That is, smart city development should prioritize citizen needs, inclusivity, and urban challenges.

4.2.2 Smart Grid and Renewable Energy

To address climate change and safeguard the environment for future generations, Malaysia must shift toward a low-carbon economy [116]. A sustainable energy policy is essential to meet growing energy needs, protect the environment, and boost economic growth. The country is well-positioned to develop a sustainable economic framework through renewable energy, which can foster economic expansion, mitigate climate change, and support long-term sustainability goals [117]. In Malaysia, smart grid and renewable energy integration showcases technological progress enabled by government support and incentives [75]. Smart energy systems leverage IoT and AI to minimize carbon footprints, with smart grid technology improving energy efficiency by 10–15% [118]. The Malaysian government also advocates for green building practices, with Green Building Index (GBI)-certified structures using 20–30% less energy than conventional buildings [119]. By adopting smart grids, green buildings, and smart lighting systems, cities help enhance energy efficiency, optimize consumption, and reduce carbon emissions.

Table 4 Smart City Policy and Major Development Areas

Smart City Policy Level	Smart City Component
Malaysia Smart City Framework (National—Malaysia)	Smart Economy, Smart People, Smart Government, Smart Mobility, Smart Living, Smart Digital Infrastructure
Smart Selangor Blueprint (State—Selangor)	Smart Governance, Smart Disaster Management, Smart Building, Smart Safety and Security, Smart Food and Agro, Smart Energy, Smart Water Management, Smart Digital Infrastructure, Smart Transport and Mobility, Smart Waste Management, Smart Healthcare and Wellbeing, Smart Education
Penang 2030 (State—Penang)	Smart Community, Smart Environment, Smart Government, Smart Mobility, Smart Economy
Sarawak Digital Economy Strategy (State—Sarawak)	Digital Government, Digital Health, E-Commerce, Smart City, Cyber Security, Digital Innovation and Entrepreneurship, Social, R&D, Digital and Data, Digital Skills and Talent Development, Digital Inclusivity, Digital Infrastructure, Agriculture, Manufacturing IR 4.0, Tourism
Cyberjaya Smart and Low Carbon City (City—Cyberjaya)	Smart Mobility, Walkability, Compact Development, Integration of Nature into Urban Areas, Efficient and Effective Resource Management, Smart and Green Buildings, Smart Community
Smart City Iskandar Malaysia Framework (Regional—Iskandar Malaysia)	Smart Governance, Smart Living, Smart People, Smart Mobility, Smart Environment, Smart Economy

Adapted from Bakhtiar and Samsudin [115]

4.3 Egypt

Egypt, a developing nation in the North Africa region, has experienced serve environmental challenges due to its demographic growth, urbanization, and economic development. The region is characterized by its predominantly arid climate, leading to significant water scarcity and vulnerability to climate change impacts. Egypt is highly vulnerable to climate change and suffers from stresses such as rising temperatures, sea-level increases, frequent droughts, severe water scarcity, and worsening air pollution, as highlighted in a World Bank report [120]. Darwish [121] examined the escalating energy and environmental challenges in the Middle East and Gulf countries resulting from rapid population growth, economic development, and modernization, all of which drive increased energy consumption and carbon emissions. The authors recommend integrating smart technologies with green building practices to address these challenges, reduce energy consumption, and promote healthier living environments. They further emphasize that combining smart technologies with sustainable design principles enhances building efficiency and provides long-term environmental benefits.

4.3.1 Smart Building Projects and Smart Cities

Egypt's pursuit of smart city development has gained momentum through flagship projects such as the New Administrative Capital (NAC) and Smart Village, both of which aim to integrate advanced technologies to enhance urban living, sustainability, and economic competitiveness. The NAC is considered Egypt's first major effort to establish a smart city that incorporates many smart building projects [122]. The goals and concepts of smart building projects (SBPs) vary from country to country, influenced by differences in time, needs, and culture. Asia and Europe are leaders in the field of SBPs,Asia primarily focuses on environmental aspects [123], while Europe concentrates on strengthening its economy by achieving low-carbon goals through the adoption of smart cities and smart buildings [122]. AI can optimize energy consumption in buildings and homes, thereby significantly reducing energy use and carbon emissions [1]. Khashaba [124] proposed using intelligent building technologies to enhance sustainability in public buildings in Cairo, developing an architectural framework aligned with local user needs. Selim et al. [67] also highlight the concept of sustainable architecture and its implementation in Egypt's flagship New Administrative Capital, featuring eco-friendly materials, energy-efficient systems, water recycling, waste reduction, green spaces, and integrated public transport. The adoption of these features showcases how sustainable urban architecture can contribute to positive environmental and social outcomes in large-scale urban developments. However, smart building projects in Egypt face challenges such as unclear definitions, missing comprehensive frameworks, vague objectives, limited expertise, and insufficient training. High upfront costs and the absence of facility management teams in early stages further hinder development and long-term sustainability [122]. Addressing these challenges is essential for ensuring that smart building initiatives can contribute effectively to Egypt's broader Vision 2030 goals of sustainability, innovation, and economic growth.

Although Egypt's per capita greenhouse gas (GHG) emissions remain below the global average, its total GHG emissions increased by more than 160% between 1990 and 2019 [125]. Given Egypt's vast arid and desert landscapes, using artificial intelligence is vital for mitigating and adapting to climate change as the country faces rising challenges. AI has the potential to help Egypt reduce GHG emissions across various sectors. In recognition of its transformative impact, Egypt launched the National Artificial Intelligence Strategy in 2021, aiming to harness AI for sustainable development, improved government efficiency, education, healthcare, and smart city initiatives. AI transforms e-government by streamlining tasks and driving data-based decisions. In smart city projects such as the New Administrative Capital, AI is being used for traffic management, intelligent surveillance, and energy optimization to advance urban sustainability [126]. In contrast to China and Malaysia, Egypt has fewer smart cities, as most of its population is concentrated in a few mega cities in the Nile Delta and Valley [44]. In the desert regions, development of smart cities faces significant challenges due to harsh environmental conditions and infrastructure limitations.

4.3.2 AI-Powered Optimization of Resource Utilization

Water scarcity is a pressing challenge that must be addressed in Egypt, as rapid urbanization places increasing demands on vital resources including energy, water and food. Egypt is projected to face severe water stress by 2030, driven by limited water resources and a population heavily concentrated in the Nile Delta and Valley, where water scarcity has already surpassed critical levels. Consequently, cities need transformative solutions to pursue sustainable growth [44]. The United Nations Sustainable Development Goal 6 (SDG 6) recognizes access to clean water as a fundamental human right. Yet this goal remains out of reach for many across Africa. The United Nations reports that one in three people on the continent still lack access to safe, clean water. To address this challenge, AI has been increasingly explored and implemented to enhance smart water management practices. In November 2023, the International Water Management Institute announced the development of *e-ReWater*, an AI-powered platform to enhance wastewater reuse in the Middle East and North Africa (MENA), supported by a $1 million grant from Google. The platform will use AI and satellite data to monitor land, sea, and atmosphere, helping estimate new water resources and improve water reuse to address water scarcity and strengthen food and water security in the region [74].

Recent studies highlight the pivotal role of the water-energy-food (WEF) nexus in optimizing resource use. ElZein et al. [44] propose a sustainable development model for the resource-scarce Rabia community in Egypt, applying the WEF nexus to respond to water, energy, and food challenges. The model offers a scalable framework for building resilient, self-sufficient communities in arid coastal regions, supporting Egypt's Vision 2030 and key Sustainable Development Goals. The authors advocate for renewable energy sources, including solar and wind, to supply power for communities and desalination plants, stating that this is expected to reduce electricity consumption by approximately 53% and lower CO_2 emissions by 54.5% compared to thermal technologies. AI could be incorporated into the above model, as one of its key benefits in energy management is the ability to integrate intermittent renewable sources by predicting demand and supply fluctuations to maximize efficiency and maintain grid stability [1]. Although the work by ElZein et al. [44] did not include AI in their proposed model, integrating AI could further enhance resource optimization and reduce CO_2 emissions. Similarly, Vahedi et al. [127] propose integrating circular economy and water-energy-food (WEF) optimization to drive near-zero CO_2 emissions in fossil-fuel industrial regions. They found that implementing circular economy principles can decrease CO_2 emissions by 41 million tons, highlighting the potential for energy and resource savings. Recognizing AI's potential to optimize resource utilization, integrating AI into the proposed framework is recommended to support cities in achieving sustainable development.

5 Discussions

This present study analyzes how AI supports smart sustainability and climate action in cities. It identifies AI applications in urban settings, including urban planning, the energy sector, the transportation sector, waste management, water management, and climate modeling and predictions, all of which contribute to urban decarbonization. For instance, AI and digital twins can track and manage real-time energy use, reducing waste and enabling actions to lower energy demand and CO_2 emissions. Additionally, smart grids and renewable energy help decrease fossil fuel consumption, further cutting greenhouse gas emissions [18, 75]. AI-powered waste management is crucial for reducing CO_2 emissions by transforming waste into renewable energy sources such as biogas (Li 2022). In urban planning, AI data-driven insights facilitate the planning and implementation of projects with minimal environmental impact, thus helping reduce CO_2 emissions in the long run.

Even though AI has great potential to support net-zero goals, its implementation requires careful oversight. Collaboration between individuals, businesses, and governments, supported by strong regulation, is key to using AI for sustainability and climate action [46]. Lack of collaboration between AI experts, urban planners, and policymakers undoubtedly hinders sustainable solutions [11]. Encouraging collaborative research and knowledge sharing among academic institutions, AI developers, urban planners, and sustainability experts is critical for advancing sustainable AI practices. Therefore, it is important to establish collaborative platforms for regular communication and knowledge sharing among stakeholders. Furthermore, it is essential to recognize that AI presents both opportunities and challenges in tackling climate change. AI helps decarbonize the economy by identifying cost-effective sustainable solutions and assessing climate risk exposure. However, the increasing energy demand of AI data centers may offset decarbonization efforts [128]. Training and deploying AI models requires high computational resources, resulting in significant carbon emissions and energy use. As demand for these technologies grows, addressing their climate impact becomes increasingly important [41]. Thus, training on AI's environmental impact is essential for urban planners and agencies to balance innovation with sustainability.

To fully leverage AI's potential in urban sustainability, several other challenges must be addressed. Technical barriers, such as the gap between AI theory and real-world implementation, may limit widespread adoption. Ethical concerns, including data privacy and cybersecurity, also require attention. Finally, addressing the digital divide is key to promoting equitable access and integrating AI into sustainability efforts.

6 Conclusions

The findings of the study reveal that countries across different regions encounter unique environmental challenges, leading to diverse approaches and strategies in pursuing sustainability goals. Nevertheless, all three cases demonstrate that AI can play a vital role in cutting greenhouse gas emissions and supporting climate adaptation through smart city development, despite the three countries being at different stages of AI adoption for climate action. For example, China benefits from a strong pool of AI talent and large AI companies to support its initiatives, while Malaysia and Egypt lack such advantages. These lessons highlight the need for proactive planning, data-driven decisions, and cross-sector collaboration to address climate challenges locally and globally.

Smart cities are central to building a sustainable future and adapting to climate change. To achieve smart city objectives, advanced technologies must be continuously integrated to enhance urban systems and urban services. China and Malaysia have established long-term strategies to harness the potential of AI. Both countries have established their long-term plans or roadmaps. For example, China's 2017 AI Development Plan and Malaysia's 2021–2025 AI Roadmap outline their respective frameworks for an AI-driven society and economy. In practice, China's smart cities follow a top-down approach, with development directed by the central government's long-term strategies. In contrast, Malaysia's smart cities involve multiple stakeholders in decision-making. China's centralized governance allows for rapid implementation of nationwide policies, supported by strong government engagement in climate initiatives and significant investments in renewable energy, electric vehicles, infrastructure resilience, and more. Moreover, diverse types of smart cities have emerged in China and Malaysia, each focusing on different dimensions. Their development requires tailored solutions, as no single model fits all contexts.

Unlike China and Malaysia, Egypt faces a different set of challenges due to its arid climate. Through initiatives such as the New Administrative Capital and Smart Village, Egypt aims to integrate advanced technologies to enhance urban living and promote sustainability. However, Egypt's megacities face increasing water stress driven by population growth and rapid urbanization. To address this, AI-powered smart resource management such as smart water management should be actively promoted. Moreover, Egypt is among the countries most vulnerable to climate change, making it crucial to leverage climate modeling and predictive analytics to enhance environmental monitoring and forecasting. AI-powered climate models process vast real-time data to accurately predict extreme weather events, facilitating quick preparedness and response to reduce the negative impacts on vulnerable communities and ecosystems [1]. Furthermore, transitioning to smart, sustainable cities demands substantial investment, with international funding playing a vital role. Cities can leverage financial instruments such as green bonds, climate funds, and smart infrastructure grants, all of which are closely aligned with sustainability objectives. Among these, green bonds are particularly important in financing renewable energy, sustainable mobility, and environmental projects [129]. For Egypt, international collaboration is crucial in supporting its green transition. Access to global technology, financial resources, and shared best practices can help build robust monitoring and response systems. This support will enable Egypt to better protect its marginalized communities and implement well-informed policies to confront climate change. [130]. Thus, countries in arid and semi-arid regions may need to follow Egypt's example in navigating their path toward smart, sustainable development and tackling climate change.

7 Recommendations

Despite the progress made in China, the country still faces significant challenges related to climate change, such as flooding and typhoons in its southern regions during the summer months. Therefore, it is advisable that China expand its use of advanced climate modeling, satellite monitoring, and big data analytics to track environmental changes in real time. By integrating these tools into both national and local policy planning, China can strengthen disaster preparedness, promote sustainable urban development, and support a low-carbon transition. Malaysia is also vulnerable to climate change, as shown in Table 3. To effectively address these risks, it is recommended that Malaysia utilize AI-powered climate modeling and predictive analytics to monitor extreme weather events, strengthen urban preparedness, and reduce the adverse impacts of climate change.

Unlike China and Malaysia, Egypt, situated in an arid region, not only faces the challenge of reducing GHG emissions but is also vulnerable to the impacts of climate change. Beyond reducing the energy consumptions supported by smart technologies, it is recommended that Egypt and countries in arid and semi-arid regions leverage digital twins and AI-based climate modeling and analytics to tackle climate change, ultimately contributing to the reduction of GHG emissions.

To achieve sustainable development, cities should be guided by the SDGs established by the United Nations to assess their specific needs and priorities. For example, SDG 6 emphasizes access to clean water and sanitation as a fundamental human right. Thus, cities in arid and semi-arid regions may need to prioritize SDG 6 to ensure reliable access to clean freshwaters for all citizens, as these areas are often prone to water stress due to their climatic conditions. China and Malaysia face their own climate-related challenges, each shaped by their unique economic structures, levels of development, and governance capacities. All three countries should align with the United Nations' Sustainable Development Goals (SDGs) to assess their strengths and weaknesses and develop strategic plans that leverage AI and other advanced technologies in building smart cities that promote sustainable development. Particular attention should be given to SDG 6, which emphasizes clean water and proper sanitation; SDG 7, which promotes affordable and reliable clean energy; SDG 11, which aims for inclusive and sustainable urban development; SDG 12, which encourages responsible consumption and production; and SDG 13, which focuses on proactive measures to combat climate change.

Finally, for all countries, key challenges such as AI-induced energy consumption, data limitations, and the need for collaboration, must be addressed to fully unlock AI's potential in combating climate change and advancing effective climate action.

References

1. Mehendale P (2022) AI for climate change mitigation: predictive analytics and sustainable solutions. Eur J Adv Eng Technol 9(5):140–144. Available online at www.ejaet.com
2. Muñoz-Pizza DM, Sanchez-Rodriguez RA, Gonzalez-Manzano E (2023) Linking climate change to urban planning through vulnerability assessment: the case of two cities at the Mexico-US border. Urban Climate 51:101674. https://doi.org/10.1016/j.uclim.2023.101674
3. Abduljabbar R, Dia H, Liyanage S, Asadi Bagloee S (2019) Applications of artificial intelligence in transport: an overview. Sustainability 11(1):189. https://doi.org/10.3390/su11010189
4. Fan W, Guo Z, Wang Y, Sun Y (2024) The future of AI in transportation: challenges and opportunities. International Journal of Transportation Science and Technology. https://doi.org/10.1016/j.ijtst.2024.09.005
5. Kabir M, Ekici S (2024) Energy-agriculture nexus: exploring the future of artificial intelligence applications. Energy Nexus 13:100264. https://doi.org/10.1016/j.nexus.2024.100264
6. Park C, No W, Choi J, Kim Y (2023) Development of an AI advisor for conceptual land use planning. Cities 138:104371. https://doi.org/10.1016/j.cities.2023.104371
7. Tseng C-J, Lin S-Y (2024) Role of artificial intelligence in carbon cost reduction of firms. J Clean Prod 247:119121. https://doi.org/10.1016/j.jclepro.2019.119121
8. Bajwa J, Munir U, Nori A, Williams B (2021) Artificial intelligence in healthcare: transforming the practice of medicine. Future Healthc J 8(2):e188–e194. https://doi.org/10.7861/fhj.2021-0095
9. Wu T, Zhang T (2022) Multi-objective optimization of energy, visual, and thermal performance for building envelopes in China's hot summer and cold winter climate zone. J Build Eng 52:105034
10. Adinkrah J, Kemausuor F, Tchao ET, Nunoo-Mensah H, Agbemenu AS, Adu-Poku A, Kponyo JJ (2025) Artificial intelligence-based strategies for sustainable energy planning and electricity demand estimation: a systematic review. Renew Sustain Energy Rev 210:15161. ISSN 1364-0321
11. Son TH, Weedon Z, Yigitcanlar T, Sanchez T, Corchado JM, Mehmood R (2023) Algorithmic urban planning for smart and sustainable development: systematic review of the literature. Sustain Cities Soc 94:104562. https://doi.org/10.1016/j.scs.2023.104562
12. Zheng X, Ma Z, Yuang Z (2024) Urban design and pollution using AI: implications for urban development in China. Heliyon 10:Article e37735. https://doi.org/10.1016/j.heliyon.2024.e37735
13. Ramyar R, Ackerman A, Johnston DM (2021) Adapting cities for climate change through urban green infrastructure planning. Cities 117:103316. https://doi.org/10.1016/j.cities.2021.103316
14. Fu G, Jin Y, Sun S, Yuan Z, Butler D (2022) The role of deep learning in urban water management: a critical review. Water Res 223:118973
15. Haider S, Rashid M, Tariq MAUR, Nadeem A (2024) The role of artificial intelligence (AI) and ChatGPT in water resources, including its potential benefits and associated challenges. Discover Water

16. Yang L, Driscol J, Sarigai S, Wu Q, Lippitt CD, Morgan M (2022) Towards synoptic water monitoring systems: a review of AI methods for automating water body detection and water quality monitoring using remote sensing. Sensors 22(6):2416. https://doi.org/10.3390/s22062416.
17. Parks D (2019) Energy efficiency left behind? Policy assemblages in Sweden's most climate-smart city. Eur Plan Stud 27(2):318–335
18. Wang D, Peng D, Huang D (2025) Application and prospects of large AI models in virtual power plants. Electr Power Syst Res 241:111403. https://doi.org/10.1016/j.epsr.2024.111403
19. Jain H, Dhupper R, Shrivastava A, Kumar D, Kumari M (2023) AI-enabled strategies for climate change adaptation: Protecting communities, infrastructure, and businesses from the impacts of climate change. Comput Urban Sci 3(25). https://doi.org/10.1007/s43762-023-00100-2
20. Abdallah M, Abu Talib M, Feroz S, Nasir Q, Abdalla H, Mahfood B (2020) Artificial intelligence applications in solid waste management: a systematic research review. Waste Manage 109:231–246
21. IEA, Buildings (2022) https://www.iea.org/reports/buildings
22. Sanchez TW, Shumway H, Gordner T, Lim T (2022) The prospects of artificial intelligence in urban planning. Int J Urban Sci. https://doi.org/10.1080/12265934.2022.2102538
23. Al-Raeei M (2024) The smart future for sustainable development: artificial intelligence solutions for sustainable urbanization. Sustain Dev 32(1):1–10. https://doi.org/10.1002/sd.3131
24. Zhang B, Li CY, Kikumoto H, Niu J, Tse TKT (2024) Smart urban windcatcher: conception of an AI-empowered wind-channeling system for real-time enhancement of urban wind environment. Build Environ 253:111357. https://doi.org/10.1016/j.buildenv.2024.111357
25. Bibri SE, Huang J, Jagatheesaperumal SK, Krogstie J (2024) The synergistic interplay of artificial intelligence and digital twin in environmentally planning sustainable smart cities: a comprehensive systematic review. Environ Sci Ecotechnol 20:100433. https://doi.org/10.1016/j.ese.2024.100433
26. Cugurullo F (2020) Urban artificial intelligence: from automation to autonomy in the smart city. Front Sustain Cities 2:38
27. Cugurullo F (2021) Frankenstein urbanism: eco, smart and autonomous cities, artificial intelligence and the end of the city. Routledge, London
28. Konya A, Nematizadeh P (2024) Recent applications of AI to environmental disciplines: a review. Sci Total Environ 906:167026
29. Ansari SPMA, Odongo JO, Nomani MZM, Salahuddin G, Hasan MF, Pallathadka LK (2022) Evaluating the role of environment management systems based on artificial intelligence. Mater Today Proc 56:2240–2244
30. Candrian C, Scherer A (2022) Rise of the machines: delegating decisions to autonomous AI. Comput Hum Behav 134:107308. https://doi.org/10.1016/j.chb.2022.107308
31. Schrage M (2017) 4 Models for using AI to make decisions. Harvard Business Review. Retrieved from https://hbr.org/2017/01/4-models-for-using-ai-to-make-decisions
32. Elallid BB, Benamar N, Hafid AS, Rachidi T, Mrani N (2022) A comprehensive survey on the application of deep and reinforcement learning approaches in autonomous driving. J King Saud Univ Comput Inf Sci 34(9):7366–7390
33. Linaza MT, Posada J, Bund J, Eisert P, Quartulli M, Döllner J, Pagani A, Olaizola IG, Barriguinha A, Moysiadis T (2021) Data-driven artificial intelligence applications for sustainable precision agriculture. Agronomy 11(1):1227
34. Tanic E (1986) Urban planning and artificial intelligence: the Urbyss system. Comput Environ Urban Syst 10(3–4):135–146. https://doi.org/10.1016/0198-9715(86)90002-5
35. Lazaroiu G, Harrison A (2021) Internet of things sensing infrastructures and data-driven planning technologies in smart sustainable city governance and management. Geopolitics Hist Int Relat 13(2):23–36
36. Wu P, Zhang Z, Peng X, Wang R (2024) Deep learning solutions for smart city challenges in urban development. Sci Reports 14:Article 5176. https://doi.org/10.1038/s41598-024-55928-3

37. Othengrafen F, Sievers L, Reinecke E (2025) From vision to reality: the use of artificial intelligence in different urban planning phases. Urban Plann 10:8576
38. Quan SJ, Park J, Economou A, Lee S (2019) Artificial intelligence-aided design: smart design for sustainable city development. Smart Sustain Built Environ 46(2019):1581–1599. Retrieved from https://ideas.repec.org/a/sae/envirb/v46y2019i9p1581-1599.html
39. Luz Castro Pena M, Carballal A, Rodríguez-Fernández N, Santos I, Romero J (2021) Artificial intelligence applied to conceptual design: a review of its use in architecture. Autom Constr 124:103550. https://doi.org/10.1016/j.autcon.2021.103550
40. Bibri SE (2025) Artificial intelligence and its generative power for advancing the sustainable smart city digital twin: a novel framework for data-driven environmental planning and design. In: Artificial intelligence of things for smarter eco-cities: pioneering the environmental synergies of urban brain, digital twin, metabolic circularity, and platform. CRC/Taylor & Francis Group. https://doi.org/10.1201/9781003536420
41. Huang J, Bibri SE Keel P (2025) Generative spatial artificial intelligence for sustainable smart cities: a pioneering large flow model for urban digital twin. Environmental Science and Ecotechnology. Advance online publication. https://doi.org/10.1016/j.ese.2025.100526
42. Alassery F, Alzahrani A, Khan AI, Irshad K, Kshirsagar SR (2022) An artificial intelligence-based solar radiation prophesy model for green energy utilization in energy management system. Sustain Energy Technol Assess 52:Article 102060. https://doi.org/10.1016/j.seta.2022.102060
43. Himeur Y, Ghanem K, Alsalemi A, Bensaali F, Amira A (2021) Artificial intelligence-based anomaly detection of energy consumption in buildings: a review, current trends, and new perspectives. Appl Energy 287:116601. https://doi.org/10.1016/j.apenergy.2021.116601
44. ElZein Z, Milad NA, Mohamed AS, Mahmoud N, Abdo N, Abdel-Ghafar HM (2024) Sustainable development on the basis of the WEF nexus in arid coastal areas for climate change mitigation: Case study of Rabia community in Matrouh Egypt. Energy Nexus 14:100299. https://doi.org/10.1016/j.nexus.2024.100299
45. Argyropoulos SA, Mitoulis S, Chatzi E, Baker JW, Brilakis I, Gkoumas K, Linkov I (2022) Digital technologies can enhance climate resilience of critical infrastructure. Risk Anal 43(2):103087
46. Olawade DB, Wada OZ, David-Olawade AC, Fapohunda O, Ige AO, Ling J (2024) Artificial intelligence potential for net zero sustainability: current evidence and prospects. Next Sustain 4:100041. https://doi.org/10.1016/j.nextsus.2024.100041
47. Ding C, Ke J, Levine M, Zhou N (2024) Potential of artificial intelligence in reducing energy and carbon emissions of commercial buildings at scale. Nat Commun. https://doi.org/10.1038/s41467-024-50088-4
48. Chen L, Chen Z, Zhang Y, Liu Y, Osman AI, Farghali M, Hua J, Al-Fatesh A, Ihara I, Rooney DW, Yap P-S (2023) Artificial intelligence-based solutions for climate change: a review. Environ Chem Lett 21:2525–2557. https://doi.org/10.1007/s10311-023-01617-y
49. Vinuesa R, Azizpour H, Leite I, Balaam M, Dignum V, Domisch S, Felländer A, Langhans SD, Tegmark M, Nerini FF (2020) The role of artificial intelligence in achieving the Sustainable Development Goals. Nat Commun 11:233. https://doi.org/10.1038/s41467-019-14108-y
50. Ajayi F, Ademola OM, Amuda KF, Alade B (2024) AI-driven decarbonization of buildings: leveraging predictive analytics and automation for sustainable energy management. World J Adv Res Rev 24(01):061–079. https://doi.org/10.30574/wjarr.2024.24.1.2997
51. Sandalow D, McCormick C, Kuckelbir A, Friedmann J, Nachmany M, Lee H, Loehr D et al (2024) Artificial intelligence for climate change mitigation roadmap, 2nd edn. ICEF Innovation Roadmap Project. Retrieved from https://www.icef.go.jp/roadmap/
52. Kezunovic M, Pinson P, Obradovic Z, Grijalva S, Hong T, Bessa R (2020) Big data analytics for future electricity grids. Electr Power Syst Res 189:106788. https://doi.org/10.1016/j.epsr.2020.106788
53. Abdalla AN, Nazir MS, Tao H, Cao S, Ji R, Jiang M, Yao L (2021) Integration of energy storage system and renewable energy sources based on artificial intelligence: an overview. J Energy Storage 40:102811. https://doi.org/10.1016/j.est.2021.102811

54. Rutenberg I, Gwagwa A, Omino M (2021) Use and impact of artificial intelligence on climate change adaptation in Africa. In: Oguge N, Ayal D, Adeleke L, da Silva I (eds), African handbook of climate change adaptation. Springer International Publishing, Cham
55. Haluza D, Jungwirth D (2023) Artificial intelligence and ten societal megatrends: an exploratory study using GPT-3. Systems 11(3):Article 120. https://doi.org/10.3390/systems11030120
56. United Nation (2020). https://www.un.org/en/climatechange/climate-solutions/cities-pollution?utm_source=chatgpt.com
57. Wei T, Wu J, Chen S (2021) Keeping track of greenhouse gas emission reduction progress and targets in 167 cities worldwide. Front Sustain Cities 3:696381. https://doi.org/10.3389/frsc.2021.696381
58. Crane M, Lloyd S, Haines A, Ding D, Hutchinson E, Belesova K, Davies M, Osrin D, Zimmermann N, Capon A, Wilkinson P, Turcu C (2021) Transforming cities for sustainability: a health perspective. Environ Int 147:106366. https://doi.org/10.1016/j.envint.2020.106366
59. Ahvenniemi H, Huovila A, Pinto-Seppä I, Airaksinen, M (2017) What are the differences between sustainable and smart cities? Cities 60(A):234–245
60. Berquier R, Gibassier D (2019) Governing the "good citizen" and shaping the "model city" to tackle climate change: materiality, economic discourse and exemplarity. Sustain Acc Manag Policy J 10(4):710–744
61. Garnett R, Adams MD (2018) LiDAR—a technology to assist with smart cities and climate change resilience: a case study in an urban metropolis. ISPRS Int J Geo Inf 7(5):161. https://doi.org/10.3390/ijgi7050161
62. Kutty AA, Kucukvar M, Abdella GM, Bulak ME, Onat NC (2022) Sustainability performance of European smart cities: a novel DEA approach with double frontiers. Sustain Cities Soc 81:Article 103777
63. Karatzimas S (2024) Smart cities' actions, performance and reporting practices on climate change challenges: an exploratory analysis in a sample of awarded smart cities. Cities 153:105270
64. Bibri SE (2021) Data-driven smart eco-cities and sustainable integrated districts: a best-evidence synthesis approach to an extensive literature review. Eur J Futures Res 9(1):1–43
65. Marvin S, While A, Chen B, Kovacic M (2022) Urban AI in China: social control or hyper-capitalist development in the post-smart city? Front Sustain Cities 4:1030318
66. Baucells Aletà N, Moreno Alonso C, Arce Ruiz RM (2017) Smart mobility and smart environment in Spanish cities. Transp Res Procedia 24:163–170
67. Selim HS, Mayhoub MS, Abuzaid A (2024) A comprehensive model to assess sustainable architecture in emerged megacities: a closer look at Cairo's New Administrative Capital (NAC). Sustainability 16(12):5046. https://doi.org/10.3390/su16125046
68. Gu Q, Sing MCP, Jefferies M, Kanjanabootra S (2025) Bridging the gap between smart cities and sustainability: current practices and future trends. Cities 159:105799
69. Bibri S, Krogstie J, Kaboli A, Alahi A (2023) Smarter eco-cities and their leading-edge artificial intelligence of things solutions for environmental sustainability: a comprehensive systemic review. Environ Sci Ecotechnol 19
70. Yigitcanlar T, Cugurullo F (2020) The sustainability of artificial intelligence: an urbanistic viewpoint from the lens of smart and sustainable cities. Sustainability 12(20):8548
71. Evans J, Karvonen A, Luque-Ayala A, Martin C, McCormick K, Raven R, Palgan YV (2019) Smart and sustainable cities? Pipedreams, practicalities, and possibilities. Local Environ 24(7):557–564
72. Herath HMKKMB, Mittal M (2022) Adoption of artificial intelligence in smart cities: a comprehensive review. Int J Inf Manag Data Insights 2:100076. https://doi.org/10.1016/j.jjimei.2022.100076
73. Eldrandaly KA, Abdel-Basset M, Abdel-Fatah L (2019) PTZ-Surveillance coverage based on artificial intelligence for smart cities. Int J Inf Manage 49:520–532. https://doi.org/10.1016/j.ijinfomgt.2019.04.017

74. IWMI (2023, November 27) International Water Management Institute (IWMI) announces development of a new, Google-supported tool that will harness AI and satellite data for water reuse in the Middle East and North Africa. International Water Management Institute. https://www.iwmi.org/news/new-google-supported-tool-will-harness-ai-and-satellite-data-for-water-reuse-in-mena/
75. Goh KC, Kurniawan TA, Zainordin N, Arifah IDC, Abas MA, Masrom MAN, Mohamed S, Omar R, Khoo SL, Gui HC, Toh TC, Onn CW (2024) Expediting decarbonization in energy, waste, and water sector through digitalization in sustainable smart cities (SSC): case studies in Malaysia and China based on Industry 5.0 paradigm. Sustain Cities Soc 117:105969
76. Kannammal A, Chandia S (2023) Applications of AI and IoT for smart cities. https://doi.org/10.2174/9789815136449123010014
77. Balas V, Kumar R, Srivastava R (2020) Recent trends and advances in artificial intelligence and Internet of Things. https://doi.org/10.1007/978-3-030-32644-9
78. Marengo A (2024) Navigating the nexus of AI and IoT: a comprehensive review of data analytics and privacy paradigms. Internet of Things 27:100318
79. Lim C, Kim KJ, Maglio PP (2018) Smart cities with big data: reference models, challenges, and considerations. Cities 82:86–99
80. Anastasiu I (2019) Unpacking the smart city through the lens of the right to the city: a taxonomy as a way forward in participatory city-making. In: de Waal M, de Lange M (eds) The hackable city: digital media and collaborative city-making in the network society. Springer, pp 239–260
81. Zhang D, Pee L, Pan S, Liu W (2022a) Orchestrating artificial intelligence for urban sustainability. Gov Inf Q 39(4):101720. https://doi.org/10.1016/j.giq.2022.101720
82. Zhang J, Jia R, Yang H, Dong K (2022b) Does electric vehicle promotion in the public sector contribute to urban transport carbon emissions reduction? Transp Policy 125:151–163. https://doi.org/10.1016/j.tranpol.2022.05.014
83. Zaugg J (2019) China's data centers emit as much carbon as 21 million cars. CNN Business
84. Ni W, Hu X, Du H, Kang Y, Ju Y, Wang Q (2022) CO_2 emission-mitigation pathways for China's data centers. Resour Conserv Recycl 202:107383
85. Yin RK (2009) Case study research: design and method, 4th edn. Sage Publications, Thousand Oaks, CA, USA
86. Pahl-Wostl C, Basurto X, Villamayor-Tomas S (2021). Comparative case study analysis. Routledge, pp 282–294. https://doi.org/10.4324/9781003021339-24
87. Azizalrahman H, Hasyimi V (2019) A model for urban sector drivers of carbon emissions. Sustain Cities Soc 44:46–55. https://doi.org/10.1016/j.scs.2018.09.035
88. Dong M, Wang G, Han X (2024) Impacts of artificial intelligence on carbon emissions in China: in terms of artificial intelligence type and regional differences. Sustain Cities Soc 113:105682
89. Global Carbon Project (2024, November 13) Territorial carbon dioxide (CO_2) emissions from fossil fuel combustion and industrial processes in China from 1960 to 2023 (in million metric tons) [Graph]. In Statista. Retrieved March 6, 2025, from https://www.statista.com/statistics/239093/co2-emissions-in-china/
90. Liengpunsakul S (2021) Artificial intelligence and sustainable development in China. Chin Econ 54(4):235–248. https://doi.org/10.1080/10971475.2021.1922914
91. Takach GS (2024) Cold War 2.0: Artificial intelligence in the new war between China, Russia, and America. Pegasus Books. ISBN: 9781639365630
92. Chu Z, Chen P, Zhang Z, Chen Z (2024) Other's shoes also fit well: AI technologies contribute to China's blue skies as well as carbon reduction. J Environ Manage 353:120171. https://doi.org/10.1016/j.jenvman.2024.120171
93. Wolniak R, Stecula K (2024) Artificial intelligence in smart cities—applications, barriers, and future directions: a review. Smart Cities 7(3):1346–1389. https://doi.org/10.3390/smartcities7030057
94. Zhu J, Gianoli A, Noori N, de Jong M, Edelenbos J (2024) How different can smart cities be? A typology of smart cities in China. Cities 149:104992

95. Guo QB, Wang Y, Dong XB (2022) Effects of smart city construction on energy saving and CO_2 emission reduction: evidence from China. Appl Energy 313:118879
96. Chu Z, Cheng MW, Yu NN (2021) A smart city is a less polluted city. Technol Forecast Soc Change 172:121037
97. U.S.-China Economic and Security Review Commission (2020) China's smart cities development. https://www.uscc.gov/research/chinas-smart-cities-development
98. Yu Y, Yang J, Chen B (2012) The smart grids in China—a review. Energies 5(5):1321–1338
99. Nyangon J (2024) Climate-proofing critical energy infrastructure: Smart grids, artificial intelligence, and machine learning for power system resilience against extreme weather events. J Infrastructure Syst 30(1)
100. Golinska-Dawson P, Sethanan K (2023) Sustainable urban freight for energy-efficient smart cities—systematic literature review. Energies 16(6):6161. https://doi.org/10.3390/en16062617
101. Mukherjee AG, Wanjari UR, Chakraborty R, Renu K, Vellingiri B, George A, Rajan SCR, Gopalakrishnan AV (2021) A review on modern and smart technologies for efficient waste disposal and management. J Environ Manage 297:113347
102. Aghaie S (2023) Sustainable urban planning: stakeholder perspectives on resource allocation challenges. J Resource Manag Decision Eng 2(1):30–36
103. Chachuli FSM, Mat S, Ludin NA, Sopian K (2021) Performance evaluation of renewable energy R&D activities in Malaysia. Renew Energy 163:544–560
104. Melo FPL, Parry L, Brancalion PHS, Pinto SRR, Freitas J, Manhães AP, Meli P, Ganade G, Chazdon RL (2021) Adding forests to the water–energy–food nexus. Nat Sustain 4(2):85–92
105. Du Plessis A, du Plessis A (2019) Current and future water scarcity and stress. In: Water as an inescapable risk: current global water availability, quality, and risks with a specific focus on South Africa. Springer, pp 13–25
106. Sabri S, Alexandridis K, Koohikamali M, Zhang S, Ozkaya HE (2023) Designing a spatially-explicit urban digital twin framework for smart water infrastructure and flood management. 2023 IEEE 3rd International Conference on Digital Twins and Parallel Intelligence (DTPI), 1–9
107. Wang WW, Zhang M, Zhou M (2011) Using LMDI method to analyze transport sector CO_2 emissions in China. Energy 36(10):5909–5915
108. Nieuwenhuijsen MJ (2020) Urban and transport planning pathways to carbon-neutral, liveable, and healthy cities: a review of the current evidence. Environ Int 140:105661
109. Brzeziński Ł (2024) Social aspects of smart urban mobility. Encyclopedia 4(2):864–873. https://doi.org/10.3390/encyclopedia4020055
110. Energy Institute (2024, June 20) Carbon dioxide (CO_2) emissions from energy consumption in Malaysia from 2014 to 2023 (in million metric tons). In Statista. Retrieved March 6, 2025, from https://www.statista.com/statistics/1394260/malaysia-co2-emissions-from-energy-use/
111. Ipsos (2019, April 19) Most concerning environmental issues according to citizens across Malaysia as of March 2019 [Graph]. In: Statista. Retrieved March 06, 2025, from https://www.statista.com/statistics/1000112/malaysia-leading-environmental-issues/
112. Kumar P, Singh RK, Kumar V (2021) Managing supply chains for sustainable operations in the era of Industry 4.0 and circular economy: analysis of barriers. Resour Conserv Recycl 164:105215
113. Lim SB, Malek JA, Yussoff MFYM, Yigitcanlar T (2021) Understanding and acceptance of smart city policies: practitioners' perspectives on the Malaysian Smart City Framework. Sustainability 13(17):9559. https://doi.org/10.3390/su13179559
114. Samsudin NA, Rosley MSF, Lai LY, Omar SR, Rashid MF, Najihah Mohd Hanifi NS, Bakhtiar IS (2022) A comparative study of smart city initiatives in Malaysia: Putrajaya and Iskandar Puteri. Plann Malays J Malays Inst Planners 20(5):14–28
115. Bakhtiar IS, Samsudin NA (2023) Comparative review of Smart City from an urban planning perspective in Johor Bahru City and Petaling Jaya City. IOP Conf Ser Earth Environ Sci 1274(1):012016

116. Fernandez MI, Go YL, Wong MLD, Früh W-G (2024) Malaysia's energy transition and readiness towards attaining net zero: review of the potential, constraints, and enablers. Renew Energy Focus 51:100640. https://doi.org/10.1016/j.ref.2024.100640
117. Adomako AB, Jamshidi EJ, Yusup Y, Elsebakhi E, Jafar MH, Ishak MIS, Lim HS, Ahmad MI (2024) Deep learning approaches for bias correction in WRF model outputs for enhanced solar and wind energy estimation: a case study in East and West Malaysia. Eco Inform 84:102898. https://doi.org/10.1016/j.ecoinf.2024.102898
118. Aldhshan SR, Abdul Maulud KN, Wan Mohd Jaafar WS, Karim OA, Pradhan B (2021) Energy consumption and spatial assessment of renewable energy penetration and building energy efficiency in Malaysia: a review. Sustainability 13(16):9244. https://doi.org/10.3390/su13169244
119. Ha CY, Khoo TJ, Loh JX (2023) Barriers to green building implementation in Malaysia: a systematic review. Progress in Energy and Environment: 11–21. https://doi.org/10.37934/progee.24.1.1121
120. World Bank (2024, November 25) Climate action: A catalyst for Egypt's prosperity and sustainable growth. https://www.worldbank.org/en/news/feature/2024/11/25/climate-action-a-catalyst-for-egypt-s-prosperity-and-sustainable-growth
121. Darwish AS (2016) Sustainable green smart buildings: future energy survivor. ISESCO J Sci Technol 12(21):35–42
122. El-Motasem S, Khodeir LM, Eid AF (2021) Analysis of challenges facing smart buildings projects in Egypt. Ain Shams Eng J 12(3):3317–3329. https://doi.org/10.1016/j.asej.2020.09.028
123. So ATP, Wong KC (2002) On the quantitative assessment of intelligent buildings. Facilities 20(13/14):368–375. https://doi.org/10.1108/02632770210435206
124. Khashaba S (2014, October) The use of intelligent buildings to achieve sustainability through an architectural proposal for public buildings in Cairo. In: Proceedings of the World SB14 Conference. Barcelona, Spain
125. European Parliament (2022) Egypt's climate change policies. European Parliament. https://www.europarl.europa.eu/BRIE/2022
126. OECD (2024) OECD artificial intelligence review of Egypt. OECD Publishing, Paris. https://doi.org/10.1787/2a282726-en
127. Vahedi R, Bararzadeh Ledari M, Fani M (2025) Advancing circular economy in the water-energy-food nexus: a framework for sustainable resource use. J Environ Manage 376:124324. https://doi.org/10.1016/j.jenvman.2024.124324
128. Cowls J, Tsamados A, Taddeo M, Floridi L (2023) The AI gambit: leveraging artificial intelligence to combat climate change—opportunities, challenges, and recommendations. AI Soc 38:283–307
129. Meng J, Ye Z, Wang Y (2024) Financing and investing in sustainable infrastructure: a review and research agenda. Sustain Futures 8:100312. https://doi.org/10.1016/j.sftr.2024.100312
130. Hamzawy A, Al-Mailam M, Arkeh J (2023, October 26) Climate change in Egypt: opportunities and obstacles. https://carnegieendowment.org/research/2023/10/climate-change-in-egypt-opportunities-and-obstacles?lang=en
131. Egypt Vision 2030 (2023) Egypt's sustainable development strategy. https://sdsegypt2030.com

Resilience in Architecture and Planning

Architecture and Resilience: A Comprehensive Overview of Purpose, Evolution and Methodologies

Aya Hany El-Senoussy

Abstract Human was created with the instinct of creation, building, and innovation. Throughout the years, the essence of architecture witnessed a change from one era to another, with the changes that occurred to man himself and the environment around him reflected in architecture. This chapter will briefly trace the evolution of the essence of architecture that we sometimes ignore and summarize it in just some redundant visuals. "The essence of architecture is not just in its form, but in its ability to respond to and shape the lives of the people who inhabit it." Gamal Bakri. Starting with defining the term "resilience" and what are the different forms of Resilient Architecture that respond to environmental issues such as climate change and man-made crises such as wars/conflicts. Success stories and case studies will be highlighted, showcasing the achievements of individuals and groups who have effectively utilized architecture and the built environment to address their needs. Emphasizing the significance of community engagement in driving change, we can harness our available resources for our benefit, even when they initially appear to be challenging.

Keywords Resilience · Adaptation · Mitigation · Vulnerability · Community engagement

1 Introduction

Human as well as most of the natural inhabitants including animals were inherently created with an embedded instinct in their nature to "build" as if it is a sign of existence. Initially, they all tended to explore and practice the building process to provide shelter and protection from the surrounding environment and any threat they could face [1, 2]. Shaping the meaning of architecture throughout history through exploration, practice, and gaining accumulated knowledge about different building

A. H. El-Senoussy (✉)
Community Architect at Megawra | Built Environment Collective, Cairo, Egypt
e-mail: ayaelsenoussy@gmail.com

Z. ElZein and A. Negm (eds.), *Recent Approaches of Sustainable Architecture in Arid and Semi-arid Cities*, Urban Sustainability,
https://doi.org/10.1007/978-981-95-1645-2_3

materials and technologies. To let architecture be developed beyond being mere structures. It transcended its practical purpose to be a powerful symbol of civilization. A mix of a form of art, function, hidden meanings, and concepts behind each building. To be more than just a shelter or a building. By some means, architecture was a sign of dominance, in other times it was used for creating a hierarchy in communities, and by all means, it acts as a reflection of human behavior, priorities, challenges, and the state of awareness and knowledge of each civilization, not by building high and massive constructions but by understanding the methodologies and the way a certain building was built. Hence, we can consider architecture as a form of human/humanity narrative [3].

By tracing the architectural evolution, we can understand important insights about our evolution (humans), too. "At the end, What Remains is the Architectural building." By Reem Basiouny (an Egyptian novelist).

1.1 Architecture and Resilience

The chapter examines the word "resilience" and its critical relevance in Architecture. As today, resilience in architecture has become a crucial topic as societies face increasing environmental crises and conflicts that threaten the built environment. The ability of architecture to withstand and adapt to natural and manmade challenges/threats is essential for ensuring the sustainability and longevity of human habitats. Understanding how architecture can respond to these challenges is vital for creating sustainable urban spaces that support both human and ecological well-being.

Resilience is not a novel concept in architecture; historical architects devised innovative solutions to combat natural challenges, such as developing primitive cooling systems and creative natural lighting designs to mitigate intense summer heat. Ideas still relevant today and ripe for modern adaptation.

The chapter will begin with an Introduction to the holistic meaning of Architecture, how its role developed, and how it cannot be separated from our daily practices and being humanized by showing its intersection with many other aspects of our lives, such as economy, environmental issues, anthropology, sociology, and politics [4, 5].

1.2 Climate Change and Architecture

This discussion will be categorized into mainly two primary sections, the first one is architectural resilience in the face of natural disasters and climate change. It will start by defining the problem of climate change as a common denominator between natural and manmade disasters [6]. It will address and analyze its underlying causes and consider Architecture's dual role as both a contributor to the climatic change problem and potentially a crucial part of the solution to these challenges.

Through this, the chapter will lead us to an exploration of how communities can be engaged in creating sustainable solutions that fit their needs and traditions, along with vulnerability assessments to investigate the effects of climate change on specific regions In addition to exploring strategies to mitigate the impact of climate change, such as adaptive reuse of buildings to reduce energy consumption and resource depletion, this study examines the role of greening in enhancing human well-being, supporting ecosystems, and improving environmental sustainability. Furthermore, it highlights water management strategies, including the use of groundwater for greening and urban farming, as essential components of climate resilience [7–9].

1.3 Architecture in the Context of Conflict and Displacement

The chapter then shifts to the second section, including discussing architectural resilience in the case of wars, conflicts, and displacements. Highlighting how Architecture is vital enough to be considered a target in wars. And how architecture could be resilient in the face of these critical and brutal actions. "Resisting to exist" reflects the will and determination of its people, symbolizing their perseverance and resistance in the face of oppression [10] and [11]. By studying examples from the case of Palestine and Syria. Shows how communities mitigate the impacts of wars/conflicts and humanitarian crises by using their knowledge and available materials in their contexts to reshape and rebuild their lives presented in a home, school, or shelter. It also links the effects of wars/conflicts on climate change.

The chapter is considered a rethinking process that discusses a redefinition of architecture by examining the meaning of resilient architecture, through an overview of how it all started from a simple fundamental need to have a complex role that is interconnected with almost everything in our lives. It discusses how architecture affects our daily life practices and how it also affects the whole environment and planet, in both positive and negative ways depending on the way that we decide to perceive architecture from it. It simply studies and shows how architecture is not only about construction and it is not meant to be only done by architects. Showing different examples of challenges that face architecture and how they could be solved in a resilient and sustainable way depending on nature as the mother inspiration source, past experiences in history that could be adapted to our modern era, including communities inherited knowledge.

1.4 Chapter Objectives and Scope

This chapter aims to redefine architecture's role beyond construction by examining its resilience in the face of environmental and societal challenges. Specifically, it will explore:

- **The intersection of architecture with key societal aspects**, including economics, environmental issues, anthropology, sociology, and politics.
- **Resilience in architecture is not a novel concept**, mentioning examples from nature and history showing how to make the best use of our environmental factors to mitigate our challenges.
- **Resilience in the face of climate change in historic settings**, with a focus on adaptive reuse, greening, and sustainable water management strategies.
- **The role of community knowledge and engagement**, in shaping sustainable and resilient architectural practices.
- **Architectural resilience in contexts with limited resources, particularly in situations of war and conflict**, exploring how communities fight for life. Reflected on how they rephrased and rebuilt their architecture with the available knowledge and resources in such difficult situations as an inspiration for architectural adaptation (Fig. 1).

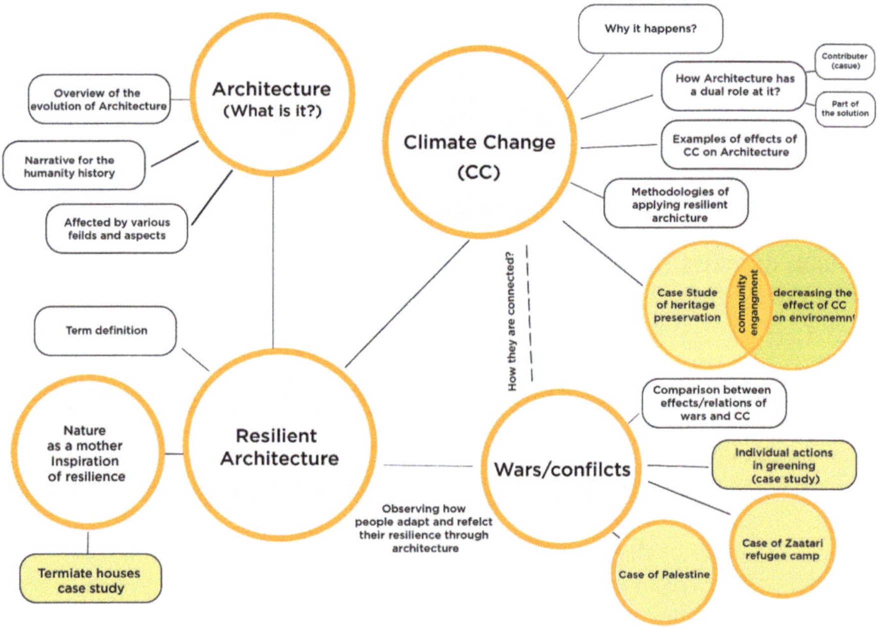

Fig. 1 Sketched mind-map illustrating the main topics of the chapter outlined with an orange stroke, with sub-titles represented by rectangles branching out. Solid lines indicate direct relationships, while dotted lines represent indirect connections between topics (the author)

2 Methods

This chapter employs a cross-sectional study design to examine the evolution, practices, effects, and roles of architecture throughout history. It begins with a comprehensive overview of the historical development of architectural concepts, utilizing comparative analysis to investigate how different architectural schools and eras have responded to their unique sociocultural and environmental contexts.

This historical context leads to the critical question of how contemporary architecture should address current challenges, particularly the threat of climate change. To explore this, the chapter draws on various resources, including scholarly articles, case studies, and design frameworks, to define architecture's role in mitigating the impacts of climate change.

Initially, the chapter defines the challenge of climate change from various references such as [12] explaining its causes and implications and examines architecture's contribution to this issue, referencing sources like [12–14]. Key insights about the phases of resilient design will be distinguished from sustainable design, drawing from. Resilience Design Toolkit, Resilience Design Integration for Architectural Projects (AIA, The American Institute of Architects, HKS, Harwood K. Smith, Inc.aia.org/resiliencedesigntoolkit), and presenting case studies with diagrams that illustrate practical strategies for implementing resilient architecture in response to predicted threats. The discussion will also highlight the positive impacts of urban greening and green rooftops on environmental quality, community health, and economic sustainability [15, 16]. As Egypt is witnessing a deterioration in its green spot area [17], and how green architecture could function not only to minimize the effect of climate change but also to enhance the quality of heritage and the community around it [9]. The chapter provides empirical observations from the field of heritage management and conservation, attaching the example of applying a successful water management strategy in a historic sites by reusing groundwater in urban farming to affect the microclimate, satisfy some of the basic nutrition needs of the neighborhood, and minimize the threat on the monument [8]. In addition to focusing on the engagement of community throughout the whole process. In the second part, the chapter discusses case narratives such as urban framing initiatives on rooftops that UN Women Palestine documented in a short film [18].

Furthermore, from this point, the chapter will explore how communities can inspire resilience in architecture, particularly in contexts of war and conflict. This will include references to works such as the books "The Architecture of Resilience" and "From Shelters to Dwellings" [10, 11, 19–21].

By synthesizing historical perspectives with contemporary practices, the chapter aims to provide a nuanced understanding of architecture's evolving role and its potential to address pressing global challenges.

3 The Evolution of the Concept of Architecture

Architecture has evolved from a fundamental instinct for shelter, into a powerful symbol of civilization, shaped by human needs, technological advancements, and cultural influences.

Initially, inspired by nature, early structures prioritized function, gradually incorporating symbolic and aesthetic elements. For example ancient Egyptian architecture, with its monumental structures and advanced techniques, set a high standard, followed by Greek and Roman contributions emphasizing beauty, proportion, and innovation. Over time, shifts in philosophy, such as Functionalism and Romanticism, reflected changing societal values. The Industrial Revolution introduced new materials, enabling modern skyscrapers [22–24].

The dynamic evolution of architecture shows how it acts as a response, influencing and being influenced by the contexts socially, industrially, environmentally, economically and politically. After this brief overview, we will explore together what should our architectural vision looks like today? And what should it respond to?

4 Resilient Architecture

Resilience is a powerful word, and resilient Architecture is a broad term that has more than one significance, definition, and shape.

"**Resilience**" is defined as the ability of a system and its component parts to anticipate, absorb, accommodate, or recover from the effects of a hazardous event in a timely and efficient manner, including through ensuring the preservation, restoration, or improvements of its essential basic structures and functions (IPCC, AIA) (as defined by Intergovernmental Panel on Climate Change, American institution of Architects).

Resilience is a term that refers to the ability to withstand and adapt to changes during tough moments. There are many types of resilience, such as, social resilience, infrastructure, economic and environmental resilience that relates to various fields and aspects, and in a way or another interconnects all of them. For example, we cannot achieve social resilience without tackling the environment [12].

Architecture has both influenced and been influenced by its time's cultural, social, and environmental context. As we move forward, the challenge remains to balance innovation with sustainability, ensuring that architectural practices contribute positively to both society and the planet.

Resilient Architecture refers to designing adaptable resilient structures that can absorb certain crises, withstand, and adapt to them. Preserving the safety of people living in the area. Currently, we are facing more than one threat that affects the livability and the quality of lives of humans, which could lead us to think about how designing resilient architecture could play a vital role in enhancing situations. One of these threats is Climate Change. That leads to catastrophes with major losses that take years to fix its effects without even being safe from the same event being repeated with the same losses or even more. In the following lines, we will explore how we could build and achieve strategies to mitigate and limit climate change's effects.

4.1 Climate Change

Change is the nature of our planet. Everything on our planet, including ourselves, has changed (physically and emotionally) since the very beginning of the circle of life on Earth.

Nature is created with a range of flexibility and adaptability to cope with the changes that occur within varied limits. The same as how our body resists illness for example to a certain change (a virus or an infection for example) and after its capability limit finishes, maybe it will need an interruption whether with medicine or by actually building a healthy routine to resist such type of infections in the future.

One of the changes that happened and will remain happening to Earth but need our interruption now to minimize its effects on our environment, is "climatic change". The consequences of climate change appeared in the nineteenth century. After 1980th, many countries tried to understand and document the impact of the climatic change on their cities and tried to devise solutions to deal with it. Ultimately, we are planning our cities for climate change and caring for the quality of life of its inhabitants [25].

The key indicator of climate change in our daily lives is the average increase in the temperature. That can easily be observed by communities giving us an alarm for more consequences that could take place later such as shortage of water (due to the more energy and water consumption that will happen as a result of the high temperatures) and the threat that could affect land quality, agriculture and the whole quality of life in our cities.

4.2 Drivers of Climate Change

The Earth's atmosphere is composed of a delicate balance of gases that act as a protective membrane, shielding the planet from harmful solar radiation. This atmospheric composition is crucial in maintaining a livable temperature on Earth, allowing life to thrive. The key gases responsible for this natural greenhouse effect are carbon dioxide (CO_2) and methane (CH_4), which are collectively known as greenhouse gases [26].

These greenhouse gases have the unique ability to trap a portion of the heat energy that radiates from the Earth's surface, preventing it from escaping into the vast expanse of space. This trapping effect is similar to the way that a greenhouse functions, where the glass or plastic structure retains warmth, allowing plants to grow even in cooler seasons [26].

The natural greenhouse effect is what has historically kept the Earth's temperature within a range that supports life. However, the problem arises when the concentration of these greenhouse gases, particularly CO_2 and CH_4, increases significantly beyond their natural levels. This is the primary driver of climate change. Activities such as the burning of fossil fuels, deforestation, and certain agricultural practices have led to a rapid increase in the levels of greenhouse gases in the atmosphere. As a result, more heat energy is trapped, causing the overall global temperature to rise. This temperature increase is the main indicator of climate change, and it then triggers a cascade of other consequences, including more frequent and severe natural disasters, rising sea levels, water and energy scarcity, and various other environmental and societal impacts [26].

The Earth's natural greenhouse effect is essential for maintaining a livable temperature, but human-induced changes to the atmospheric composition, primarily through the increased emission of greenhouse gases, are the root cause of the ongoing climate change crisis [26].

4.3 Architecture and Climate Change

Architecture plays a vital role in climate change. Buildings cause 36% of the global energy expended. 22% from residential buildings, 8% from non-residential and 6% from construction industry [26].

A sample of man-made actions that cause or accelerate the effect of environmental actions is present in **the case of Balat, at Dakhleh Oasis in Egypt**, where the wrong system of irrigation and water management causes a sudden collapse to the structures of the buildings and its almost the only place in Egypt that faces this phenomenon of sudden collapses. Due to the random way of irrigation water gets collected in some areas passing by the buildings built with material which absorbs water. Due to the high change in the temperature ranges from summer to winter, the water absorbed in the clay shrinks and evaporates causing these sudden collapses. Aside from the change in the temperature, a key issue in this problem, is our wrong practices. To study and solve such issues, the architect should study and understand many aspects to be able to offer solutions that are compatible with the social contextual culture [27].

Another example is the MENA (Middle East and North Africa) region **urban and city planning**, in some cases, architects who don't study the climate before planning, locate industrial cities in the wrong locations according to the wind direction, where all the fumes coming out from them go in the wind direction to the city causing environmental problems [27].

Also, **Rapid Urbanization** causes UHI (Urban Heat Island effect), a key meteorological phenomenon that speeds up the effect of climate change in urban areas, causing complicated impacts that need complex solutions in return [25]. "The increase of highly vulnerable urban systems and communities in many countries to a high risk of climate change is mainly due to rapid urbanization" [28].

4.4 How Architecture Could Play a Role in Mitigating Climate Change Impacts?

The previous examples show how our actions impact the whole system and hence, could hopefully affect it oppositely as well. What is meant by that is to study how our built environments could be more resilient while facing the threats and changes that could face it. To do that we should start by studying the **vulnerability** of our region. As the effect from one region to another (even if they both suffer from the same problem) varies due to different factors and parameters.

It is useful to tackle a **participatory approach** while assessing as the **Indigenous communities** are always aware of aspects that professionals could miss. In the "Handbook of climate change" [29]. It was mentioned that the participatory approach began in 1960s with "Activist participatory research" Afterwards many other approaches started to take place and develop as "field research farming systems", " Participatory research PR", "Rapid Rural Appraisal", "Agro-ecosystem Analysis". Others took place in the 1980s like "Applied Anthropology" [29].

The late 20th and early twenty-first centuries witnessed a shift towards sustainable architecture in response to the increase in global environmental issues. Concepts like **green roofs**, **passive solar design**, and the **use of renewable materials** became prominent. For example, The Leadership in Energy and Environmental Design (LEED) certification emerged as a standard for environmentally responsible building practices. New vocabularies, such as resilient and sustainable architecture, entered the architecture dictionary.

4.5 Resilient Design

Resilient architecture is about designing to face predictable changes and hazards that could occur in a specific area while adapting to it and preserving the safety and the livability of its communities, it is like designing to cope with the problem.

> The architect anticipates change and hazards throughout the building's life and provides design features to mitigate risk and vulnerability. [12]

There is **no one-size-fits-all solution**, as each case differs depending on its level of vulnerability, level of awareness, type of threat, economic state, politics, social relations, and power dynamics. However, there is common ground that can be used to meet all the processes.

4.5.1 Phases of Applying Resilient Design

The first phase is the resilience scope which focuses on understanding the need to apply resilient design at this project or place, type of hazard that could occur (no project or place is immune from hazards). It is the Architects duty to research well and to investigate on different levels to achieve the best results that could last for the future and the benefit of the stakeholders), and most importantly interact with communities and stakeholders as a main part of the process and the design.

First of all, the architect should observe and understand the client's level of awareness. Whether they are unaware, exploring, evaluating, or embedded [12]. After deciding to work on a specific site, a **climate profile** for the area should be done to understand its history with climatic change and the major threat factor on this site. The climate profile studies some main factors such as temperature, humidity, wind, rain, flash floods, and Sea level rise. It helps us understand and predict the climatic future of a certain region [27]. After that and according to the case of the catastrophe and the ability to intervene to a certain level. We should choose the **type of mitigation** that fits the situation the best, whether it's going to be a **preventive**, **corrective**, or **conservative action** [27].

4.5.2 Examples Showing How Resilient Design Could Look on Different Scales

A case study example was given in the Design Resilience Toolkit [12]. This was an example of how we can make design decisions for a house in the mountains region that is vulnerable to wildfire and winter weather. The architects and users needed to design multi-functional mitigation plans for both challenges. The goal is to design a house that is durable and sustainable while affording its occupants safety and livability in the case of hazards. The design process took into consideration the material selection, structure design, energy efficiency, and surrounding site design.

Starting from the site, a buffer zone should be achieved to prevent the house from being too near to the vegetation to limit and decelerate the fire spread to the house. That gives firefighters the chance to have space to do their work and save the situation during emergencies. Designing more than one access and exit facilitates, for the occupants' escape route and evacuation process. Including renewable energy sources and water supply to ensure the functionality of the space. Also Choosing a reinforced roof structure to be strong enough to withstand the snow pressure and accumulation [19].

Figure 2 shows how the designer started by understanding the hazard and the context, thinking of design solutions on macro and micro scale. From infrastructure and the space surrounding the building to thinking of the building's structure, form, and design. While taking into consideration the scenarios and the needs that the occupants will be looking for in the case of facing a hazrad/threat [12, 19].

Another given example on a larger scale is a coastal hotel, that could be affected by sea level rise and storm surge in addition to other contextual characteristics. In Figs. 3 and 4, the designer points out the challenges that surround the building like, the wind load, hydrodynamic pressure from the water waves and movements, sea level rise risk, and the intense heat waves that will affect the top of the building. Figure 2 shows how the regular design could be shifted with initial mitigation strategies that lift up

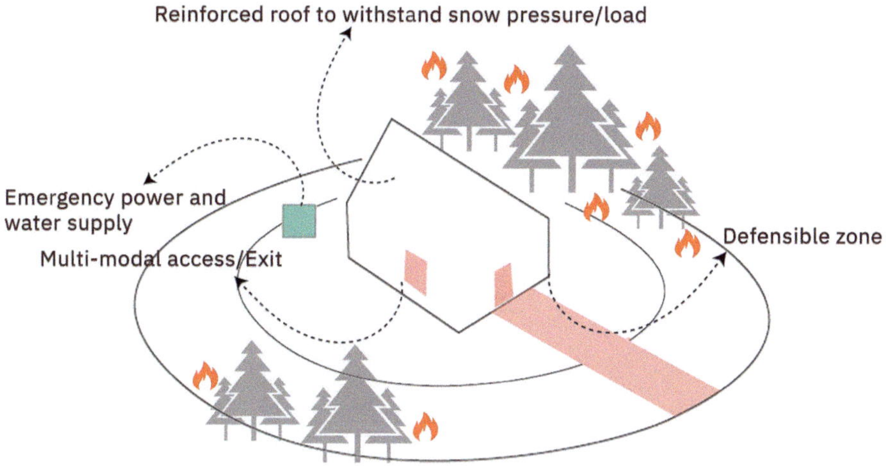

Fig. 2 A study sketch shows the design decisions taken for mitigating the environmental hazards that could face the building. Based on the Resilience Design Toolkit 2023 [12]

the design thinking process, making it more applicable and resilient to its users, and able to fit and adapt to its context conditions.

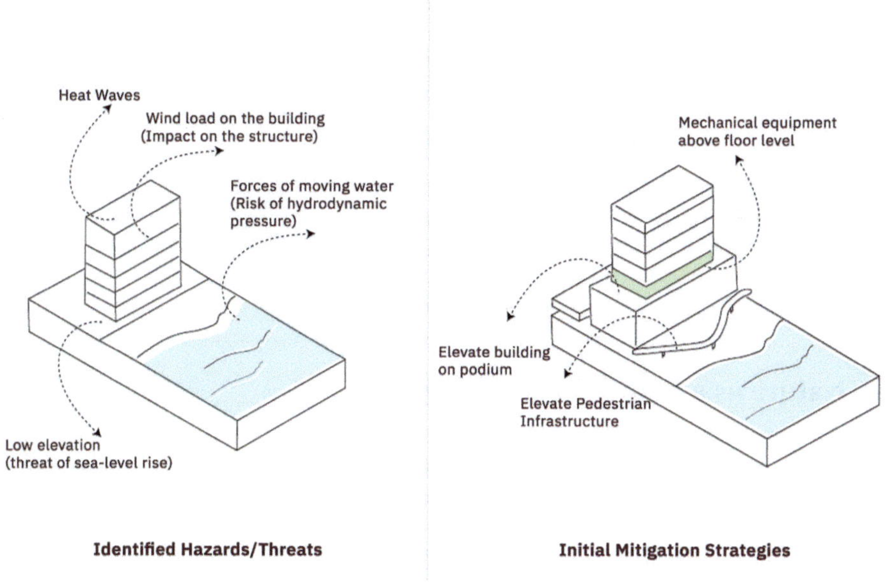

Fig. 3 A study sketch shows the environmental threats versus the proposed mitigation solutions, Based on the Resilience Design Toolkit 2023 [12]

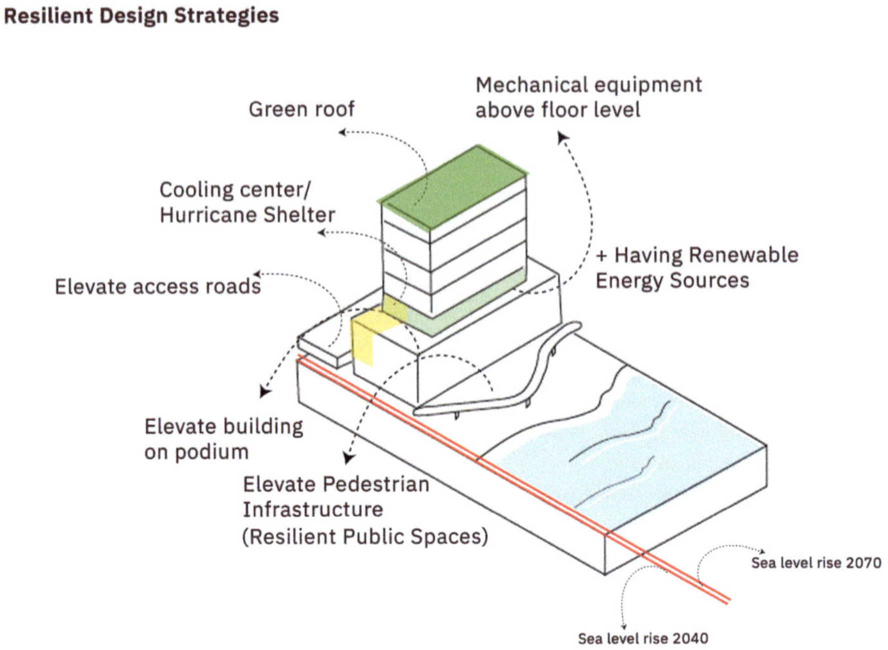

Fig. 4 A sketch shows the final design decisions. Based on the Resilience Design Toolkit 2023 [12]

In the coastal hotel example, for the sea level rise problem, the architect proposed to elevate the pedestrian level, the access, and the podium of the building. To avoid in the cases of flood damage to mechanical equipment, it is planned to be placed above the predicted flood level. And to minimize the heat waves on the roof, green roofing was addressed as a good idea for cooling and insulating the level below it. In addition to having a renewable energy source and a safe shelter area in critical cases [12].

5 Nature as a Source of Inspiration

The concepts of resilient architecture as well as sustainable architecture are not a new trend in architecture, The concepts of resilient and sustainable architecture have deep historical roots, reflecting an age-old awareness and adaptation to environmental and social challenges. One notable example is the traditional Islamic house in Cairo, which showcases design principles to mitigate heat and enhance circulation, demonstrating an early form of sustainable architecture [30, 31].

Beyond human examples, nature itself offers remarkable instances of sustainable and resilient architecture, particularly in the structures built by animals. Nature always offers a great source of inspiration as it automatically has a built-in response and adaptation to all the changes that occur on Earth. In examining these natural construction methods, we can see that animals build for a wide range of purposes—from simple shelter and protection to elaborate structures for mating, nesting, and climate control. Some species engage in daily construction activities, while others create enduring edifices that can last for generations. The process of natural construction often follows a systematic approach, from carefully surveying a suitable site to selecting appropriate building materials and crafting the final form, both externally and sometimes even internally [2, 32].

This wealth of expertise embedded within the natural world is the foundation of our knowledge. One of the most compelling examples of biomimicry in action is the use of termite mound structures to inspire passive cooling systems in buildings.

5.1 A Case Study of Termite Mounds as an Efficient and Resilient Structure

> It is easy to marvel at the forms of nature seen from the outside, but this is not enough. As soon as we pay closer attention, its processes and functionalities become fascinating. [33]

Termite mounds can endure for thousands to millions of years and demonstrate remarkable resilience against environmental conditions and natural disasters. These mounds maintain their passive cooling systems and efficient air regulation throughout their lifespan. Architects can draw inspiration from these structures to create self-sustaining buildings with lower energy consumption and greater efficiency.

The material used by termites to construct their mounds includes soil, termite saliva, and dung, all of which possess inherent thermal insulation properties. These mounds are typically tall and slender, often reaching heights of up to six meters. Their porous structure, characterized by numerous holes, facilitates effective airflow. Utilizing the stack effect, these mounds feature a central chimney connected to an extensive network of tunnels, which termites diligently maintain. This system allows hot air to escape through the central chimney while cooler air settles in the lower parts of the mound, ensuring a stable internal environment [32, 34].

By studying these natural structures, architects can develop building designs that mimic these efficient, low-energy systems, promoting sustainability and resilience in modern architecture. Certain buildings in Zimbabwe and Kenya have drawn inspiration from the passive cooling systems of termite mounds. These innovative designs result in buildings that use up to 90% less energy compared to conventional structures, demonstrating the efficacy of nature-inspired solutions [35] (Fig. 5).

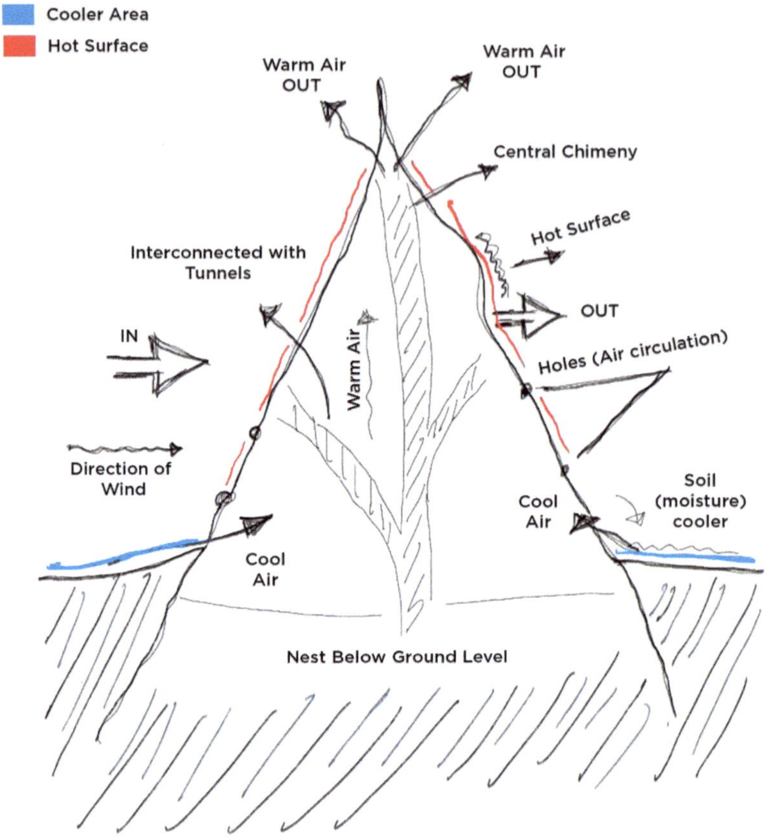

Fig. 5 A study sketch for the case of termite mounds by the author based on [1]

6 Case Studies of Building Resilient Communities in Cairo Through Participatory Work

6.1 "Tree/Greening" as a Key Element for Impactful Solutions

In a world that suffers from water shortage, air pollution, and food insecurity, the tree is the only organism that absorbs carbon dioxide and releases oxygen. It comes as a key element In front of all these problems as a fundamental, primitive, simple yet effective, and always a smart idea to implement even if it cannot solve all the environmental problems alone, they remain a crucial element to consider for social, economic, and environmental sustainability. Despite their importance, the past decades have seen a decline in green spaces, as if planting trees is a luxury rather than a necessity. This neglect overlooks the significant benefits green areas provide to communities, including improved air quality, enhanced mental health, and increased urban resilience. It is essential to recognize and harness the power of trees to support the development of resilient cities (Fig. 6).

Table 1 shows the shrinkage of green spaces in the city of Cairo in three years and its effect per each individual [36, 37].

The upcoming case studies in Al-Khalifa district offer an example of how we can try to build resilient communities by preserving our history represented in the architectural monuments while greening and applying trials of rewilding our city in the hope of achieving healthier environments. Through implementing water management techniques and methodologies that vary from one site to another based on specific circumstances, constraints, and opportunities.

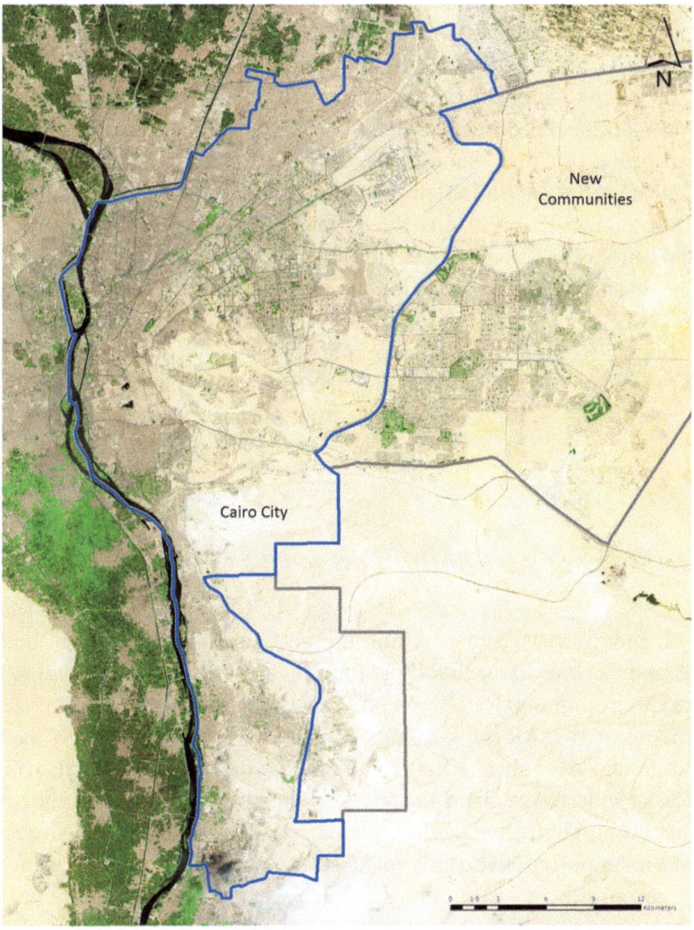

Fig. 6 Satellite image showing the green spots in Cairo in 2019 from public green space quantity and distribution in Cairo [17]. The blue line set the boundaries of Cairo

Table 1 The decline in green spaces over time compared to the increase in population [17]

Year	Population number	Green spaces area m^2	Individuals share of green spaces. M^2/ individual
2017	7,834,388	9,030,128	0.87
2020	6,923,494	9,415,766	0.74

6.2 *Heritage as a Driver for Change, Case Studies from Historic Cairo, from Megawra—Built Environment Collective Framework*

Historic Cairo faces significant groundwater issues that affect many of its buildings, especially the older and historical structures, which are more vulnerable due to to the properties of their construction materials and their lower ground level. The humidity caused by the groundwater threatens the structural integrity of these buildings.

Megawra—Built Environment Collective, a dual entity comprising Megawra—an architectural office in Cairo—and the Built Environment Collective —an NGO— is leading heritage conservation projects through its Athar Lina Initiative. More recently, these conservation efforts have aimed to preserve historical buildings by reducing and mitigating the effects of groundwater. Their strategy presents a win-win solution: conserving historical buildings in the Al-Khalifa district while reusing the groundwater for the surrounding community's benefit.

This includes greening initiatives to enhance the environment and support food security, especially as the neighborhood struggles with rising costs of food. In the upcoming case studies, it is clear how the community is a vibrant factor in the project [8, 38].

6.3 *Case Study of Al-Khalifa Park*

In 2016, "Megawra" started researching the problem of groundwater, and detected monuments and old buildings already getting by groundwater. The land occupied by "Al-Khalifa Environmental Park" was previously used as a debris dumpsite. Located directly in front of two Mamluk domes are flooded by groundwater, the domes of Fatima Khatun and Al-Ashraf Khalil. The land layout is longitudinal, which makes the land take a wide range from the view of the street and on a higher level from the street level, lying between Zynhoum area and Al-Ashraf Street. All these factors encouraged the team to convert this misused plot into a green heaven in al-khalifa, that lacks green spaces. By using piezometers, water was transferred from the corresponding domes to the park. At first, the team did their studies to understand the chemistry of water. The major challenge in the groundwater in this area was not its pollution but its salination. As it is salinated with a considerable amount of salt that not all plants can withstand. Instead of treating the water, the decision was to choose

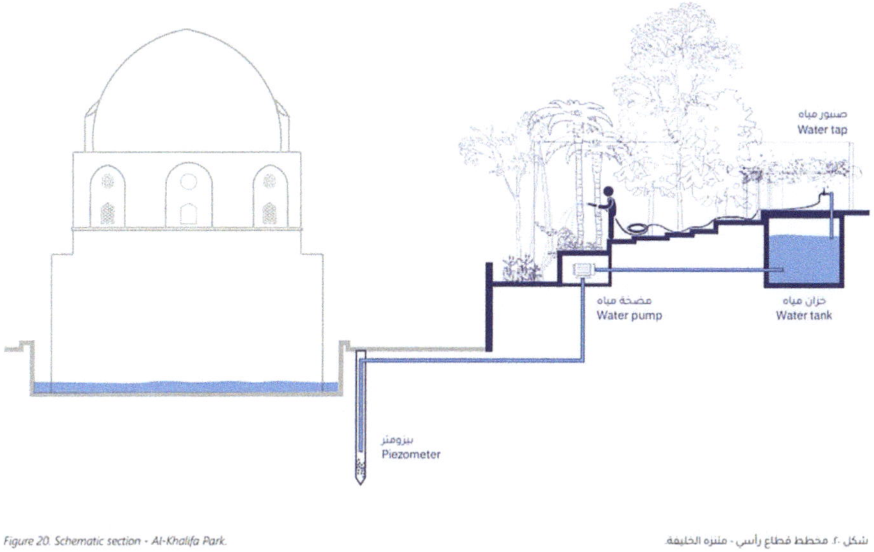

Fig. 7 Showing the process of drawing groundwater using a piezometer at Al-Ashraf Khalil Dome and transferred to Al-Khalifa Park [8]

suitable plants that could withstand salty water. it took time till all the plants got irrigated from the groundwater. At first, all the plants were watered by pure water, then by a mix of pure and underground water, then by a mix but with the increase of the percentage of the groundwater. And then finally all non-edible plants are being watered by 100% groundwater [8, 39, 40] (Figs. 7 and 8).

6.4 *Case Study of Safiyy Al-Din Jawhar Dome*

A similar methodology is being applied in the area surrounding Safiyy Al-Din Jawhar Dome (SDJ). The 2022–2024 conservation intervention by Megawra—BEC includes the implementation of a dewatering system. It collecting groundwater to fulfill the same main two purposes: preserving the monument and water reuse. In this scenario, it is used for greening of urban pockets in the street, specifically two neighbouring alleys, and of rooftops [41].

What is worth considering in megawra initiative framework, is that all its work aims to be from and to the community. By studying and understanding their needs, challenges, and hopes, and trying to fit them into the solutions. Then the initiative involves them through the whole process of achieving a participatory design method, that got achieved through conducting ethnographic and anthropological research before any project, to build considerable knowledge about the community and build

Fig. 8 (A) Photo before the Al-Khalifa Park intervention. (B) Photo after the Al-Khalifa Park intervention. Image courtesy of Megawra—Built Environment Collective/Athar Lina [40]

relations with them that encourage them to be part of the process and believe in the ability to create a change.

An example of this in greening interventions, a needs assessment and food security survey were conducted. Participatory research that fed into these studies included building a mind map to understand the community's patterns of food, what they like/ use on a daily basis. Which crops do they prefer to grow? If they have any previous knowledge of greening or farming?

Understanding that social and power dynamics will affect the processing exchange between people of the community in the future.

That is the paving way to involving them in the design process through conducting community meetings as shown in Fig. 9 sharing and getting feedback on designs and interventions. Suitable presentation tools were used to help them understand the architectural interventions create informed opinions on what is being presented. Even after executing the project, the connection was not cut with the community, as they were involved in the [39] organizing activities for children and women to build a sense of ownership, which Is one of the key factors of preserving the spaces, keeping it alive and [8, 40, 41] (Table 2 and Fig. 10).

7 From Architecture Enabling Communities to Be Resilient, to Communities Inspiring Resilience in Architecture

Usually, when we search for resilient architecture, it is focused mainly on how to create buildings, urban spaces, and cities that are resilient while facing hazards caused by the natural environment and climate. However, it is very inspiring and worth studying to observe and analyze how people stand resilient in the face of more brutal and violent hazards caused by "humans". The losses that wars, armed forces, and conflicts cause are sometimes more effective and catastrophic, carrying massive emotional and materialistic losses that are very hard to heal. We tend to calculate and anticipate when a specific environmental hazard could occur after years and start to apply plans years before to avoid losses that could be caused by it. But how can we predict and plan for insane wars that could happen over one night causing millions of people to be homeless, demolishing their history, and opening many questions about an uncertain future?

How architecture got affected and how people are being resilient in such situations, is a great source of inspiration, to understand how a group of people could fight and have their right to exist, live and appropriate in some situations that seem hopeless at first sight [19, 42, 43].

Wars have been a constant throughout history, influencing architectural design significantly. This is evident in the development of defensive architecture, Certain building typologies occurred for defense and protection, such as forts and castles that were usually built on higher levels to overview the whole city. On the urban scale, the streets were designed to be narrow (so that any stranger could be detected easily) and each neighborhood had a door that could be closed during threat times, while on the city scale, cities were surrounded and encircled by walls and fences. Even on a residential level, the home was designed to afford privacy and a high sense of security, making it hard for a stranger to know about the home's interior or access it easily. Starting from the narrow and small openings on the exterior façade, moving to the inclined entrance doesn't allow an outsider to see and understand the

Fig. 9 A and B from a community meeting around Safiyy Al-Din Jawhar Dome, in one of the targeted alleys for greening intervention with the neighborhood, to share with them the designs as part of communit engagement and pariticpation process. Images by Salma Shehatta, courtesy of Megawra—BuiltEnvironment Collective/Athar Lina (www.atharlina.com)

spatial arrangement of the interior. Now, all of that is not applicable in front of the advancement of weapons technology that could destroy a whole city in a couple of seconds [10, 19, 42] (Table 3).

As an illustration of climate change in a city like Gaza, Fig. 11 demonstrates the rising temperatures over the years. In this figure, darker bars indicate hotter temperatures, while lighter colors represent cooler weather (Fig. 12).

Table 2 Defining the stages and ways of community involvement through the whole process

Stages and ways of **community involvement** throughout the whole process	**Before** the project	**During** the project	**After** the project
	• Ethnographic research work. (observatory/ interviews in groups and individuals) Therefore that leads to the next phase of designing and taking decisions that respond to community needs • Awareness sessions In case the community is not aware of the bigger scope of the project or doesn't know who are we and what we are aiming to do in their neighborhood	• Follow the ethnographic work development (observatory/ interviews in groups and individuals) • Participatory design through community meetings for consultation • Receiving and observing the ongoing impressions/ feedback from the community on the work/intervention	• Share in the project management plan • Conducting community events/ events, for kids and adults to raise their ownership feeling towards the space/monument • Evaluating the results and feedback

However, the community envolvment process is not linear, this representation meant to show examples of the community involvement through various stages of a project, which could change and intersect according to each project and its affecting factors. By author(s) based on practical experience in the field work of Megawra and Athar-Lina [39] and Athar-Lina [40, 41]

So, what are the types of resilient architecture that people tend to apply in the face of wars and armed forces?

We can categorize them into two types.

8 Resilient Strategies for Everyday Life During the Occupation

8.1 Individual Greening Initiative in the Land of Gaza

Example one: In situations where individuals are trapped in their country under occupation, with their government lacking the capacity to enact meaningful change, we often observe the emergence of innovative and resilient strategies aimed at improving living conditions. This phenomenon is particularly evident in Gaza, Palestine, where the ongoing blockade, restrictions, and rapid urbanization prioritize the construction of residential buildings to accommodate a growing population, resulting in a significant scarcity of green spaces.

Fig. 10 A photo collage from a story designed for kids (as part of the awareness actions with the community before and during the project) to create a simple narrative for telling the children the story of the monument of Yahya Al-Shabih and how it suffers from the groundwater in a simple way that allows us to create a narrative for the kids to be aware of their context potentials and challenges. As well as being part of the process. By author(s)

A compelling case study is that of Suzan Atallah, an architect who faced the limited job opportunities like many of her contemporaries. She took the initiative to address this pressing issue through her project, "Veranda." By studying green economy and climate change, she incorporated sustainable practices into her work, focusing on creating green spaces on rooftops and balconies. This initiative enriches the urban landscape with greenery and contributes to food security, exemplified by her efforts to cultivate apple trees and other crops in such unconventional spaces. Among her notable achievements is the establishment of a children's garden in Deir Al-Balah, funded by a Swiss grant, showcasing the potential of individual resilience in architecture to foster community well-being amidst challenging circumstances [18].

Table 3 A comparison between Climate Change and man-made disasters represented in Wars/conflicts

Point of comparison	Climate change	Wars/conflicts
Crisis can be predicted/anticipated	**Yes** (by studying the climatic history of a specific area we can predict that a specific hazard could occur after number of years, like an earthquake, sea-level rise..etc.)	**Not often** (As a country can decide over one night to drop a bomb on another city, it's not a natural phenomenon that we could trace, it's a human decision to devastation)
Upon the prediction, we can design resilient architecture to mitigate it	**Yes** (as given in the previous examples in Figs. 2 and 3)	**Not often**
Reconstruction of buildings can be done after the hazard	**Yes**	**Yes** (It can happen but depends on when we decide brutal actions to stop. It can continue for months and years with more daily losses)
Affects the economy, emotions, and politics	**Yes**	**Yes**
Does it have an influence on the other type of crisis?	**Not often** (climate change is not a normal cause for wars)	**Yes** (Wars do affect the environment and climate change) Figs. 12 and 13
Needs a study and working hand in hand with the community	**Yes**	**Yes**

By author(s) based on United Nations Development Programme Programme of Assistance to the Palestinian People [44], Fry [45], and Kiyala [46]

8.2 *Everyday Use of Architecture as a Form of Resilience*

The interconnected perception of architectural elements and spaces has shifted significantly during the occupation. The presence of danger and control redefines spaces and structures that once felt safe and familiar.

Walls, windows, and doorways are no longer just structural features; they become potential shields or threats. This shift changes how people navigate and interact with their environment, emphasizing the complex relationship between occupation and its impact on architectural perception. Where streets and communal areas are no longer open shared spaces for people to communicate, they shift to be fire and battle zones. Markets are being targeted areas and Windows are no longer framing the beauty of the view, instead, it's a way for the gunshot to penetrate you [49]. People had to invent ways to mitigate such brutal actions that could occur at any moment every day.

For example, having escape spaces to shelter them from sudden attacks is a form of resilience and survival. Figure 14 shows how in a primary school in 1980, kids were taught how to act and where to go in the case of "threat". Shelters have various

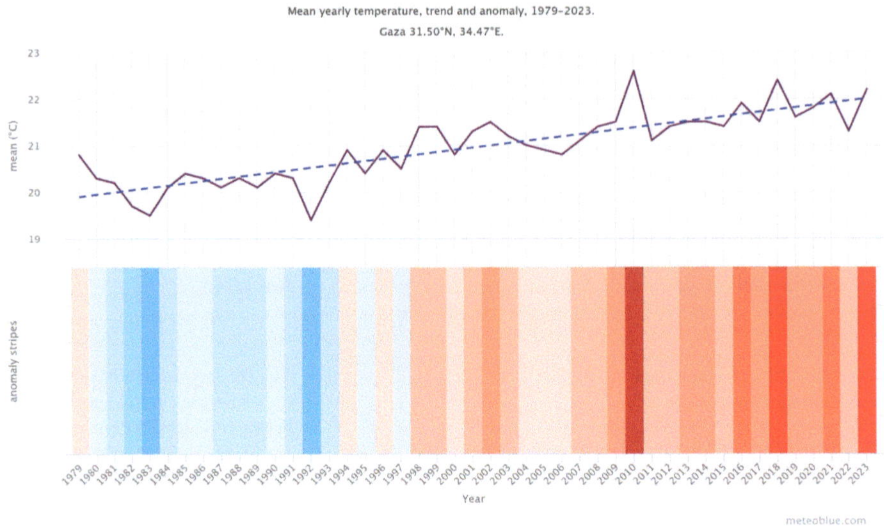

Fig. 11 Graph showing the climate change effect on the city of Gaza from 1979 till 2023, indicating how it's getting warmer, and having the warmer year in 2010 [47]

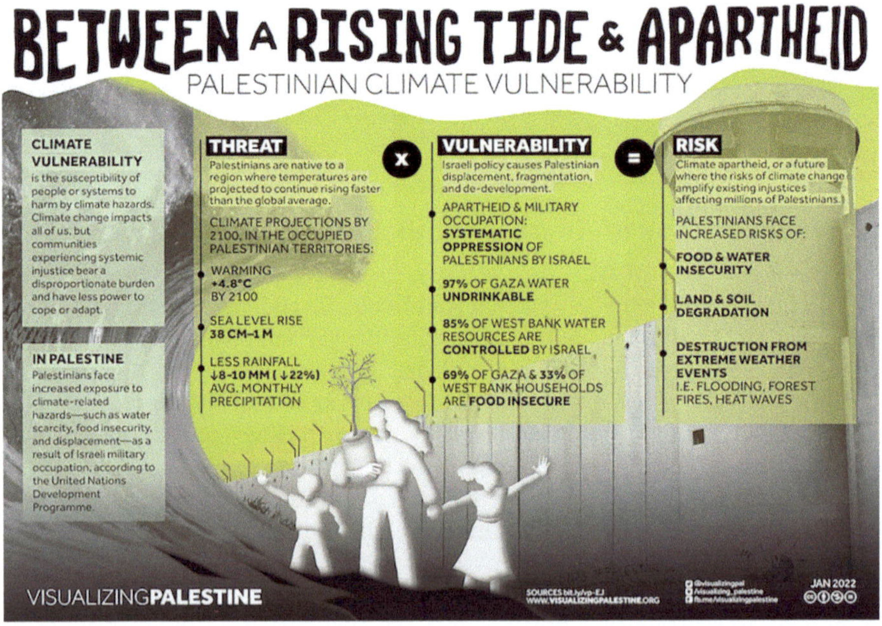

Fig. 12 Showing the effect of war on the climate vulnerability of the city of Gaza [48]

Fig. 13 Photo by Thomas Billhardt, from his photo book "Where is Palestine" Showing how kids were taught at schools where to escape during attacks [50]

types, whether underground, creating this duality of home and shelter under it. even if it was not built or designed to shelter citizens, people reuse spaces like basements and garages to do this function.. or abandoned buildings that are not targeted by the armed forces [10] (Fig. 13).

8.3 A Shift in the Uses of Architectural Spaces and Architectural Dictionary as a Tactic for the Daily Resilience Actions, Case of Kasbah, Nablus

"The backbone of this matrix of "*sumoud*" is the specific architecture of the kasbah and its urban tissue" [49]. In the paper "*Architecture as an Agency of Resilience in urban armed conflicts*", writers showed how in the case of Nabuls, the inhabitants created their new architectural dictionary or matrix to resist the armed forces and find a way of survival in their everyday lives. Citizens perceive their city through main types of spaces including, nodes (as the spaces) and sneaks (as the non-exposed routes to reach the spaces):

Nodes: They are the places where they can shelter (as mentioned before in basements for example) or apply for emergency medical care in a safe area away from the fire zones, sometimes in mosques. Other times nodes had the function of storing basic needs and necessities as food, fuel, and water [42].

Sneaks are "Alternative circulation patterns" where regular citizens can reach them through various types of "Junctions".

Junctions are the connection points that lead you to the escape route or the "sneak". A junction could be a back door or window [42].

9 Post-War Resilient Actions

9.1 Case Study of Al-Khan Al-Ahmar School, Palestine

An inspiring example of how good architecture can emerge even in the most challenging and hopeless conditions is the School of Al-Khan Al-Ahmar.

Al-Khan Al-Ahmar is home to a Bedouin community that was forcibly displaced from other areas. However, this location falls within Zone C (Oslo II division of the West Bank under full Israeli control), a highly contested area where both Israelis and Palestinians coexist under complex and restrictive conditions. The community faces numerous obstacles, prohibitions, and limitations that make daily life incredibly difficult. One such restriction is the prohibition of constructing any permanent building and not allowing to use of cement in their construction [51–53].

Despite such political, economic, and environmental challenges (hot and dry weather in summer and occasional rainfalls in winter), when Al-Khan Al-Ahmar School was built, the community turned to the most readily available materials in their surroundings, which are "tires" to be the main building material for construing the only school available till now for the people of Al-khan Al-Ahmar that consists of sixty families with their children. The school walls were constructed primarily from recycled tires, packed with sand, an abundant and unrestricted resource that also acts as thermal mass to stabilize indoor temperatures. Tires absorb heat during the day and release it at night, helping to regulate indoor temperatures naturally. For the roofs, sandwich panels supported by **wooden beams** were used, with an extended overhang offering shade and sloped to facilitate rainwater drainage during the winter to prevent accumulation and potential damage with openings locations that allow natural cross-ventilation within the classrooms [51–53].

After the school became actively used, a new need for expansion of spaces evolved, however, they were not allowed to increase the volume of the school by any means. They added spaces for the community using "straw" as a smart building material in such a context where the Bedouins work mainly in herding sheep and camels, so straws will be used from their production, an easy material to find around the context and have properties as it has high acoustic and thermal insulation which fits perfectly with the high temperature weather during summer at Al-khan Al-Ahmar where people mention that they cannot stay inside their shelters that are built from corrugated sheets, due the high temperature gained by the absorption of the corrugated sheets. In addition to its breathability, it allows air and moisture to pass through the indoors naturally [51–53] (Figs. 14, 15, 16, and 17).

Fig. 14 A photo through the process of building Al-khan Al-Ahmar school from the tires [51, 52]

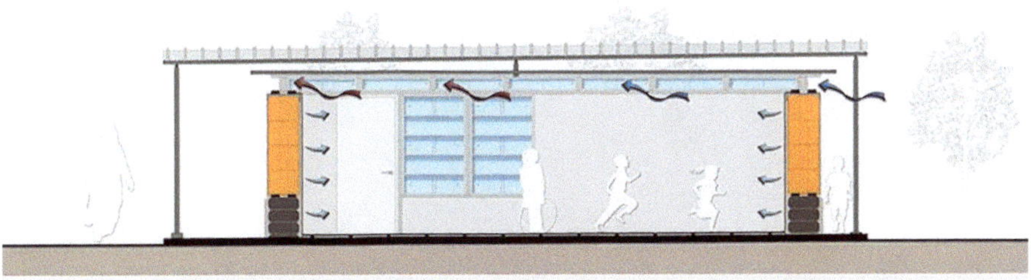

Fig. 15 The diagram shows air circulation using arrows, Cool air (blue arrows) enters from the lower vents on both sides of the building, passing through the porous straw walls. While hot air (red arrows) rises and exits through upper openings, creating natural cross-ventilation. The diagram also shows that the base of the wall is from recycled tires, acting as a stable foundation to place the straw walls above it which acts as a thermal insulator letting cool air pass through it to the interior. The extended-sloped overhang offers shade and facilitates rainwater drainage during the winter [51, 52]

Fig. 16 A photo through the process of the new phase of adding spaces to Al-khan Al-Ahmar school using straw as an easily available material to find in the bedouin camp helps the farmers by using their production and breathable with thermal and acoustic insulation [51, 52]

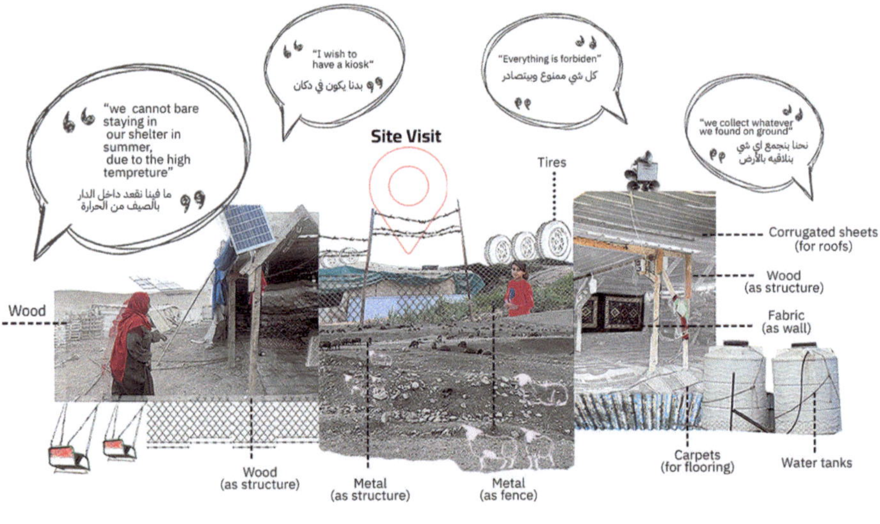

Fig. 17 A collage of the available material in the context of Al-khan Al-ahmar with pop-up words from the Bedouin community. Photos by: Sarah Jamjoum, collage by (the author)

9.2 The Architecture of Refugee Camps, a Case Study of "Zataari Refugee Camp"

After displacements, the architecture of Refugee camps was established, passing by many stages. One of the Refugee camps that is worth studying, is "The Zataari refugee camp". The largest Syrian refugee camp in the world. located in Jordan, in the northern region of Mafraq. Almost 20 percent of Jordan's population is in Zataari camp, occupying 1.4 million Syrian refugees. The process of creation of Zataari camp started in 2013 when some Syrians crossed the borders during the civil war in Syria to reach Jordan and started to dwell and create their basic living needs from scratch. Then, as more families are settled in the area, it gradually developed to be a city within the city. With markets, shops, and communal kitchens. It turned from being a temporary stay to being a home for an unknown number of years of an uncertain future. Through these years people change, some get married, others give birth and life continues. As well as practicing change in their living spaces to respond to these changes, social needs including gender, family relations, privacy needs, and hierarchy, in addition to many other aspects and dynamics that affect these processes, let one caravan arrangement be different from the other.

The process developed to have certain typologies for refugee shelters. Starting from a separate tent to a tent attached to a caravan and then to only having the caravan form, adding, adjusting, and subtracting from it, creating corridors, shared areas, and spatial spaces of each family house according to their own needs. It is worth looking to experience how ordinary people generally understand architecture, practice place-making, and problem-solving processes, and apply an example that architects should take a deep look and re-think then what should our role be. "considering refugees as architects can lead us to question the limitations and potential of architecture as a practice" [11]. With this massive increase in the number of refugees. The government started thinking about shifting this camp from being an emergency shelter to representing a humanized place and offering a better way of living for people there. Here it came to the role of the government and organizations to enhance the whole state of the refugees at Zataari camp by trying to build infrastructure, connect water supplies, and supply electricity by using solar panel sheets instead of sodium street lights. it according to their requirements. Zaatari Camp offers a clear example of the following three points: how conflicts have affected people and architecture for years, how people respond and mitigate spontaneously on all levels including architecture, how architects should study and learn from people's responses and solutions, by building on it to offer them the least good living state of life they deserve to have as human beings [20].

10 Conclusion

In conclusion, the chapter rethinks the meaning of architecture through the lens of resilience. The term resilience is highlighted by its different perceptions and means.

Which is already embedded in people who practice it every day. Shows that building, constructing, and dwelling is not only meant by architects; it's already done and happen every day by people to cope with circumstances they find themselves obligated and forced to face. Being resilient is an instinct to live and exist. How it is translated into architecture is a powerful source of inspiration for professionals to rethink what is the role of architecture, how it can serve humanity and the environment, how people interact with it in their lives, and how architecture is not about rigid masses and high buildings. It's a visual language that is interconnected with various fields and circumstances. Crafting that being an architect, means that you should be an observer first, observe threats, opportunities, natural patterns, and primitive solutions, as well as observe people's challenges and hopes, engaging them in the process of creating a participatory work from their needs. As discussed, architecture is influenced by various fields, including both the tangible and intangible aspects of its surrounding environment. This places a responsibility on us to be keen observers, aware of future threats such as climate change. We are empowered by theory and practice to enhance our situation by drawing from previous mitigation and adaptation actions. Additionally, we can learn from indigenous knowledge, engage with communities. Furthermore, sometimes we can just turn some difficulties into stepping stones for solutions, leveraging our local resources and transforming challenges into opportunities for innovative approaches in a collaborative mindset with the local communities.

11 Relevant SDGs

This chapter aligns with global efforts to address sustainability and resilience in architecture, particularly in the context of climate change, urban planning, and social justice. It directly relates to the United Nations Sustainable Development Goals (SDGs), including SDG 6 (Clean Water and Sanitation), SDG 9 (Industry, Innovation, and Infrastructure), SDG 11 (Sustainable Cities and Communities), SDG 13 (Climate Action), SDG 15 (Life on Land), and SDG 16 (Peace, Justice, and Strong Institutions). Each section will explore how architecture can contribute to these goals through innovative and adaptive design strategies.

- **SDG 6 (Clean Water and Sanitation)**: Water management and greening initiatives in urban environments.
- **SDG 9 (Industry, Innovation, and Infrastructure)**: Nature-inspired and biomimetic design strategies for resilient buildings.
- **SDG 11 (Sustainable Cities and Communities)**: Sustainable urban planning, adaptive reuse, and resilient architecture in historic and modern settings.

- **SDG 13 (Climate Action)**: Mitigating climate change through architecture, green spaces, and energy-efficient design.
- **SDG 15 (Life on Land)**: The role of greening in enhancing urban sustainability and biodiversity.
- **SDG 16 (Peace, Justice, and Strong Institutions)**: Architecture as a tool for resilience in conflict zones and post-disaster reconstruction.

Throughout your chapter, explicitly mention SDGs where they apply. For example:

Climate Change and Architecture (SDG 11, SDG 13, SDG 9)

The impact of climate change on architecture and urban planning is a major challenge, directly linked to SDG 11 (Sustainable Cities and Communities) and SDG 13 (Climate Action). By integrating climate-responsive architecture, adaptive reuse, and sustainable materials, we can create cities that are resilient to environmental stressors while reducing their carbon footprint (SDG 9: Industry, Innovation, and Infrastructure).

Water Management in Historic Cairo (SDG 6, SDG 11)

The reuse of groundwater for greening in Old Cairo presents an innovative approach to sustainable water management (SDG 6: Clean Water and Sanitation). By addressing the rising groundwater levels that threaten heritage sites, this initiative also contributes to SDG 11 (Sustainable Cities and Communities) by preserving cultural heritage and improving urban resilience.

Architecture in Conflict Zones (SDG 16, SDG 11)

Resilient architecture in war-affected regions, such as Palestine and Syria, plays a critical role in supporting displaced communities (SDG 16: Peace, Justice, and Strong Institutions). From emergency shelters to long-term urban strategies, architectural solutions can provide stability and preserve cultural identity in post-conflict recovery (SDG 11: Sustainable Cities and Communities).

12 Recommendations

12.1 The Architect as an Observer and Learner

A good architect is first and foremost a keen observer. This chapter highlights the importance of studying and learning from the surrounding environment to develop innovative and context-sensitive solutions. Inspiration can be drawn from multiple sources:

- **Nature**: Observing biological resilience and adaptive systems, such as termite mounds inspiring passive cooling techniques in architecture.

- **Historical and Traditional Knowledge**: Integrating time-tested architectural solutions, like Islamic courtyard houses in Cairo, with contemporary technologies to meet modern needs.
- **Community Practices**: Understanding how people naturally adapt to challenges in their daily lives, as seen in the Kasbah of Nablus, where residents have developed alternative circulation patterns for safety under occupation. As well as the case study of Al-Khan Al-Ahmar School, which was built with the minimum resources, however, it fills the need and responds as well to climate and the surrounding context constraints.

Actionable Steps:

- Architects should incorporate biomimicry into their designs, learning from how nature adapts to environmental conditions.
- Historic architecture should be studied not only for conservation but also for reinterpreting traditional methods in modern, sustainable ways.
- Community engagement starting from firstly understanding them through Ethnographic research and community mapping can document how people engage with their built environment, ensuring architecture responds to real needs and that they will be encouraged to be part of the whole process of design and implementation.

12.2 *Community Engagement as a Foundation for Resilient Design*

Designing and building hand in hand with communities is crucial for creating spaces that are functional, meaningful, and sustainable. When architects involve local people in the design and construction processes:

- They gain deeper insight into local knowledge, traditions, and climate-adaptive strategies.
- The community benefits from skill development, ownership, and a strengthened sense of ownership towards their built environment.
- Trust is established, ensuring that projects remain aligned with community aspirations and long-term well-being.

Case Study Example:

Megawra's community-based conservation work in Historic Cairo, where local residents were actively involved in preserving monuments and transforming groundwater issues into greening and urban farming solutions.

Actionable Steps:

- Architects should conduct participatory workshops where community members contribute ideas and insights.

- Engaging local artisans and builders ensures that traditional techniques are preserved and valued.
- Developing long-term maintenance plans with community input fosters sustainability beyond initial construction.

12.3 The Architect's Expansive Role in Society

Architecture is not just about constructing buildings; it is deeply intertwined with social, economic, and environmental challenges. Architects should recognize their critical role in shaping communities by:

- Designing solutions that respond to social justice issues, such as displacement and economic inequality.
- Acting as mediators between communities, policymakers, and stakeholders, ensuring equitable urban development.

Case Study Example:

Zaatari Refugee Camp in Jordan evolved from a temporary emergency shelter to a self-sustaining city through community-led modifications, reflecting how displaced people actively shape their built environment.

Actionable Steps:

Architects should collaborate with policymakers to integrate resilience into urban planning.

- Architectural education should incorporate social impact training, emphasizing the role of design in addressing real-world challenges.
- Encouraging architects to work with NGOs and humanitarian agencies can expand architecture's humanitarian role.

12.4 Turning Challenges into Opportunities

Problems can sometimes turn into solutions when approached with resourcefulness. A key example discussed in the chapter is the **groundwater issue in Historic Cairo**, where an environmental challenge was transformed into an opportunity by reusing groundwater for **greening projects and urban farming**. This approach demonstrates that:

- Environmental threats can be mitigated through adaptive reuse strategies.
- Local resources can be leveraged for broader community benefits, such as improving public spaces and food security.
- Architecture, when thoughtfully applied, can turn constraints into catalysts for innovation and regeneration.

Case Study Example:

Al-Khalifa Park in Cairo, where groundwater threatening historic monuments was repurposed for irrigation, transforming an abandoned space into a thriving green community hub.

Actionable Steps:

- Architects should conduct environmental impact assessments to identify hidden opportunities within challenges.
- Developing low-cost, high-impact solutions can make sustainability more accessible in underserved communities.
- Strengthening collaborations with environmental experts and urban ecologists ensures solutions are scientifically sound and replicable.

References

1. PBS (2011) The Incredible Termite Mound. [Online]. Available at: https://www.pbs.org/wnet/nature/the-animal-house-the-incredible-termite-mound/7222/
2. Vrcelj Z, Jamei E (2021) Biomimicry and the built environment, learning from nature's solutions. [Online]. Available at: https://www.mdpi.com/2076-3417/11/16/7514
3. Shah V (2023) The Role of Architecture in Humaity's story. [Online]. Available at: https://thoughteconomics.com/the-role-of-architecture-in-humanitys-story/
4. McIntyre MH (2006) A literature review of the social,economic and environmental impact of architecture and design. s.n., s.l.
5. Cutieru A (2021) How social sciences shape the built environment. [Online]. Available at: https://www.archdaily.com/962538/how-social-sciences-shape-the-built-environment
6. United Nations (2021) Human activity the common link between disasters around the world. [Online]. Available at: https://www.ungeneva.org/en/news-media/news/2021/09/69141/human-activity-common-link-between-disasters-around-world
7. Sherif S (2022) Cairo's Green Sprawl: The Move of Urban Green Space towards Exclusivity. [Online]. Available at: https://aps.aucegypt.edu/en/articles/947/cairos-green-sprawl-the-move-of-urban-green-space-towards-exclusivity
8. Al-Ibrashy M et al (2022) Athar Lina toolkit for groundwater solutions in historic settings-treatment and reuse of water. Megawra Papers, s.l.
9. Zahran O, Micklewright J (2024) How The past meets the future: greening strategies with pocket parks in historic Cairo
10. Okasha A (2011) The Architecture of Resilience. Ahmed Bahaa Al-Din assosiation, s.l.
11. Dalal A (2022) From Shelters to Dewllings. s.n., s.l.
12. Hsu S et al (2023) Resilience Design toolkit. In: Resilience Design toolkit. AIA, s.l.
13. Crook L (2021) Ten ways in which architecture is addressing climate change. [Online]. Available at: https://www.dezeen.com/2021/04/22/architecture-climate-change-earth-day/
14. Jungudo MM (2022) The Impact of Climate Change in Egypt
15. Ragheb A, El Shimy H, Ragheb G (2015) Green Architecture, a concept of sustainability
16. Ugai T (2015) Evaluation of sustainable roof from various aspects and benefits
17. Dimitrijevic B, Aly D (2022) Public green space quantity and distribution in Cairo, Egypt. [Online]. Available at: https://jeas.springeropen.com/articles/10.1186/s44147-021-00067-z
18. UN women Palestine_Green spaces (2022) [Film]. UN women Palestine, s.l.
19. Burayidi MA, Allen A, Twigg J, Wamsler C (2020) Handbook of urban resilience. Routledge, s.l.

20. Carlisle L (2022) The UN Refugee Agency. [Online]. Available at: https://www.unhcr.org/news/stories/jordans-zaatari-refugee-camp-10-facts-10-years
21. Ketana A (2022) "From every neighborhood, house and street": on the role of architecture in resisting colonialism. February
22. Tapalaga A (2023) History of Yesterday. [Online]. Available at: https://historyofyesterday.com/the-role-of-ancient-egypt-in-shaping-modern-architecture/
23. Huyssteen Jv (2023) Art in Context. [Online]. Available at: https://artincontext.org/history-of-architecture/
24. Nickles T (2024) Scientific Revolutions. [Online]. Available at: https://plato.stanford.edu/archives/spr2024/entries/scientific-revolutions/
25. Ghoneem MYM (2015) Planning for climate change, why does it matter?
26. Architecture 2030 (2023). [Online]. Available at: https://www.architecture2030.org/. Accessed 2024
27. Medhat A (2024) ARCE Egyptian cultural-natural heritage and climate change. s.l., s.n.
28. Al-Ghamdi SG, Al-Humaiqani M (2022) Exploring the significance of resilience qualities in the context of the middle east built environment. In: Environment and Sustainable development. Springer, s.l., pp 319–330
29. Aliou Diouf ATG (2015) A methodological framework for building an index for vulnerability assessment in rainfed agriculture. In: Filho WL (ed) Handbook of climate change adaptation. Springer, Sénégal, pp 3–15
30. Sepehri H et al (2023) Traditional dwellings in four middle eastern cities: adaptation strategies to harsh climate in privacy. In: Climate change and cooling cities. Springer, s.l.
31. Petrucci C (2023) Thermal Comfort in traditional and sustainable buildings: comparative analysis of thermal perception in the Middle East and Southeast Asian buildings. Università Ca' Foscari Venezia, 14 July
32. Laidre ME (2021) Animal architecture. [Online]. Available at: https://www.cell.com/current-biology/fulltext/S0960-9822(21)01350-6
33. Estevez AT (2021) Sustainable nature-inspired architecture. In: Sustainable engineering technologies and architectures. s.n., s.l.
34. Korb J (2010) Termite mound architecture, from function to construction. In: Bignell DE, Roisin Y, Lo N (eds) Biology of termites: a modern synthesis. Springer, Dordrecht
35. Wood, J (2024) 5 ways giant termite mounds can help architects create more efficient and sustainable buildings. [Online]. Available at: https://www.weforum.org/agenda/2024/06/termite-mounds-sustainable-architecture/
36. Lotspeich J (2023) Why planting and growing trees matters for climate, people and biodiversity. [Online]. Available at: https://www.weforum.org/stories/authors/john-lotspeich/
37. Kirschbaum MU (2024) Is tree planting an effective strategy for climate change mitigation? Sciencedirect, 20 January
38. Coppola M, Di Benedetto F, Pecchioni E, Santo AP (2020) Groundwater damages on the historic buildings of Cairo. [Online]. Available at: 346581171_Groundwater_damages_on_the_historic_buildings_of_Cairo_the_case_of_the_medieval_walls_of_Mokattam_limestone_Groundwater_damages_on_the_historic_buildings_of_Cairo_the_case_of_the_medieval_walls_of_Mok
39. Athar-Lina (2018) Athar-Lina. [Online]. Available at: https://atharlina.com/ar/projects/athar-lina-groundwater-project/
40. Athar-Lina (2022) Athar-Lina. [Online]. Available at: https://atharlina.com/ar/projects/alkhalifa-park/
41. Athar-Lina (2023) Athar-Lina. [Online]. Available at: https://atharlina.com/projects/erth-athar-lina-heritage-and-climate-project/
42. Kittana AM, De Meulder B (2019) Architecture as an agency of resilience in urban armed conflicts
43. Mir JE (2024) Planning for resilience in and after conflict—a world away. Wilson Center, 3 June

44. United Nations Development Programme Programme of Assistance to the Palestinian People (2010) Climate change adaptation strategy and programme of action for the Palestinian authority. [Online]. Available at: https://eprints.lse.ac.uk/30777/1/PA-UNDP_climate_change.pdf. Accessed 2021
45. Fry T (2019) Climate, change and war. In: Unstaging war, confronting conflict and peace. Springer, s.l.
46. Kiyala JCK (2024) Climate change and climate wars in the anthropocene. In: Kiyala JCK, Chivasa N (eds) Climate change and socio-political violence in Sub-Saharan Africa in the anthropocene. Springer, s.l.
47. meteoblue (2024) meteoblue. [Online]. Available at: https://www.meteoblue.com/en/climate-change/gaza_palestine_281133
48. Visualizing Palestine (2022) Visualizing Plaestine. [Online]. Available at: https://visualizingpalestine.org/visual/between-rising-tide-and-apartheid/
49. Kittana AM, Leuven B, De Meulder B (2019) Architecture as an agency of resilience in urban armed conflicts. Archnet-IJAR
50. Billhardt T (1988) Where is Palestine. [Online]. Available at: https://www.reddit.com/r/Palestine/comments/18h8lq9/wo_liegt_pal%C3%A4stina_eng_where_is_palestine_a_photo/?rdt=38382
51. ARCò - Architecture & Cooperation (2014a) Architecture in development.org. [Online]. Available at: https://architectureindevelopment.org/project/357
52. ARCò - Architecture & Cooperation (2014b) Architecture in development.org. [Online]. Available at: https://architectureindevelopment.org/project/446
53. Firrone T (2024) IRIS UNiPA. [Online]. Available at: https://iris.unipa.it/handle/10447/641949
54. Segal C (2016) PBS News. [Online]. Available at: https://www.pbs.org/newshour/arts/zaatari-refugee-camp-syrian-photographer

Developing a Framework for the Design of a Resilient Social-Ecological System in the Arab Region: An Overview on Arid and Semi-arid Ecosystems

Noha Essam Khamis

Abstract The Principles for Social-Ecological Urbanism, which was released in 2013, served as the conceptual foundation for the context of the Social-Ecological Urban field. It highlights how thoughtful institution and urban form design—both influenced by urban discourse—can address resilience in interconnected social and ecological urban systems. Globally, cities are realizing more and more that they must strengthen their resilience in order to handle converging challenges, including natural disasters, deteriorating environments, urban development, and climate change. In the last several years, the Arabian countries have witnessed profound transformations of their social urban paradigms from the conventional approaches of generous built-up areas and urban sprawl designs to more environmentally, socially, and economically sustainable and resilient social buildings. Through a critical, fact-based analysis of the decision-making processes and urban-architectural products that drove its development, this paper investigates the state of recent social urban projects in a few Arab cities. The aim of this Research Topic is to highlight promising, recent, and innovative research in the subject of social-ecological urbanism. In order to assess the local and global sustainability agendas put into action, mixed-methods data were created using archival research, semi-structured interviews, and field observations. The study also provides evidence of legitimate environmental practices, local urban practices, and community well-being. In order to enlighten upcoming social urban developments, it concludes with suggestions for overcoming problems with the current procedures surrounding the design and implementation of sustainable urban forms.

Keywords Resilient cities · Social-ecological urbanism · Urban development · Sustainable developments goals

N. E. Khamis (✉)
Design Department, College of Architecture & Design, Effat University, Jeddah, Saudi Arabia
e-mail: Noha.e.khamis@gmail.com; nkhamis@effatuniversity.edu.sa

Z. ElZein and A. Negm (eds.), *Recent Approaches of Sustainable Architecture in Arid and Semi-arid Cities*, Urban Sustainability,
https://doi.org/10.1007/978-981-95-1645-2_4

1 Introduction

The study presents a review of the most recent developments in the dynamic shift of arid and semi-arid ecosystems in the Arab region. According to recent research, the Arab region's arid land area has been growing over the past several decades and is expected to continue throughout the upcoming decades as a result of a larger warming effect than in other regions. The productivity, biodiversity, and stability of the arid ecosystem in the Arab region have all suffered as a result of the warming trend and increased human activity. The rate of land degradation and desertification was accelerated by population growth, overgrazing, and grazing abandonment.

Limited water resources, significant interannual rainfall variability, a high frequency of extreme climate events, substantial hydrologic variability, and spatial heterogeneity are the characteristics of ecosystems of arid lands [1]. Arid regions are, therefore, among the most vulnerable to the effects of climate change and human activity [2], which could have a permanent effect on the ecosystem of arid regions [3]. According to scenarios, the world's arid is warming more quickly than other regions. For instance, Huang et al. discovered that arid temperatures may climb by 3.2–4.0 °C for every 2 °C increase in the world temperature. In turn, a rise in temperature may heighten the likelihood of a drought, prolong its duration, diminish surface runoff, and lower grain yield and vegetation covering [4].

Due to the rising of the global warming, the arid regions are generally more susceptible to anthropogenic perturbations and climate dynamics. According to a prior study [5], the pace of land deterioration in dryland areas is almost 12 million hectares annually. Land degradation will lead to social and political instability, a decrease in the delivery of dryland services, and an increase in the danger of food shortages. The Arab region is a typical and representative arid region, of which hyper-arid, arid, semi-arid, and arid account for 48.76%, 13.44%, 18.75% and 5.03%, respectively.

Spatially, North Africa and West Asia in the Mediterranean region are dominated by hyper-arid and semi-arid drylands, while the coastal areas of Morocco and Algeria are dominated by semi-arid lands. The Mediterranean region is a biodiversity hotspot in the world with up to 25,000 species of plants, which makes it a typical and representative region for the study of dryland ecosystems.

The Middle East and North Africa (MENA) countries have experienced a significant population increase and is expected to double in size between 2025 and 2080 due to its role as the primary host of semi-arid and arid regions in the Mediterranean region. This region is under threat from rapidly rising food and water prices, land degradation, and biodiversity loss [6] because of population growth and climate change. Therefore, greater attention is needed to this region in order to manage dryland ecosystems sustainably.

The ecosystems found in arid regions are delicate and susceptible to both human activity and climate change, which can quickly result in a host of environmental issues [7].

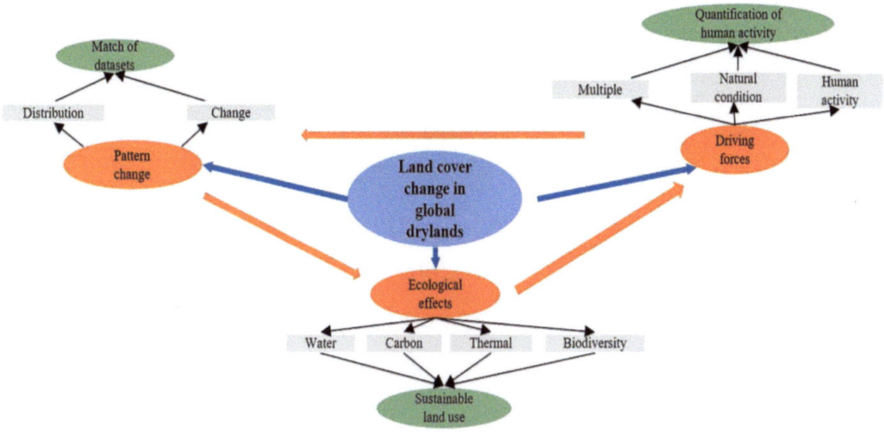

Fig. 1 Land cover change aspects in global drylands. *Source* Wang et al. [10]

Human activities such as urbanization, cropland reclamation, and overgrazing have altered the patterns and processes of the dryland landscape [8], which in turn has affected biodiversity, thermal processes, the carbon cycle, and water cycles. As a result, researching changes in land cover in arid regions has emerged as a crucial strategy for comprehending environmental change worldwide. However, the existing research results are inconsistent due to a variety of data sources and analysis techniques, which means that our knowledge of the distribution and evolution of land cover in drylands is still incomplete [9]. In order to improve the livelihoods and climate change resilience of populations living in drylands, regular monitoring can help evaluate the effects of various governance frameworks, laws, and policies pertaining to land use.

Drylands are hotspots for global change due to their sensitivity and vulnerability. Numerous studies have examined the ecological effects, driving forces, and spatial and temporal characteristics of land cover change in arid regions [7]. Nonetheless, disparities in data and methodology have resulted in an uneven comprehension of the alteration of land cover in drylands (Fig. 1).

1.1 Key Ecosystem in the Arab Region

Arid lands, defined as arid, semiarid, are among the world's most fragile lands, caused by continuous drought and increasing overexploitation. Nearly 30% of the earth's land area is made up of these areas (Fig. 2). One of the primary causes of soil deterioration and desertification in arid and semiarid regions is overgrazing, which overutilizes trees [11]. According to the quantity of precipitation that occurs each year, the soil is classified as ultra-dry (less than 100 mm), meaning that it contains virtually no water, dry (100–250 mm), and semiarid (200–500 mm). Other definitions

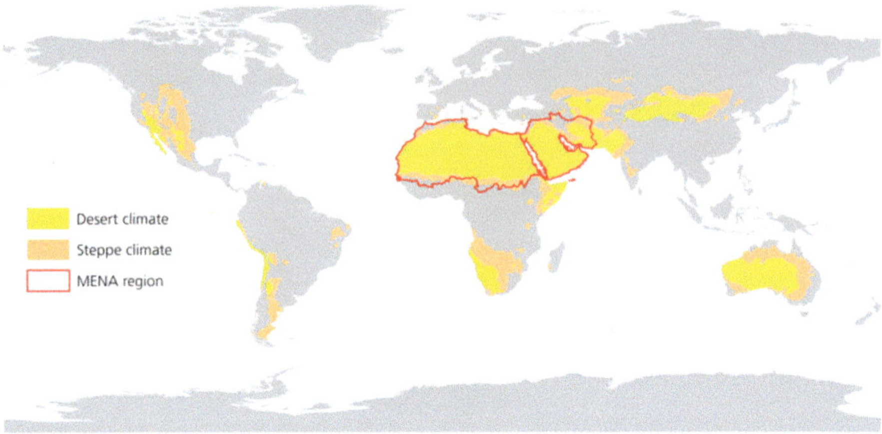

Fig. 2 The Köppen-Geiger climatic classification divides the world into arid and semi-arid zones. The MENA region, which has a great solar potential, is represented by the color red. *Source* Kottek et al. [13]

and criteria are used to classify the soils [12]. The definition of arid ecosystems is a lack of water availability, generally due to low precipitation. Evapotranspiration is the term used to describe the evaporative loss of water from plants as well as transpiration and evaporation, which can both remove water from the landscape.

The countries that make up the Arab region are those that reach from the Mediterranean Sea in the north to the Horn of Africa and the Indian Ocean in the southeast, and from the Atlantic Ocean in the west to the Gulf in the east [9]. The Arab region, which spans North Africa and Western Asia and has a combined population of 358 million, is roughly 14.2 million km. About 90% of the Arab continent is composed of desert, semi-arid, and dry subhumid regions [14]. The fruitful Crescent, which stretches from Iraq through Syria, Lebanon, and Palestine, the High Atlas Mountains, and the Nile Valley are among its many fruitful areas. Southern Arabia and Sudan both have deep woods.

In the Arab region, coastal and marine habitats, as well as agricultural systems, drylands, forests, mountains, and urban areas, are all significant ecosystems. The section that follows provides an explanation of these ecosystems along with their salient characteristics and roles.

1.1.1 Dryland Ecosystems

Drylands are defined as regions with an aridity index value of less than 0.65 by the World Atlas of Desertification [15]. Drylands are those where the yearly precipitation is less than two-thirds of the possible evapotranspiration, according to the UN Convention to Combat Desertification. Water scarcity, limited soil moisture, extreme rainfall variability, frequent but unpredictably occurring droughts, high temperatures, high rates of evaporation, low soil fertility, high salt, overgrazing, and fires are characteristics of desert ecosystems [16]. The great bulk of the land in the Arab world is made up of drylands. They rule over regions such as Saudi Arabia, the Arabian Peninsula's Empty Quarter, and North Africa's Great Sahara. Many places, such as Aarsal, a mountainous region (36,000 ha) in Lebanon, where around 5000 ha are seriously at risk of land degradation, have semi-arid systems [17]. A prolonged deterioration in the ability of dryland ecosystems to deliver products and services is thought to be the cause of land degradation, or desertification, which affects dryland areas generally and the Arab region specifically. The primary causes of this decrease in dryland land productivity are salinization, compaction and crusting of the soil, wind and water erosion, pollution, acidification, alkalinization, and water logging [18]. Dryland ecosystems are frequently linked to suffering, destitution, and hostilities [19].

1.1.2 Ecosystems in Forests

In Arab countries, forest ecosystems cover around 89.64 million hectares of land [20]. Compared to the global forest cover, which accounts for 30.3% of the total land area, this only makes up 7.2% of the land area [21]. Approximately 83% of the Arab region's forests are thought to be in North Africa, Sudan, and Somalia, with the remaining portion being in the Arabian Peninsula and the Mashriq countries. Syria and Lebanon both have areas of forest cover. Forests dominate the Atlas Mountains from Algeria to Morocco and the Green Mountains in Libya (see Table 1 for the latest recent FAO data regarding forest cover in Arab countries).

1.1.3 Ecosystems in Mountains

Mountains provide various ecosystem services, such as the production of plants and animals, clean water, pharmaceuticals, medicinal herbs, and non-timber forest products.

Moreover, they supply regulating and supporting services, like soil fertility, biodiversity, and climate regulation. Additionally, they play a significant role in providing cultural services, primarily in the areas of diversity and recreation [23]. Furthermore, mountains have inherent aesthetic and spiritual significance [24]. In order to promote sustainable production systems, mountain inhabitants have developed a wide variety of cultures, dialects, and traditional agricultural knowledge. Generally speaking, mountainous areas have greater levels of poverty and ethnic diversity, and its residents are frequently more vulnerable than those in other areas. Developing nations are home to 90% of the world's mountain inhabitants. According to Körner and Ohsawa [23], 90 million mountain people are thought to be impoverished and at risk of food insecurity.

Table 1 Forest area by country in the Arab region

Country	Forest area (1000 Hectares)	Percent of total land area
Sudan	69,949	29
Lebanon	137	13
Somalia	6747	11
Morocco	5131	11
Tunisia	1006	6
UAE	317	4
Syria	491	3
Iraq	825	2
Palestinian Terriority	9	2
Algeria	1492	1
Yemen	549	1
Jordan	98	1
Bahrain	1	1
Saudi Arabia	977	Not significant
Mouritania	242	Not significant
Libya	217	Not significant
Egypt	70	Not significant
Dijibouti	6	Not significant
Kuwait	6	Not significant
Oman	2	Not significant
Qatar	0	0

Source FAO [22]. Global Forests Resources Assessment

1.1.4 Cultivated Ecosystems

The cultivated systems in the Arab world are primarily found in Sudan and in countries that are directly exposed to the Mediterranean Sea, such as Morocco, Tunisia, Syria, and Egypt. In Egypt, new technologies have been introduced to increase wheat production and water productivity [25].

1.1.5 Coastal Ecosystems

According to Agardy [26], coastal habitats are some of the most abundant yet delicate ecosystems on the planet. They are extremely prolific, important to the environment, and valuable to humans. More services pertaining to human well-being are produced by coastal ecosystems than by the majority of other systems [26]. However, the environment is changing quickly in many environments. Due to the increased demands of expanding populations and the ongoing quest for greater productivity, human dependence on coastal systems has grown over the past few centuries. There are 16.6 thousand kilometers of rivers, freshwater lakes, and semi-fresh water lakes in the Arab world, along with 22.4 thousand kilometers of shoreline [27].

1.1.6 Marine Ecosystems

In the Arab region, marine ecosystems are found in all countries of the region, such as Egypt, Lebanon, Syria, Yemen, Sudan, Libya, Algeria, Morocco, Tunisia, Oman, KSA, and other Gulf countries. Poor coastal Communities in this area rely on the fishing industry for employment, revenue production, and sustenance. Furthermore, fisheries are a significant export industry for a number of Maghreb nations. Numerous marine ecosystems in the Arab world, particularly those in the Gulf and several international tourist destinations, exhibit signs of stress as a result of invasive species, unplanned coastal development, pollution from large amounts of desalination plant effluents, urban coastal runoff, and waste and sewage disposal. The overexploitation of fish populations, pollution from oil exploration, and industrial pollutants are other factors contributing to the decline of marine biodiversity [28].

2 The Resilient Social-Ecological System: A New Paradigm

Ambiguous definitions of resilience make it difficult to define goals accurately enough to guide management. The complexity of system behavior can be broken down into easily understood parts using multi-indicator frameworks [29, 30], where indicators serve as stand-ins for mechanisms that are hard to see or measure. However, single measures are unlikely to adequately capture the dimensions of resilience [31, 32]. When making decisions about river protection, it is preferable to take into account how conservation (from resource extraction or alteration) provides the functional ability to preserve the system's state or states in the face of climate change. This implies that indicators should ideally be made to be thorough, evaluating the present characteristics of systems that will allow for resilience in the face of future climatic patterns while taking historical circumstances into consideration. Biophysical diversity, for instance, can preserve a variety of critical functions, interactions, and disturbance responses, as well as increase the availability of organisms that can tolerate changed environmental conditions [33–35]. A number of obstacles must be overcome in order to achieve these important conservation objectives. For example, regions identified as crucial for preserving endangered species and reducing the effects of climate change may not always be found adjacent to existing protected areas [36]. In order to carry out a global conservation agenda, international, national, and regional conservation bodies must coordinate in order to, for instance, determine which places should get the majority of protection efforts [37, 38] (Table 2).

Table 2 Examples of biological, physicochemical, and social indicators of resilience

Resilience indicator	Components	Examples of indicators
Biological	Diversity	Species richness; evenness; community types; trait or functional diversity; functional redundancy; intra-taxon genetic diversity; diversity measured at multiple levels, local, regional, and global diversity
	Functional	Drying resistance traits; reproductive flexibility; community association; life history strategies favoring high variability
	Connectivity	Dispersal ability; metapopulation size and structure
Physicochemical	Diversity	Runoff source diversity; neighboring (free-flowing) stream classification typologies (up and downstream); flow magnitude
	Functional	Hydrologic class robustness
	Connectivity	Stream Dendritic Connectivity Index; fragment size
Social	Diversity	Number of vested stakeholder groups and institutions; nonmarket and market economic uses; number and scale of primary, secondary, and tertiary economic uses
	Connectivity	Public access to river systems; recreational use; cultural uses; ecotourism value; proximity to communities; trade

Source by author

3 Resilience Conditions, Trends and Impacts

Conditions, trends, and impacts should ideally relate to the state of ecosystems and the services they provide, both now and in the past. The misuse and exploitation of these services by humans can have detrimental effects on the environment and increase the strain on delicate ecosystems. Approximately 60% of ecosystem services directly related to human well-being are in decline globally [39]. Different places have very different anthropogenic and environmental factors that influence ecosystem changes and, consequently, changes in ecosystem services. As a result, it is challenging to establish broad claims of causation and to establish a direct correlation between specific motivators and alterations in ecosystems and the benefits they provide [40].

The Millennium Ecosystem Assessment (M.E.A) is regarded as one of the most extensive initiatives to date to evaluate ecological change and how it affects human well-being. The MA comprises a number of sub-global assessments at various spatial scales in addition to the global assessment of ecosystem changes. In contrast to other international studies that only concentrate on global or regional scales, these evaluations are unique in that they have offered insightful information about how scale and knowledge systems affect the intricate relationship between ecosystems and human well-being [41].

3.1 Land Cover Change

Land cover has changed, particularly due to conversion to farmland, as a result of population development and urbanization, which have also altered global consumption patterns and raised demands for food and ecosystem services. Other ecosystem services deteriorate as a result of rising demands for provisioning services. The services that wood formerly offered are diminished when they are turned into farmed areas. Approximately 30% of the Earth's surface is currently covered by cultivated lands [42]. There are several instances of land cover change in the Arab world, which is mostly caused by overgrazing, deforestation and forest degradation, urbanization, agricultural intensification, and other man-made and natural factors. The region is characterized by drylands.

3.2 Climate Change Influence

On a global scale, climate change is thought to be one of the main causes of ecosystem change. Significant changes in the global climate system would result from continued greenhouse gas emissions at or above current rates, causing temperatures to rise steadily [43]. Climate change is one of the additional stresses that ecosystems in the arid, water-scarce Arab world are facing. Large increases in greenhouse gas emissions are predicted to occur in the Arab world over the next 20 to 50 years if current development activities continue [27]. As a result, air quality becomes a crucial environmental factor that influences sustainable development both locally and globally.

3.3 Land Degradation

Desertification, or the degradation of land in arid regions, is another worldwide issue. Desertification is defined as "land degradation in arid, semi-arid, and dry sub-humid areas resulting from various factors, including climatic variations and human activities by the United Nations Convention to Combat Desertification" [44]. Drylands continue to reduce providing services in tandem with desertification. It is anticipated that factors including population growth, changes in land use and cover, and climate change will hasten the scarcity of water and deterioration of arid regions [45].

3.4 Loss of Biodiversity

Biodiversity is essential to ecosystem function. "The variability among living organisms from all sources, including terrestrial, marine, and other aquatic ecosystems and the ecological complexes of which they are a part", including diversity within and among species and diversity within and among ecosystems" is how the Convention on biodiversity [42]. The sub-global evaluations focused on certain aspects of biodiversity, and this term is broad.

4 Methods

The systematic literature review procedures used in this review-based paper have demonstrated the literature review's integrity as a rigorous framework. The methodology of this study was based on examining the status of research on the issue, which is based on "Global Climate Change Impacts, adaptation, social and ecological resilience systems, arid and semi-arid ecosystems" by carefully examining prior published and unpublished academic work. Additionally, highlighting previous research on the same subject and make recommendations for further research on the same subject. Specifically, the study methodology was firstly, to organize publications to make them easy and quick to find. Secondly, to explore issues in this area, propose an outline of research for future work and to synthesize the previous literature on climate change, various sectors, and their mitigation measurement, resilience framework, ecological and social system. Reports used to search for multiple keywords such as "Climate Change," "Mitigation and Adaptation, Resilient Cities, Social-Ecological Urbanism, Biodiversity, Urban Development, Sustainable Developments Goals, etc.; in summary, keyword list and full text have been made. Initially, the search for keywords yielded a large amount of literature.

5 Findings

The evaluation's findings show that arid lands are productive landscapes with significant potential for both economic growth and environmental preservation rather than being wastelands. They emphasize the value of funding dryland forests and their resources, including trees that are not part of forests. The results can serve as a starting point for highlighting important new risks that arid and their inhabitants face on a regional and global scale, including as issues with food security, biodiversity, and climate change. As a result, they can help focus and prioritize the investments required for arid lands restoration.

In order to assess the effects of climate change and human activity, the outcomes of adaptation and mitigation efforts, and the progress made towards attaining regional targets for land degradation neutrality, the evaluation should be the first step toward regular monitoring of changes in dry landscapes. To more effectively improve the livelihoods and climate change resilience of dryland populations, regular monitoring can help analyze the effects of various governance frameworks, policies, and legislation connected to land use.

6 Discussion

Adjustments to natural, social, financial, physical, and human capitals in response to actual or anticipated climatic events and their implications are referred to as adaptation in this study. Examples from the three case studies are used in this section to illustrate the various forms of adaptation to climatic variability and change. Activities that reflect modifications in certain characteristics of one or more livelihood capitals and are directly linked to lessening susceptibility to climate variability and change are the kinds of adaptations that are discussed here. In order to discover environmental management practices that serve as adaptation to climate hazards, including drought, community members took part in focus groups and consultations in each of the case studies. Rising temperatures are expected to have an effect on the continent's interior regions [25]. Many planted species are under risk of going extinct as a result of changing weather patterns brought on by a lack of natural resources (water), increased glacier melting, and rising mercury. The coastal environment, however, is in danger of being destroyed [46]. A significant change in rainfall patterns and a 2–3% increase in mercury by 2050 could have detrimental effects. Globally, natural and environmental disasters resulted in significant losses, including reduced agricultural productivity, system restoration, and the reconstruction of essential technologies. In addition, smog-related eye and skin conditions, as well as an increase in traffic accidents because of poor visibility, have been plaguing the world for the past three or four years (Fig. 3).

7 Conclusions

Progress towards sustainable development is vitally dependent upon sound human decision-making regarding ecosystems. With its emphasis on the human dimension of ecosystem change, this chapter can be considered as a contribution for the search for a sustainable development for the Arab region. It is the central message of this chapter to show that the resolution of environmental problems involves real changes in the behaviors and decisions of people towards their natural surroundings. Ecosystems and environmental resources are becoming overused and polluted. Empowering local communities through public awareness and community participation will provide more opportunities towards sustainable development policies. Although the role of government and regulations is needed to manage environmental problems,

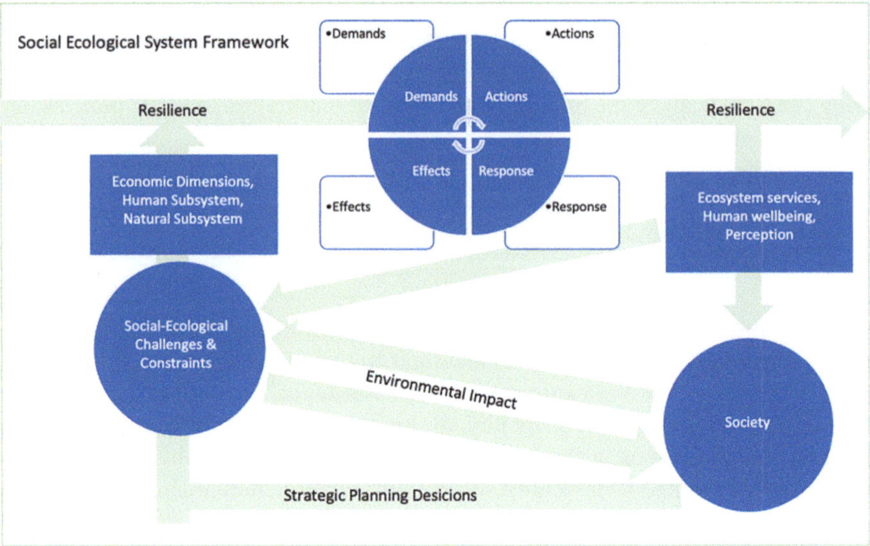

Fig. 3 A model of the framework of the design of a resilient social-ecological system. *Source* by author(s)

the role of individuals and societies is central. The link between public awareness, policy structure, and institutional framework needs to be recognized. Approaches to dealing with sustainable development issues are often complex due to information imprecision. Despite this complexity, a political will as well as the will of the local populations can help bring about win-win solutions. Finally, the chapter clearly demonstrates that human aspiration for better wellbeing is usually the primary factor that explains trends and impacts of ecosystem change. Nevertheless, the interactions of modern human activities with ecosystems have contributed to increasing human vulnerability and to the impact of ecosystem changes on well-being. Proper ecosystem management remains an important tool to enhance human well-being. Climate change, access to clean water, food and energy are key factors that can impact the health, stability and security of individuals and communities. Today, Arab countries realize the dependence of local communities on their ecosystems as well as the profound association between ecosystem conditions and development. Thus, policy makers in the Arab region are beginning to integrate environmental factors into development decisions. Despite the limitations of data in the Arab region as a whole, sufficient knowledge exists in most countries to apply sound ecosystem management practices and sustainable development strategies in order to provide opportunities for better human well-being.

References

1. Aghsaei H, Mobarghaee Dinan N, Moridi A et al (2020) Effects of dynamic land use/land cover change on water resources and sediment yield in the Anzali wetland catchment, Gilan, Iran. Sci Total Environ 712:136449. https://doi.org/10.1016/j.scitotenv.2019.136449. PMID: 32050376
2. Ban N, Schmidli J, Schär C (2014) Evaluation of the convection-resolving regional climate modeling approach in decade-long simulations. J Geophys Res Atmos 119:7889–7907. https://doi.org/10.1002/2014JD021478
3. Dirmeyer P, Cash B, Kinter J, Jung T, Marx L, Satoh M, Stan C, Tomita H, Towers P, Wedi N, Achuthavarier D, Adams J, Altshuler E, Huang B, Jin E, Manganello J (2011) Simulating the diurnal cycle of rainfall in global climate models: resolution versus parameterization. Climate Dynamics - CLIM DYNAM 39:1–20. https://doi.org/10.1007/s00382-011-1127-9
4. Huang G-Q, Li S (2019) Review of the genus Falsotrachystola Breuning, 1950 (Coleoptera: Cer- ambycidae: Lamiinae: Morimopsini). Zootaxa 4555(1):45–55. https://doi.org/10.11646/zootaxa.4555.1.3
5. Dong Z, Xu L, Yang T (2022) Transient response characteristics analysis of high-power piezoelectric transducers. Micromachines 13:1638
6. Dimobe K, Tondoh J, Weber JC, Bayala J, Ouédraogo K, Greenough KM (2018) Farmers' preferred tree species and their potential carbon stocks in southern burkina faso: implications for biocarbon initiatives. PLoS One 13(12):e0199488. https://doi.org/10.1371/journal.pone.0199488
7. Huang G, Liu B, Gouverneur X (2015) Note on the genus Neacanista Gressitt, 1940 (Coleoptera: Cerambycidae: Lamiinae: Acanthocinini). Zootaxa 3981(4):553–564. https://doi.org/10.11646/zootaxa.3981.4.6
8. Li G, Hui Z, Zhao Y, Hu (2019) The bivariate distribution characteristics of spatial structure in natural Korean pine broad-leaved forest. J Veg Sci 23(6):1180–1190. https://doi.org/10.1111/j.1654-11032012.01431
9. Verburg P, Alexander P, Evans T, Magliocca N, Malek Ž, Rounsevell M, van Vliet J (2019) Beyond land cover change: towards a new generation of land use models. Curr Opin Environ Sustain 38:77–85. https://doi.org/10.1016/j.cosust.2019.05.002
10. Wang H, Liu Y, Wang Y, Yao Y, Wang Ch (2023) Land cover change in global drylands: a review. Sci Total Environ 863:160943. https://doi.org/10.1016/j.scitotenv.2022.160943
11. Malagnoux M, Sène EH, Atzmon N (2007) Forests, trees and water in arid lands: a delicate balance. Unasylva 58:24–29
12. Verheye WH (2009) Agro-climate-based land evaluation systems. In: Verheye WH (ed) Land use, land cover and soil sciences, vol. 4. Encyclopedia of Life Support Systems (EOLSS), Oxford, United Kingdom, pp 130–145
13. Kottek M, Grieser J, Beck C, Rudolf B, Rubel F (2006) World Map of the Köppen-Geiger climate classification updated. Meteorol Z 15:259–263. https://doi.org/10.1127/0941-2948/2006/0130
14. Abdulhussain A, Abdu ASh, Al-Zubari WK, El-Deen NA, Abdul-Raheem M (2021, August) Desertification in the Arab Region: analysis of current status and trends. J Arid Environ 51(4):521–545. http://www.sciencedirect.com
15. Middleton M, Midgley C (1997) Avoiding the demonstration of lack of ability: an under-explored aspect of goal theory. J Educ Psychol 89:710–718. https://doi.org/10.1037/0022-0663.89.4.710
16. Levy M, Babu S, Hamilton K (2015) Ecosystem Conditions and Human Well-being. Chapter 5 in: Millennium ecosystem assessment. current state and trends: findings of the current states and trends working group. Ecosystems and Human Well-being, vol.1, Island Press, Washington DC
17. Zurayk R, Al-Awar F, Hamadeh S, Talhouk S, Sayegh C, Chehab A, Al Shab K (2000) Using indigenous knowledge in land use investigations: a participatory study in a semi-arid mountainous region of Lebanon

18. Lal R (2003) Soil erosion and the global carbon budget. Environ Int 29:437–450. https://doi.org/10.1016/S0160-4120(02)00192-7
19. ICARDA (2007) ICARDA Annual Report 2006. Aleppo, Syrian Arab Republic: International Center for Agricultural Research in the Dry Areas (ICARDA). https://hdl.handle.net/20.500.11766/7623
20. Wen H, Li Si, Chen X, Qin, C, Li, L (2024) Amplified production and export of dissolved inorganic carbon during hot and wet subtropical monsoon. Water Resour Res 60
21. UNDP (2009) Human Development Report 2009. Oxford University Press, New York
22. FAO (2023) Global Forests Resources Assessment (FRA 2023): Main Report
23. Körner C, Ohsawa M (2015). Mountain systems. Chapter 24 in: Millennium ecosystem assessment. current state and trends: findings of the current states and trends working group. Ecosystems and Human Well-being, vol.1. Island Press, Washington DC
24. Sponsel LE (2003) The social life of nature [Review of Nature in the Global South: Environmental Projects in South and Southeast Asia, by P R Greenough & A L Tsing]. Anthropol Q 76(4):775–787. http://www.jstor.org/stable/3318290
25. Abbass K, Qasim MZ, Song H (2022) A review of the global climate change impacts, adaptation, and sustainable mitigation measures. Environ Sci Pollut Res 29:42539–42559. https://doi.org/10.1007/s11356-022-19718-6
26. Agardy T (2005) Global marine conservation policy versus site-level implementation: the mismatch of scale and its implications. Marine Ecology-Progress Series - MAR ECOL-PROGR SER. 300:242–248. https://doi.org/10.3354/meps300242
27. UNEP (2019) Air Quality and Atmospheric Pollution in the Arab Region. Joint Technical Secretariat: League of Arab States, United Nations Economic and Social Commission for Western Asia, and United Nations Environment Programme, Regional Office for West Asia
28. UN (2020) The Third Arab Report on the Millennium Development Goals 2010 and the Impact on the Global Economic Crisis, New York
29. Richter BD, Baumgartner JV, Powell J, Braun DP (1996) A method for assessing hydrologic alteration within ecosystems. Conserv Biol 10:1163–1174. https://doi.org/10.1046/j.1523-1739.1996.10041163.x
30. Saxena A, Huneeus C (2022, June) Establishing evidence for resilience: a case of monsoon flood-affected communities in the Gangetic Plains of South Asia. Mitig Adapt Strat Glob Change 27(5):1–27
31. Allen CR, Cumming GS, Garmestani AS, Taylor PD, Walker BH (2011) Managing for resilience. Wildl Biol 17:337–349. https://doi.org/10.2981/10-084
32. Folke C (2016) Resilience (republished). Ecol Soc 21:44. https://doi.org/10.1093/acrefore/9780199389414.013.8
33. Duffy JE (2009) Why biodiversity is important to the functioning of real-world ecosystems. Front Ecol Environ 7:437–444. https://doi.org/10.1890/070195
34. Mori AS, Furukawa T, Sasaki T (2013) Response diversity determines the resilience of ecosystems to environmental change. Biol Rev 88(2):349–364. https://doi.org/10.1111/brv.12004
35. Soliveres S, Van Der Plas F, Manning P, Prati D, Gossner MM, Renner SC, Alt F, Arndt H, Baumgartner V, Binkenstein J, Birkhofer K, Blaser S, Blüthgen N, Boch S, Böhm S, Börschig C, Buscot F, Diekötter T, Heinze J et al (2016) Biodiversity at multiple trophic levels is needed for ecosystem multifunctionality. Nature 536(7617):456–459. https://doi.org/10.1038/nature19092
36. Dreiss LM, Malcolm JW (2022) Identifying key federal, state, and private lands strategies for achieving 30 × 30 in the United States. Conserv Lett 15:e12849. https://doi.org/10.1111/conl.12849
37. Eckert I, Brown A, Caron D, Riva F, Pollock LJ (2022) 30 × 30 biodiversity gains rely on national coordination. Nat Commun 14:7113. https://doi.org/10.1101/2022.12.16.520787
38. Fernandez VM, Teixeira MTA (2024) Life and death without telomerase: the saccharomyces cerevisiae model. Cold Spring Harb Perspect Biology
39. SCBD (2006) Secretariat of CBD - ITS/DC/MC/54802, letter from SCBD 24th of May 2006

40. Janetos A (2015) Developing better indicators to track climate impacts. Front Ecol Environ 13:403–403. https://doi.org/10.1890/1540-9295-13.8.403
41. Figliuolo G (2014, April 25) Addressing biodiversity conservation methods with fagus sylvatica genetic indicators. Open J Genet 4(2)
42. UNEP (2023) Millennium ecosystems assessment: ecosystems and human well-being: synthesis report. Island Press, Washington DC, p 57
43. IPCC (2022) Climate Change 2022: Mitigation of Climate Change. Contribution of Working Group III to the Sixth Assessment Report of the Intergovernmental Panel on Climate Change [Shukla PR, Skea J, Slade R, Al Khourdajie A, van Diemen R, McCollum D, Pathak M, Some S, Vyas P, Fradera R, Belkacemi M, Hasija A, Lisboa G, Luz S, Malley J (eds)]. Cambridge University Press, Cambridge, UK and New York, NY, USA. https://doi.org/10.1017/9781009157926
44. UNCCD (United Nations Convention to Combat Desertification) (1994) Convention to Combat Desertification in Those Countries Experiencing Serious Drought and/or Desertification, Particularly in Africa. Part I, Introduction, Article 1(a). UNDP (2019)
45. Safriel U, Adeel Z, Niemeijer D, Puigdefabregas J, White R, Lal R, Winslow M, Ziedler J, Prince S, Archer E, King C (2005) Dryland systems. Ecosystems and human well-being: current state and trends
46. Schuurmans C (2021) The world heat budget: expected changes. Climate Change. CRC Press, pp 1–15

Resilient Spaces Through Urban Agriculture Practices

Mai Mahmoud Mourad

Abstract Cities and their public spaces are the earliest places where people congregate and gather to exchange their various commercial trades, social, and other activities. The aim of this chapter is to examine the potential of urban agriculture as a sustainable and resilient solution to the challenges faced by public spaces in rapidly urbanizing cities. Specifically, this study seeks to investigate how integrating urban agriculture into public spaces can enhance the resilience of cities in the context of rapid urbanization, global climate change, and political instability. This chapter adopts a reviewing methodology to explore the design of resilient urban spaces through urban agriculture practices. The approach begins with a thorough literature review, which aims to identify the key Resilience City Dimensions (RCDs) outlined in the Rockefeller Foundation's City Resilience Index. This review helps establish the foundational framework for understanding the various factors contributing to urban resilience, including social, economic, environmental, and governance aspects. By combining theoretical frameworks, urban agriculture examples, and practical applications. In conclusion, this chapter offers valuable insights into how urban agriculture can serve as a tool for creating more resilient, sustainable, and adaptive urban environments, urban agriculture has the potential to transform public spaces into vital components of a city's environmental and social fabric. However, successful implementation requires collaboration between government authorities, urban planners, and local communities, as well as the development of policies that support the integration of agriculture into the urban landscape.

1 Introduction

Resilience is a term used by disciplines to describe how a system reacts to a disruption of any sort [1]. The majority of disciplines also distinguish between resilience, which is positive, and resilience, which is negative. Positive resilience improves a person's

---

M. M. Mourad (✉)
Arab Academy for Science, Technology & Maritime Transport, Cairo, Egypt
e-mail: maimmourad13@gmail.com

Z. ElZein and A. Negm (eds.), *Recent Approaches of Sustainable Architecture in Arid and Semi-arid Cities*, Urban Sustainability,
https://doi.org/10.1007/978-981-95-1645-2_5

capacity to handle stress, whereas poor resilience does so with unfavorable externalities and prolonged. Efforts to bolster resilience across different disciplines emphasize positive adaptation despite adversity [2]. Many distinguishing traits may be extracted and combined from the literature to provide a suitable definition of urban resilience in fragile cities, even if it may be difficult to come up with a definitive description [3].

City resilience is a concept that spans multiple dimensions and involves various systems and actors [4–6]. Since urban areas are complex, dynamic socio-ecological systems, Resilient cities are proposed to be reflective, resourceful, flexible, redundant, robust, and integrated, characteristics which make them safe to fail rather than fail-safe in the face of challenges [7]. The urban resilience concept can applied to two main difficulties. The first one is when cities and areas are robust to sudden and sporadic shocks, such as natural catastrophes -the most well-known examples being the Asian tsunami and Hurricane Katrina in New Orleans- or terrorist attacks. The second difficulty when cities and regions are generally more robust to long-term stresses such as deindustrialization, aging and declining populations, urban sprawl, and suburbanization [8] The shocks and slow burns which are common for systems going through change need to be distinguished from one another. The Slow burns may increase competition for attention, but shocks may bring people together [8].

Adaptability and adaptability are the foundations of the evolutionary method, which explains many forms of resilience. According to Dawley [9], adaptability refers to the choice to veer off of a current growth route in favor of a new, similar, or alternative trajectory, whereas adaptation represents a city or region's innate desire to better its circumstances along the path that has succeeded in the past [9]. Resilient spaces are places that feed community life where people congregate to organize activities, share knowledge, and assist one another [10]. Although developing countries are characterized by a limited number of capabilities and resources, it's shaped by numerous amount of public spaces where characterized by many innovative urban solutions that emphasize resilience in public spaces in arid regions such as Egypt. Then, urban agriculture becomes one of the urban sustainable solutions and plays a vital role as climate defense infrastructure.

2 Methods

This chapter employs a multi-faceted approach to designing resilient urban spaces through urban agriculture practices. The methodology begins with a comprehensive literature review to identify key Resilience City Dimensions (RCDs). Then the book chapter investigates how these dimensions can be implemented in public spaces through urban agriculture. Finally, it emphasizes the several aspects of urban agriculture practices depending on key Resilience City Dimensions (RCDs).

Urban agriculture practices were identified according to the main seven variables through Resilience City Dimensions (RCDs) based on the Rockefeller Foundation's city resilience index. These seven variables are Integrated, Inclusive, Reflective, resourceful, Robust, Redundant, and flexible. The chapter establishes a clear understanding of how these dimensions can contribute to the overall resilience of urban spaces, specifically in the Implementation of Urban Agriculture Practices into public spaces in arid regions, with a specific focus on Egypt.

Background on Resilience and Urban AgricultureBuilding resilience through urban agriculture has become increasingly important, especially in relation to the mounting demands of urbanization and climate change. Urban agriculture contributes to reducing food insecurity concerns, lessens the environmental effect of food transportation, and encourages the adaptive reuse of urban spaces by bringing food production into urban settings. Additionally, by promoting social cohesiveness and community involvement, urban agriculture allows people to actively participate and feel empowered. The chapter indicates how urban agriculture can enhance environmental resilience by enhancing biodiversity, mitigating the urban heat island effect, and fostering sustainable water management. In terms of the economy, it boosts regional economies by generating employment, decreasing dependency on outside food supplies, and promoting sustainable regional food systems.

2.1 City Resilience Index—The Rockefeller Foundation

According to the Rockefeller Foundation, the City Resilience Index is being developed by Arup. The Index has been designed to enable who are interested to measure and monitor the multiple factors that contribute to their resilience. Its primary purpose is to diagnose strengths and weaknesses and measure relative performance over time. This provides a holistic articulation of Resilience City Dimensions (RCDs), structured around four dimensions, 12 goals and 52 indicators that are critical for the resilience of our cities as Fig. 1 indicated.

These Resilience City Dimensions (RCDs) suggest that the resilience of a city relates to the four main dimensions, which are Health and well-being, ensuring the health and well-being of everyone living and working in the city; Economy and society, the social and financial systems that enable urban populations to live peacefully, and act collectively; infrastructure and environment, man -made and natural systems that provide critical services, protect and connect urban citizens; and leadership and strategy, the need for informed, inclusive, integrated and iterative decision making in our cities. According to RCDs there are seven variables that help cities to be more resilient which are Integrated, Inclusive, Reflective, resourceful, Robust, Redundant, and flexible [12].

2.2 Urban Agriculture Through Conflicts

Urban agriculture (UA) which is the production of food, mostly fruit and vegetables, within urban areas [13, 14] is increasingly recognized for its multiple social and

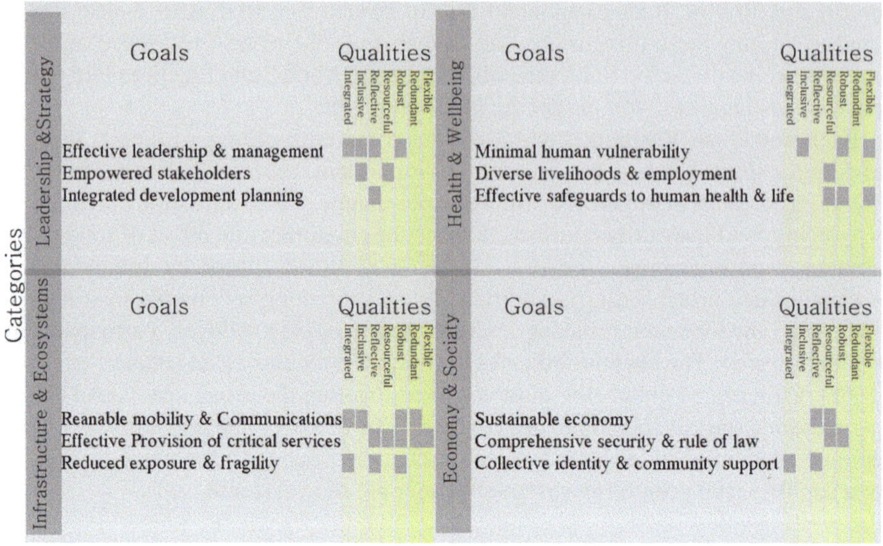

Fig. 1 Qualities of resilient systems according to the Rockefeller Foundation, Arup. *Source* Author depending on Metaxas and Psarropoulou [11]

ecological benefits by city governments around the world [15, 16]. During World War II, UA played an important role in increasing food security and boosting national morale in Britain and the USA, where household fruit and vegetable production was promoted through the Dig for Victory and Victory Garden campaigns [5, 6, 17] as Fig. 2 shown, respectively Following a period of decline in post-war years, urban agriculture is now enjoying a resurgence of interest in the global North, which has recently spiked during the Covid-19 lockdowns 17[5]. UA takes various forms, from vegetable plots in private gardens, through allotments and community farms, to rooftop gardens and edible walls using technologically advanced, soil-free cultivation methods [5,6, 17].

For urban dwellers, urban food gardening has always been vital during times of crisis, such as war or economic downturn. Urban agriculture practices have been established on urban spaces in New York City as shown in Fig. 3. Also, when city dwellers in Germany were given the option of going hungry or growing part of their

Fig. 2 Urban agriculture during the war years. *Source* Platform Wederopbouw Rotterdam [18]

Fig. 3 Shows the impact of establishing urban agriculture on urban spaces in New York City. *Source* New York City Department of Parks and Recreation [20]

own food following World War I, the Schrebergaerten movement was born. Urban land was extensively cultivated during the Second World War in Britain thanks to the Dig for Victory effort. The type of crisis is different: prolonged high unemployment rates are becoming a major issue in several places, pushing many people to adjust or take up new survival techniques, such as devoting some of their time to food gardening. Thus, in many respects, social, economic, and environmental. urban agriculture supports the sustainability of cities [19].

2.3 Urban Agriculture Classifications and Urban Spaces

Urban agriculture is classified according to its scale and size into three main levels: Subsistence, Micro-scale, and medium to sizeable-scale urban agriculture as shown in Fig. 4. Subsistence urban agriculture often occupies less than 500 square meters, while Micro-scale urban agriculture occupies areas between 500 and 1,000 square meters, while medium to sizeable-scale urban agriculture occupies more than 1,000 square meters [21].

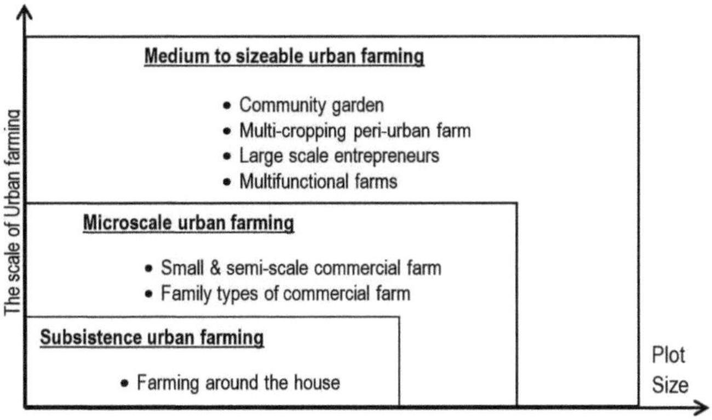

Fig. 4 Scales of the Urban Farming Production System. *Source* Hashim et al. [22]

These three scales include more than 40 applications for urban agriculture, according to the United Nations Urban Development Program for the year 1996, starting with applications at the subsistence-urban agriculture levels as kitchen garden and Allotment farming. The kitchen garden is a small- scale food production adjacent to existing building with related program activities, typically enclosed with raised beds as a residential kitchen garden and institutional garden. While Allotment farming is a plot-based food-producing, individually managed within divided plots typically enclosed with raised beds. The community garden and victory garden are some of these scale examples.

In addition, Micro-scale urban agriculture as edible landscape which is a food production integrated with existing landscape design and program, including integration with non-edible plants and non-food related programs, enclosed and unclosed examples such as edible park and edible school yard and campus. Small-scale farm where the food-producing landscape, occasionally integrated with alternative programs; enclosed and un-enclosed models exist as communally-managed urban farm and demonstration educational farm.

As well as urban agriculture that occurs on a medium to sizable scale within diverse plots of land, such as large-scale farms where food-producing landscape, typically designed for productive output, occasionally balanced with alternative programs; enclosed and unenclosed models exist. market farm, communal farm, and demonstration educational farm are examples of large-scale farm, retail distribution land, and support sites which are temporary or permanent structures housing programs related to the retail and distribution of urban-grown products. there are a various example of that scale as farmer's markets, food pantries, restaurants, grocery stores, value-added and packaging centers, distribution centers, and resource centers as shown in Fig. 5.

3 Discussion

Urban agriculture is increasingly recognized as a vital strategy for enhancing the resilience of cities, especially in arid regions like Egypt, where the **arid** or **semi-arid climate** refers to regions that receive very little rainfall and experience significant water scarcity for most of the year. These climates are characterized by high temperatures, low humidity, and limited precipitation. Urban agriculture in arid areas is not only possible but can also be highly sustainable. These practices focus on maximizing space, conserving water, and using innovative technologies to ensure that food production can continue despite challenging climatic conditions. This chapter has delved into various dimensions of urban agriculture, each contributing uniquely to urban resilience. The discussion will reflect on the key insights derived from the study and explore their broader implications for urban planning and policy. The findings of this chapter explore that Urban Agriculture functions as an integrated

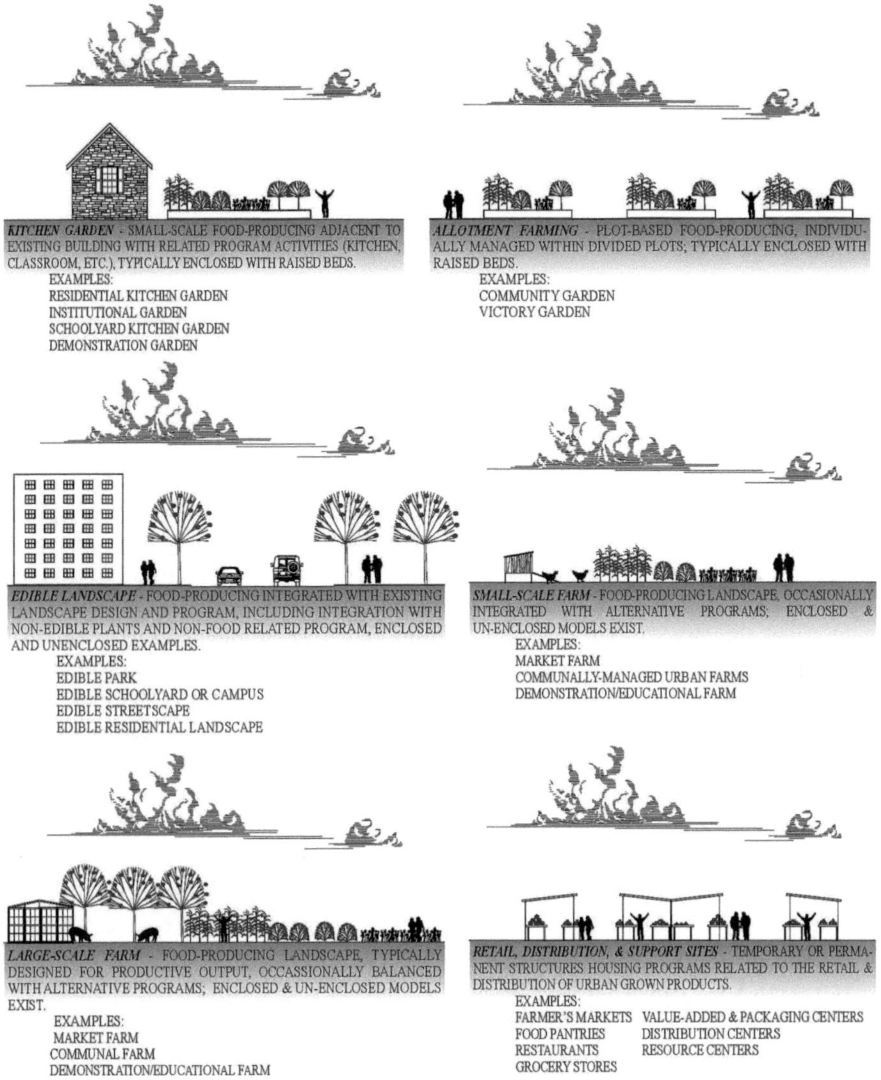

Fig. 5 Urban agriculture applications in public spaces. *Source* Napawan [23]

system, Inclusive, Reflective Practices, Resourceful, Robust, Redundant, and Flexible. These seven key Dimensions shape the resilience city characteristics' according to the Resilience City Dimensions (RCDs).

Fig. 6 Shows the impact of establishing urban agriculture on abandoned urban spaces in New York City. *Source* Argitecture [27]

3.1 Urban Agriculture as Integrated System

Urban Agriculture practices worldwide are investigating the potential to re-establish the connection between agricultural and food production through two methods; facilitating the re-establishment of the relationship between urban residents and peri-urban farms, or granting permission for food production inside city borders [24]. This increases the city's resilience to withstand through its ability to provide food, which gives the city's urban spaces great importance by implementing urban agriculture practices. There is increasing interest in city farms managed by professional farmers, often in collaboration with committed citizen organizations. These locations may include community, school, or private allotment gardens. Urban agriculture has demonstrated its ability to transform neglected spaces into cultivated spaces, which ensures the continuity and sustainability of these spaces. Urban and peri-urban agriculture is being used as a strategy to renew abandoned urban spaces [25]. New York City has adopted this strategy by having non-profit organizations and residents convert vacant city-owned sites into green spaces and community gardens. More than 135 community gardens, urban farms, and 835 school gardens have been established throughout the five boroughs, adding more than a million square feet of green space to New York City. These parks were built on vacant city-owned lots, within public housing projects, and on empty lots in churches, day care, and elder centers [26] (Fig. 6).

3.2 Inclusive Urban Agriculture

In unstable and divided societies, urban agriculture can play an active role in alleviating conflict, individualism, and isolation [28, 29]. As gathering points for different age, ethnic, and socio-economic groups where social ties and social inclusion are strengthened for disadvantaged and often discriminated against people or social groups, such as women or social groups headed by women Households, the unemployed, migrants, ethnic minorities, refugees, asylum seekers, the elderly, pensioners, disabled or indigent persons, or first-time juvenile offenders [30]. Urban gardens and farms help break down stereotypes of race, class, and gender [31]. Once this type of barrier is removed and differences are embraced, working together in these inclusive spaces can help provide neighbors with trust and positive engagement, creating a shared cultural identity that fosters a sense of solidarity and distinction, developing a sense of community, and allows them to establish social connections and build social bonds Mutual, and participation in the social fabric of society [32]. Urban agriculture can ultimately lead to building a strong, cohesive community where all voices are heard, a community that feels empowered to take back control of the neighborhood and improve it [33] (Figs. 7 and 8).

Fig. 7 Shows a beneficiary of a rooftop urban farming project tending to the roof of a building in Egypt. *Source* FAO [34]

Fig. 8 Urban farmers demonstrate in one of the small horticultural gardens at the Training and Nursery Centre, Senegal. *Source* FAO [34]

Fig. 9 It explains residents' practices in growing their home gardens and their sense of intimacy in space. *Source* Blog/Agribazaar [36]

3.3 Urban Agriculture as Reflective Practices/Create a Feeling of Intimacy in the Space

Urban agriculture can be a reflective practice by integrating continuous learning, adaptation, and community engagement, and creating a feeling of intimacy in space. Urban agriculture is one of the collective activities of residents in cities and residential neighborhoods, and intimate urban spaces represent places to contain these activities, as they are public open spaces that residents share in their use and are a suitable ground for meetings and interaction between them, and the formation of these spaces depends on the designer's creative ability to reach solutions. Distinctive beyond the physical dimensions to the social aspects that work to transform it from a mere physical vacuum into a vacuum for life and social rapprochement, and relationships are formed and strengthened between participants and organizers when practicing urban agriculture in the urban space, as it enhances positive responses from the general population to the practice of urban agriculture through backyards and lands vacant commercial offered for use [35]. These responses to public recognition influence a shift in attitudes and support from individuals and enhance their sense of belonging and intimacy during urban agriculture projects (Fig. 9).

3.4 Urban Agriculture as Resourceful System

UA can contribute to enhancing urban food security and healthy nutrition of the urban poor. Urban households that are involved in some sort of farming or gardening are more food secure, have a better and more diverse diet, and eat more vegetables than non-farming households [37]. Production of food by poor urban households can supply 0·20–0·60 of their total food consumption (especially in green vegetables, medicinal and aromatic herbs, eggs, milk and meat from small animals [14]). Involvement in UPA may also lead to better mitigation of diseases (better nutrition, home-grown medicinal plants), more physical exercise, less dependency on gifts and food aid and enhanced self-esteem [38].

UA helps in the Possibility of providing a healthy and safe environment through Easy access to safe, nutritious food, primarily fresh produce, which helps improve the health conditions of the urban poor, as urban agriculture has been linked to a range of health benefits, as does providing affordable access to fresh produce, especially in neighborhoods with few grocery stores. Supermarkets help reduce the consumption of canned foods. Maintaining a healthy diet also helps reduce the risk of chronic diseases such as diabetes among city residents. Urban agriculture contributes to the arrival of grown products with minimal processing and in a short time, which reduces nutrient deterioration and damage resulting from poor transportation. Or storage if shipped from other places [39]. The goal is to promote well-being, prevent diseases, and improve the quality of life of the population. These are the health, physical, mental, and environmental concerns of communities and populations [40].

3.5 Robust Urban Agriculture

A robust UA system incorporates diverse crop varieties, innovative growing techniques, and sustainable resource management practices [41]. By diversifying crop species and implementing climate-resilient practices, such as drought-resistant crops and efficient water use, robust UA can mitigate the impacts of extreme weather events and climate change [42]. Additionally, robust UA systems often include infrastructure designed to withstand natural disasters, such as raised beds to prevent flooding and windbreaks to protect against storms. This level of resilience is essential for maintaining food security and urban stability, particularly in regions prone to environmental uncertainties. Robust UA not only ensures a steady supply of fresh produce but also contributes to the overall health and well-being of urban communities by creating green spaces, improving air quality, and fostering community resilience [43].

3.6 Redundant Urban Agriculture

Increasing urban agriculture might be a catalyst for revolutionary change in the creation and upkeep of resilient, sustainable urban systems. However, current knowledge on the factors that influence, facilitate, and guide the expansion of urban agriculture is still incomplete and mostly isolated within several fields and industries [44]. Urban Agriculture includes a range of practices, including community soil-based, commercial rooftop as engineered soil, community hybrid, rooftop hydroponic greenhouse, and commercial vertical Aeroponic (Fig. 10).

Fig. 10 Shows patterns of urban agriculture in urban spaces in the city. *Source* Blog/Simon, 2024

Fig. 11 Shows patterns of urban agriculture in urban spaces in the city. *Source* Translated by author(s) based on Branduini et al. [46]

3.6.1 Flexible Urban Agriculture

Urban agriculture is one of the components of green urban infrastructure, which can enhance the city's ecosystem and resilience by integrating food production with the city's urban structure. Various practices exist through public gardens, rooftop gardens, community gardens, backyard food gardens, and other urban agriculture practices. This integration can take many patterns, as it comes in the form of a green belt, a green corridor, or a green connected network. Green patch is considered one of the patterns of urban agriculture present in the city, and the last pattern comes in being a metropolitan network. It connects the city's open spaces and public spaces [45] (Fig. 11).

4 Conclusions

Building resilient urban environments is made possible by urban agriculture. Cities may strengthen their resistance to a variety of threats by including inclusive, thoughtful, inventive, resilient, redundant, and adaptable urban agriculture solutions. In addition to addressing concerns of sustainability and food security, establishing urban agriculture in dry areas as Egypt helps to create resilience, and livable cities.

Urban agriculture will continue to be an essential part of resilient and prosperous cities by maintenance, which will require ongoing review and adaptation.

An integrated approach to urban agriculture ensures that these practices are seamlessly incorporated into the urban fabric, promoting sustainability and enhancing the overall ecosystem. Inclusive urban agriculture practices engage diverse community members, fostering social cohesion and ensuring that the benefits of urban farming are accessible to all. Redundancy in urban agriculture provides multiple layers of safety and security, ensuring that if one system fails, others can take over, maintaining overall stability. Flexibility in urban agriculture allows for adaptation to changing conditions and needs, ensuring long-term sustainability and resilience.

Specifically, in arid regions as Egypt, urban agriculture plays a crucial role in addressing the unique challenges posed by the environment. These practices contribute significantly to city resilience through food security and sustainability, providing fresh produce and improving the quality of urban life. The continuous evaluation and adaptation of urban agriculture practices are essential to maintaining their effectiveness and ensuring their long-term impact. This involves regular monitoring, feedback mechanisms, and the willingness to implement necessary changes based on evolving circumstances and new insights.

5 Relevant SDGs

Creating a resilience spaces by adapting urban agriculture practices is essential to achieving the Sustainable Development Goals (SDGs), especially those related to food security, sustainable cities, and climate resilience. The integration of urban agricultre into urban spaces to provide access to wholesome food in these urban settings, especially in underprivileged communities and by that the Sustainable Delopment Goals 2 and 3 (SDG2 and SDG3), which are Zero Hunger and Good Health And Wellbeing. Urban agriculture also promotes community involvement, enhances green open spaces, and convert neglegted spaces into productive one, all of which contribute to Sustainable Delopment Goal 11(SDG 11) which is Sustainable Cities and Communities. Additionally, urban agriculture supporting the Sustainable Delopment Goal 13 (SDG 13) which is climate action as urban agriculture directly links biodiversity preservation and climate change mitigation, positioning it as a required strategy for climate action in the context of sustainable urban development.

6 Recommendations

It is suggested that future research investigate at the ways that capacity-building projects and educational programs can enable communities to embrace Urban farming practices. To encourage a sense of ownership and accountability for creating resilient urban environments, this involves investigating ways to develop local knowledge and expertise in sustainable farming, climate adaptation, and urban design. Additionally, in order to give urban dwellers—particularly those residing in vulnerable areas—the knowledge and skills necessary to participate in urban agriculture.

The local governments ought to fund sustainable farming practices so that city dwellers may help make their communities more resilient. Adopting these suggestions will enable cities to beinifit from urban agriculture as a crucial instrument for creating resiliance spaces.

References

1. Vale J (2014) The politics of resilient cities: whose resilience and whose city? Build Res Inf 42(2):191–201. https://doi.org/10.1080/09613218.2014.850602
2. Fleming J, Ledogar, R (2008) Resilience, an evolving concept: a review of literature relevant to Aboriginal research. Pimatisiwin 6(2):7–23. https://pmc.ncbi.nlm.nih.gov/articles/PMC2956753/
3. Patel R, Nosal L (2016) Defining the Resilient City. United Nations, University Centre for Policy Research, Working Paper. ISBN: 978-92-808-9034-1
4. Olsson G, Kerselaers E, Kristensen LS, Primdahl J, Rogge E, Wästfelt A (2014) Peri-urban food production and its relation to urban resilience. Sustainability 8:1340. https://doi.org/10.3390/su8121340
5. Sharifi A, Yamagata Y (2014) Resilient urban planning: major principles and criteria. Energy Procedia 61:1491–1495. https://doi.org/10.1016/j.egypro.2014.12.154
6. Tendall D, Joerin J, Kopainsky B, Edwards PR, Shreck A, Le Q, Kruetli P, Grant MD, Six J (2015) Food system resilience: defining the concept. Glob Food Secur 6:17–23. https://www.researchgate.net/publication/282577265_Food_system_resilience_Defining_the_concept.
7. McMillen H, Campbell L, Svendsen E, Reynolds R (2019) Recognizing Stewardship practices as indicators of social resilience: in living memorials and in a community garden. Sustainability 8:775. https://doi.org/10.3390/su8080775
8. Schiappacasse P, Müller B (2018) One fits all?: Resilience as a multipurpose concept in regional and environmental development. Raumforschung und Raumordnung | Spat Res Plan 76. https://doi.org/10.1007/s13147-018-0520-9
9. Dawley S, Pike A, Tomaney J (2010) Towards the resilient region?: Policy activism and peripherial region development. "SERC Discussion Paper", No. 53, September. https://blogs.ncl.ac.uk/curds/files/2013/10/SERC-Discussion-Paper-No53.pdf
10. Abshirini E, Koch D (2017) Resilience, space syntax and spatial interfaces: the case of river. ITU A|Z, Vol 14, No 1, Sweden. https://doi.org/10.5505/itujfa.2017.65265
11. Metaxas T, Psarropoulou S (2021) Sustainable development and resilience: a combined analysis of the cities of Rotterdam and Thessaloniki. Urban Sci 5(4):78. https://doi.org/10.3390/urbansci5040078
12. Silva J, Morera B (2012) City Resilience Index - The Rockefeller Foundation, Arup. https://www.rockefellerfoundation.org/wp-content/uploads/CRI-Revised-Booklet1.pdf
13. FAO (2018) FAO's Role in Urban Agriculture. Food and Agriculture Organization of the United Nations: Roma, Italy. Available online: http://www.fao.org/urban-agriculture/en/
14. Smit J, Nasr J, Ratta A (2001) Urban agriculture: food, jobs and sustainable cities. United Nations Development Programme, The Urban Agriculture Network, New York, USA. http://www.jacsmit.com/book/Chap02.pdf
15. FAO (2020) Urban food systems and COVID-19: the role of cities and local governments in responding to the emergency, FAO, Rome, Italy

16. Morgan K (2015) Nourishing the city: the rise of the urban food question in the Global North. Urban Stud 52:1379–1394. https://doi.org/10.1177/0042098014534902
17. Olsson GA, Kerselaers E, Kristensen LS, Primdahl J, Rogge E, Wästfelt (2014) A peri-urban food production and Its relation to urban resilience. Sustainability 2016(8):1340
18. Platform Wederopbouw Rotterdam, Urban agriculture during the war years (2024) https://wederopbouwrotterdam.nl/en/articles/urban-agriculture-during-the-war-years
19. Deelstra T, Girardet H (2000) Urban agriculture and sustainable cities. Growing Cities, Growing Food: Urban Agriculture on the Policy Agenda, 43–65. https://www.researchgate.net/publication/284992045_Urban_agriculture_and_sustainable_cities
20. Blog/New York City Department of Parks and Recreation (2024) History of Farm Gardens in NYC Parks. New York City Parks. https://www.nycgovparks.org/about/history/community-gardens/farm-gardens
21. Phung C (2019) Implications of the circular economy and digital transition on skills and green jobs in the plastics industry, Field Actions Science Reports. The journal of field actions, Special Issue 19:100107. https://journals.openedition.org/factsreports/5498#bibliography
22. Hashim N, Hussain M, Ismail A (2020) Green roof concept analysis: a comparative study of urban farming practice in cities. Malays J Sustain Environ 7(1):115–132. https://doi.org/10.24191/myse.v7i1.8914
23. Napawan C (2015) Production places: evaluating communally-managed urban farms as public open space. Landscape J Aug 34(1):37–56. https://doi.org/10.3368/lj.34.1.37
24. Sonnino R (2009) Feeding the city: towards a new research and planning agenda. Int Plan Stud 14:425–443. https://doi.org/10.1080/13563471003642795
25. Beckie M, Bogdan E (2010) Planting roots: urban agriculture for senior immigrants. J Agric Food Syst Commun Dev 1(2):77–89. New Leaf Associates. https://doi.org/10.5304/jafscd.2010.012.004
26. Mougeot L (2000) Urban agriculture: Definition, presence, potentials and risks. Growing Cities, Growing Food: Urban Agriculture on the Policy Agenda. https://www.researchgate.net/publication/242091146_Urban_agriculture_Definition_presence_potentials_and_risks
27. Blog/Agritecture (2024) 7 ways urban agriculture can transform unused spaces. Agritecture. https://www.agritecture.com/blog/2021/10/11/7-ways-urban-agriculture-can-transform-unused-spaces
28. Levkoe CZ (2006) Learning democracy through food justice movements. Agric Hum Values 23(1):89–98. https://doi.org/10.1007/s10460-005-5871-5
29. Slater R (2001) Urban agriculture, gender and empowerment: an alternative view. Dev South Afr 18(5):635–650. https://doi.org/10.1080/03768350120097478
30. Orsini F, Kahane R, Nono-Womdim R, Gianquinto G (2013) Urban agriculture in the developing world: a review. Agron Sustain Dev 33(4):695–720. https://doi.org/10.1007/s13593-013-0143-z
31. McIlvaine-Newsad H, Porter R (2013) How does your garden grow? Environmental justice aspects of community gardens. J Ecol Anthropol 16(1)
32. Duchemin E, Wegmuller F, Legault AM (2009) Urban agriculture: multi-dimensional tools for social development in poor neighbourhoods. Field Action Sci Report 2:1–8. https://doi.org/10.5194/facts-2-1-2009. https://www.researchgate.net/publication/250396876_Urban_agriculture_Multi-dimensional_tools_for_social_development_in_poor_neighbourhoods
33. Ackerman K et al (2014) Sustainable food systems for future cities: the potential of urban agriculture. Econ Soc Rev 45(2):189–206. https://www.researchgate.net/publication/268448635_Sustainable_Food_Systems_for_Future_Cities_The_Potential_of_Urban_Agriculture
34. FAO, Rikolto and RUAF (2022) Urban and peri-urban agriculture sourcebook – From production to food systems. Rome, FAO and Rikolto. https://doi.org/10.4060/cb9722en
35. Hallett S, Hoagland L, Toner E (2016) Urban Agriculture: Environmental, Economic, and Social Perspectives. https://doi.org/10.1002/9781119281269.ch2
36. Blog/Agribazaar (2024) The rise of urban agriculture in India by 2025: Trends to watch. Medium. https://medium.com/@agribazaar/the-rise-of-urban-agriculture-in-india-by-2025-trends-to-watch-6e1ebb1ba5bb

37. Zezza A, Tasciotti L (2008) Does urban agricultureenhance dietary diversity? Empirical evidence from a sample of developing countries, 12th EAAE Congress People
38. Lock K, Dezeeuw H (2001) Mitigating the health risksassociated with urban and peri-urban agriculture. Urban Agric Mag 1:6–8. https://www.researchgate.net/publication/284056754_Mitigating_the_health_risks_associated_with_urban_and_periurban_agriculture
39. Santo R, Palmer A, Kim B (2016) Vacant lots to vibrant plots: a review of the benefits and limitations of urban agriculture. John Hopkins center for a Livable Future, May
40. Baumgartner B, Belevi H (2001) A systematic overview of urban agriculture in developing countries. EAWAG/SANDEC, Dübendorf. https://doi.org/10.1504/IJETM.2003.003382
41. Zuma M, Arthur G, Coopoosamy R, Naidoo K (2023) Incorporating cropping systems with eco-friendly strategies and solutions to mitigate the effects of climate change on crop production, J Agric Food Res 14:100722. https://doi.org/10.1016/j.jafr.2023.100722
42. Baraj B et al (2024) Climate change and resilience, adaptation, and sustainability of agriculture in India: a bibliometric review, vol 10, issue 8. https://doi.org/10.1016/j.heliyon.2024.e29586
43. Gulyas B, Edmondson J (2021) Increasing city resilience through urban agriculture: challenges and solutions in the global north. Sustainability 13(3):1465. ISSN 2071-1050. https://doi.org/10.3390/su13031465
44. Qiu J, Zhao H, Chang NB et al (2024) Scale up urban agriculture to leverage transformative food systems change, advance social–ecological resilience and improve sustainability. Nat Food 5:83–92. https://doi.org/10.1038/s43016-023-00902-x
45. Sirowy B, Ruggeri D (2024) Setting the stage: Urban agriculture, public space, and human well-being. In: Sirowy B, Ruggeri D (eds) urban agriculture in public space. GeoJournal Library, vol 132. Springer, Cham. https://doi.org/10.1007/978-3-031-41550-0_1
46. Branduini P, Giacchè G, Laviscio R, Torquati B, Scazzosi L (2016) For a systemic reading of urban agriculture. Typologies, policies, business models, spatiality and metabolism. Agriregionieuropa (Per una lettura sistemica delle agricolture urbane. Tipologie, politiche, modelli imprenditoriali, spazialità e metabolismo. Agriregionieuropa). 44. https://www.researchgate.net/publication/299341223_Per_una_lettura_sistemica_delle_agricolture_urbane_Tipologie_politiche_modelli_imprenditoriali_spazialita_e_metabolismo

Sustainable Building Materials and Circular Economy

Sustainable Environmental Architecture: Approaches for Sustainable Building Materials and Waste Management in the Era of Climate Change

Engy Farrag

Abstract Arid and semi-arid cities face increasing challenges related to climate change and resource scarcity; sustainable environmental architecture has emerged as a vital approach to address these pressing issues. This chapter explores recent approaches to sustainable environmental architecture in arid and semi-arid cities, focusing on two key aspects: sustainable building materials and sustainable waste management strategies. With plastic waste accounting for a significant portion of urban and rural waste, totaling approximately 5.4 million tons annually, addressing the environmental hazards and health risks associated with plastic pollution is imperative. This study highlights the importance of reducing and recycling plastic waste generated and how to mitigate plastic pollution by implementing low-tech and open-source plastic recycling technologies specifically designed for emerging countries. It will be through the inclusion of case studies showcasing successful implementations of these technologies in architecture and construction. The study also provides recommendations based on the findings from the literature review, data analysis, and technology evaluation.

Keywords Sustainable architecture · Climate change · Waste management · Plastic waste · Recycling

---

E. Farrag (✉)
Architechural Engineering Department, Delta University for Science and Technology, Dammam, Kingdom of Saudi Arabia
e-mail: engy.farrag@alasala.edu.sa; engy.m.farrag@gmail.com

Z. ElZein and A. Negm (eds.), *Recent Approaches of Sustainable Architecture in Arid and Semi-arid Cities*, Urban Sustainability,
https://doi.org/10.1007/978-981-95-1645-2_6

1 Introduction

Arid and semi-arid regions, characterized by their fragile ecosystems and scarce resources, are increasingly vulnerable to the intensifying impacts of climate change. Water scarcity, extreme temperatures, desertification, and rapid urbanization converge to create complex challenges demanding innovative and sustainable solutions. Sustainable environmental architecture emerges as a critical framework for addressing these issues by integrating ecological principles into the built environment.

This chapter highlights the critical role of sustainable architecture and waste management in addressing plastic pollution. The selection and utilization of building materials are essential for creating structures that are both environmentally friendly and resilient to the unique climatic conditions of these regions [1]. For example, the use of natural materials like bamboo, compressed earth blocks, and recycled plastics has proven effective in reducing both environmental footprint and construction costs. Equally important is the effective management of waste generated throughout the building lifecycle, particularly in areas with limited infrastructure. In countries like India, the implementation of low-cost, decentralized waste-to-energy solutions has provided valuable lessons in how to handle waste sustainably in areas lacking robust waste management infrastructure. A pressing challenge exacerbating these issues is the global proliferation of plastic waste, with severe environmental and health consequences. Some countries, such as Kenya, have introduced successful bans on plastic bags, while others are focusing on the development of biodegradable alternatives to reduce plastic pollution.

By exploring opportunities for plastic waste reduction, reuse, and recycling within the built environment, this study aims to contribute to the development of sustainable and resilient communities in arid and semi-arid regions. This research seeks to provide insights and recommendations for a more sustainable future in these challenging environments.

2 Methods

A multifaceted research methodology was employed to pursue the chapter's objectives and to comprehensively investigate sustainable building materials, waste management strategies, and their interplay in arid and semi-arid regions. A systematic literature review was conducted to examine existing research on sustainable architecture, building materials, waste management, and plastic pollution within these specific geographic contexts. Emphasis was placed on identifying best practices and innovative approaches to address the unique challenges of these regions. Case studies of exemplary projects were analyzed to extract practical applications and lessons learned. Moreover, an in-depth evaluation of low-tech and open-source plastic recycling technologies was undertaken to assess their suitability for arid and semi-arid environments, considering factors such as technical feasibility, economic viability, environmental impact, and social implications. By combining these research methods, this study aimed to develop a robust understanding of the opportunities and challenges associated with sustainable environmental architecture in these regions, contributing to the broader discourse on sustainable development in vulnerable ecosystems.

3 Findings

The study underscores the crucial importance of selecting environmentally friendly building materials and implementing effective waste management strategies in arid and semi-arid regions to create resilient structures that can withstand the unique climatic conditions. It highlights the urgent need to address plastic pollution through sustainable architecture and waste management practices, emphasizing opportunities for plastic waste reduction, reuse, and recycling within the built environment. By showcasing Egypt's substantial plastic waste generation as a case study, the research illuminates the potential of low-tech plastic recycling technologies in developing cost-effective building materials and fostering a circular economy, ultimately aiming to provide valuable insights and recommendations for fostering sustainability and resilience in these challenging environments.

3.1 The Threats Posed by Plastic

The proliferation of plastic waste constitutes a significant global challenge with far-reaching environmental and human health implications [2]. Since the mid-twentieth century, plastic production has surged, resulting in a staggering accumulation of plastic debris [3]. The dominance of disposable plastic products, coupled with the use of harmful additives, exacerbates the issue. Beyond environmental contamination, plastic pollution poses risks to human health through chemical leaching and the release of greenhouse gases during production [4].

From 1950 to 2017, a staggering amount of plastic, totaling 9.2 billion tones, was manufactured [5]. This quantity surpasses the weight of one ton for every individual currently residing on Earth. The majority of this plastic production comprises disposable items and packaging materials. The use of plastic presents numerous health risks. To enhance its properties, various chemicals are incorporated into the base plastic. However, these chemicals pose hazards to human health and tend to accumulate in indoor air and house dust. In addition, the production of plastics is one of the largest and fastest-growing contributors to these emissions. The production of plastics alone could generate 53.5 billion tons of carbon dioxide emissions by 2050 (Fig. 1). For these reasons, the good management of plastic waste is recycling, it is essentially for many aspects such as the economic, environmental, and social

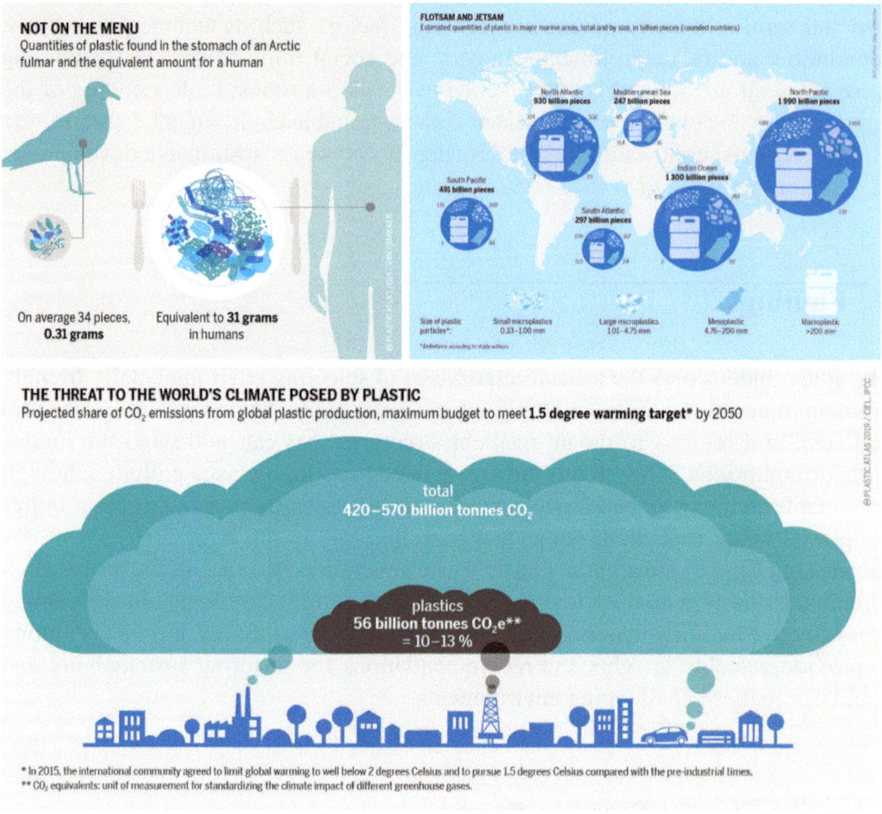

Fig. 1 The harmful impact of plastic waste. *Source* Heinrich Böll Foundation [5]

aspects. The imperative to transition towards a circular economy, prioritizing plastic waste reduction, reuse, and recycling, is evident in the face of this mounting crisis [6].

3.2 Waste Management: Circularity in the Built Environment

Effective waste management is paramount for conserving resources, safeguarding public health, and minimizing environmental degradation. The challenge of waste disposal is particularly acute in arid and semi-arid regions, characterized by scarce water resources and limited infrastructure. Adopting circular economy principles, such as waste reduction, reuse, and recycling, is essential for mitigating the environmental impacts of construction and building operations. Implementing waste prevention strategies, diverting waste from landfills through robust recycling and composting programs, and promoting the utilization of recycled materials in construction are essential steps towards a more sustainable built environment. A comprehensive waste management system, including collection, sorting, processing, and disposal facilities, is necessary to achieve this. Public awareness campaigns and education programs can also play a crucial role in fostering behavioral change and promoting waste reduction [7, 8].

3.3 Low-Tech Plastic Recycling

The integration of low-tech recycling processes into the construction sector of arid and semi-arid regions presents a promising avenue for sustainable development. By implementing effective waste collection and sorting systems, communities can divert plastic waste from landfills and recover valuable materials for reuse. Low-tech recycling methods, such as manual sorting, shredding, and washing, can be adapted to local conditions, requiring minimal technological infrastructure. Once processed, recycled plastic can be incorporated into various building components, including insulation, roofing materials, wall panels, and furniture. This approach not only contributes to waste reduction but also creates opportunities for local employment and economic growth. Moreover, the use of recycled plastic in construction can enhance building performance, providing insulation, durability, and water resistance. By adopting a circular economy approach, these regions can transform a significant environmental challenge into a resource for sustainable development [9, 10].

Figures 2 and 3 outline a typical plastic recycling process, consisting of the following key steps:

Fig. 2 The process of plastic recycling. *Source* Plastic odyssey & polystar.com

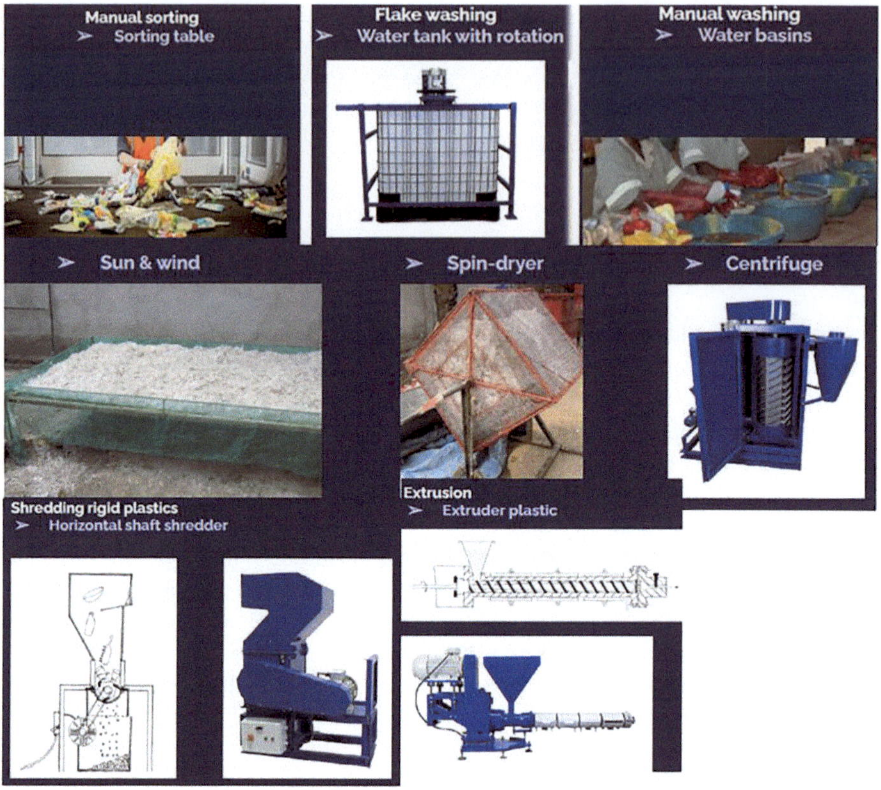

Fig. 3 Material transformation technics. *Source* Plastic odyssey [9, 10]

- **Collection:** This initial stage involves gathering plastic waste from various sources, such as households, businesses, and public spaces.
- **Sorting:** Collected plastic waste is categorized based on its type, color, and composition to ensure efficient processing.
- **Washing:** Plastic materials are cleaned to remove impurities, labels, and contaminants, preparing them for further processing.
- **Drying:** Moisture is removed from the plastic to prevent issues during the melting process.
- **Shredding:** Plastic pieces are reduced in size through shredding to facilitate handling and melting.
- **Material Preparation:** The shredded plastic undergoes further preparation, which may include additional cleaning or separation processes.
- **Heating/Mixing:** Plastic flakes are melted and mixed to create a homogenous material.
- **Shaping:** The molten plastic is molded or extruded into new products, completing the recycling process.

Table 1 An evaluation table of low-tech and open-source plastic recycling technologies for arid and semi-arid environments

Technologies	Technical feasibility	Economic viability	Environmental impact	Social & Community benefits
Manual Sorting	5	5	5	5
Shredding (Hand-powered)	4	3	5	4
Washing (Simple Filtration)	4	4	4	4
Sun-Drying	5	5	5	3

Table 1 presents an evaluation of technologies based on how feasible they are to implement, how economically viable they are, their impact on the environment, and their contribution to social or community well-being. Manual Sorting scores high across all categories, indicating it is very effective across the board. Shredding (Hand-powered) has a lower economic rating, while Sun-Drying stands out for its simplicity and low cost, though its social benefits are less pronounced. Based on Table 1, here is the evaluation of Low-Tech Plastic Recycling Solutions for Arid and Semi-Arid Regions:

A. **Technical Feasibility**

- **Manual Sorting:** Manual sorting is highly feasible for arid and semi-arid environments as it requires minimal equipment and can be done effectively regardless of climatic conditions.
- **Shredding (Hand-powered):** Hand-powered shredding machines are well-suited for these regions as they require no electricity and can be adapted to local materials. However, their processing capacity might be lower compared to motorized shredders.
- **Washing (Simple Filtration):** Simple filtration systems using locally available materials like sand or fabric can be effective for washing plastic waste in arid and semi-arid areas. However, contaminant removal efficiency may be lower than more sophisticated washing techniques.
- **Sun-Drying:** Sun-drying is an ideal technology for arid and semi-arid environments due to the abundant sunlight. It is a low-cost and energy-efficient way to remove moisture from shredded plastic.

B. **Economic Viability**

- **Manual Sorting:** Manual sorting is labor-intensive, but minimal equipment requirements make it a cost-effective solution.
- **Shredding (Hand-powered):** Hand-powered shredders are generally less expensive than motorized options. However, the lower processing capacity may affect overall productivity and economic viability.

- **Washing (Simple Filtration):** Simple filtration systems are cost-effective due to their reliance on locally available materials. However, they may require more frequent maintenance or replacement compared to more advanced washing systems.
- **Sun-Drying:** Sun-drying is a free and readily available method for drying shredded plastic, making it highly economical.

C. **Environmental Impact**

- **Manual Sorting:** Manual sorting has minimal environmental impact as it does not require any energy or generate emissions.
- **Shredding (Hand-powered):** Hand-powered shredding generates no greenhouse gas emissions or air pollution.
- **Washing (Simple Filtration):** Simple filtration systems may use water for cleaning, but the amount is typically minimal compared to industrial washing processes. However, the effectiveness of contaminant removal should be considered to ensure minimal environmental impact from residual contamination.
- **Sun-Drying:** Sun-drying is a clean and energy-efficient technology with minimal environmental impact.

D. **Social & Community Benefits**

- **Manual Sorting:** Manual sorting can create employment opportunities in arid and semi-arid regions.
- **Shredding (Hand-powered):** Hand-powered shredding can be implemented at a community level, promoting local participation in recycling.
- **Washing (Simple Filtration):** Using local skills and knowledge, simple filtration systems can be built and maintained.
- **Sun-Drying:** Sun-drying requires minimal technical expertise, making it accessible for communities in arid and semi-arid regions.

Overall, low-tech and open-source plastic recycling technologies can be suitable solutions for arid and semi-arid environments due to their low dependence on external resources, ease of operation, and minimal environmental impact. By carefully selecting and implementing these technologies, communities in these regions can effectively manage plastic waste and promote sustainable development.

3.4 Incorporation of Recycled Plastic in Construction

Recycled plastic can be utilized in various building applications. Plastic shreds can be incorporated into insulation materials to enhance thermal performance and reduce energy consumption. Recycled plastic can also be used to produce roofing materials, providing water resistance and durability. Additionally, plastic can be employed in manufacturing wall panels, furniture, and other building components [11, 12].

Securing a consistent and adequate supply of plastic feedstock is paramount for the successful operation of recycling facilities. Two primary procurement strategies emerge waste collection and the purchase of plastic scrap. While the former relies on the aggregation of post-consumer and commercial plastic waste through various collection systems, the latter involves acquiring processed plastic materials from industrial sources. Each approach presents distinct challenges and opportunities, including issues related to material quality, contamination, economic viability, and infrastructure. A comprehensive assessment of these factors is essential for optimizing plastic supply chains and maximizing recycling efficiency [13].

Two Primary Methods of Acquiring Plastic for Recycling:

a. **Collection of Plastic Waste**

   Community-based collection: Involves gathering plastic waste from households, businesses, and public spaces through organized collection programs.
   Waste management systems: Relies on municipal or private waste management services to collect and sort plastic waste.
   Informal recycling: Involves individuals or groups collecting recyclable materials, including plastic, for resale to recycling facilities.

b. **Purchasing Plastic Scrap**

   Industrial sources: Acquiring plastic scrap directly from manufacturing facilities that generate plastic waste during production processes.
   Recycling companies: Purchasing plastic scrap from other recycling companies that have already processed and sorted the material.

Both methods have their advantages and challenges (Fig. 4), and the choice of acquisition method often depends on factors such as the availability of plastic waste, the desired quality of the recycled material, and economic considerations.

The integration of recycled plastics into the construction sector presents a promising opportunity to mitigate environmental impacts and promote resource circularity. This approach provides a dual benefit of effective waste management and resource conservation by diverting plastic waste from landfills and transforming

Fig. 4 Methods of Collecting Plastic Waste: (a) Direct Collection from Trash, (b) Purchasing Recyclable Plastics. *Source* Plastic odyssey [9, 10]

it into valuable building materials. The incorporation of recycled plastics can be realized through various applications, including insulation, roofing, wall panels, and structural components.

Incorporating recycled plastics into the construction sector offers a diverse range of applications, contributing to both environmental sustainability and building performance.

b.1 Insulation Materials

Recycled plastic fibers and foams can be effectively utilized as insulation materials, enhancing energy efficiency and reducing heating and cooling demands. These materials provide excellent thermal resistance, improving indoor comfort and reducing energy consumption.

b.2 Roofing Materials

Recycled plastics can be transformed into durable and water-resistant roofing materials, offering long-term protection for buildings. Plastic-based shingles, tiles, and membranes provide cost-effective and sustainable alternatives to traditional roofing options.

b.3 Structural Components

Reinforced concrete incorporating recycled plastic fibers exhibits enhanced mechanical properties, including increased strength and ductility. This innovative approach expands the application of recycled plastics in structural engineering.

b.4 Road Construction

Recycled plastic aggregates can be used as a substitute for traditional aggregates in road construction, reducing the consumption of natural resources and improving road performance. Plastic-modified asphalt demonstrates enhanced durability and resistance to cracking.

b.5 Other Applications

Beyond these primary applications, recycled plastics can be utilized in various other construction components, such as pipes, flooring, and interior finishes. The versatility of recycled plastics offers opportunities for creative and innovative design solutions.

By exploring these diverse applications (Fig. 5), the construction industry can contribute to the circular economy and promote sustainable building practices.

4 Conclusion

The integration of sustainable principles into the built environment is imperative for addressing the complex challenges posed by climate change and resource scarcity in arid and semi-arid regions. This chapter has explored the pivotal roles of building

Fig. 5 Products from the plastic waste. *Source* Plastic odyssey [9, 10]

materials and waste management in achieving sustainable architecture. By emphasizing the use of locally sourced and environmentally friendly materials, coupled with effective waste management strategies, it is possible to create resilient and eco-conscious structures.

The escalating plastic pollution crisis necessitates innovative solutions, and the construction sector offers a promising avenue for addressing this challenge. Incorporating recycled plastics into building materials can divert waste from landfills, conserve resources, and promote a circular economy. While material quality, durability, and cost-competitiveness challenges persist, advancements in recycling technologies and supportive policies are essential to unlock the full potential of recycled plastics in construction.

To achieve a sustainable built environment, a holistic approach encompassing material selection, waste management, and integrating recycled materials is required. By embracing these principles, arid and semi-arid regions can enhance their resilience to climate change while fostering economic growth and environmental stewardship.

5 Relevant SDGs

The research outlined and related to several United Nations Sustainable Development Goals (SDGs). Below are the most relevant SDGs:

SDG 11: Sustainable Cities and Communities
This goal focuses on making cities inclusive, safe, resilient, and sustainable. The focus on sustainable building materials, waste management, and urban planning in arid regions is crucial for creating sustainable cities that can cope with the challenges of rapid urbanization, water scarcity, and plastic pollution.

SDG 12: Responsible Consumption and Production
Waste management, plastic recycling, and the circular economy are key components of this goal. Reducing, reusing, and recycling plastic waste in the built environment, as well as promoting sustainable construction practices, contribute to more sustainable consumption and production patterns.

SDG 13: Climate Action
Addressing the impacts of climate change in arid and semi-arid regions, such as extreme temperatures, desertification, and resource depletion, is integral to this goal. The research on sustainable architecture and waste management practices directly contributes to mitigating climate change by reducing carbon footprints and adapting to changing environmental conditions.

SDG 15: Life on Land
Desertification, land degradation, and the preservation of fragile ecosystems in arid regions are key concerns here. Sustainable practices in architecture, such as utilizing local materials and managing waste, can help protect ecosystems and promote land restoration efforts in these vulnerable regions.

6 Recommendations

Recommendations for Sustainable Environmental Architecture in Arid and Semi-Arid Regions:

- Prioritize locally sourced and low-embodied energy materials.
- Promote the use of bio-based and recycled materials.
- Implement a circular economy approach.
- Develop robust waste management infrastructure.
- Promote public awareness campaigns.
- Pilot and implement low-tech plastic recycling programs.
- Develop standards and guidelines for recycled plastic use.
- Promote innovation and design solutions.
- Introduce incentives for sustainable construction practices.
- Mandate the use of recycled content in construction projects.
- Strengthen regulations on plastic waste management.

By implementing these recommendations, arid and semi-arid regions can pave the way for a more sustainable future. Integrating these approaches can foster the development of resilient and resource-efficient built environments that contribute to a circular economy and mitigate the impacts of climate change.

These recommendations provide a framework for fostering sustainable environmental architecture in arid and semi-arid regions. By prioritizing material selection, waste management, and the integration of recycled plastics, these regions can enhance their resilience to climate change while promoting economic growth and environmental stewardship.

References

1. McDonough W, Braungart M (2002) Cradle to cradle: rethinking the way we make things. North Point Press.
2. Bidashimwa D, Hoke T, Huynh TB, Narkpitaks N, Priyonugroho K, Ha TT, Burns A, Weissman A (2023) Plastic pollution: how can the global health community fight the growing problem? BMJ Glob Health 8(Suppl 3):e012140. https://doi.org/10.1136/bmjgh-2023-012140
3. United Nations Environment Programme (2023) Visual Feature | Beat Plastic Pollution. Retrieved from https://www.unep.org/interactives/beat-plastic-pollution/
4. Singh N, Walker TR (2024) Plastic recycling: a panacea or environmental pollution problem. NPJ Mater Sustain. 2:17. https://doi.org/10.1038/s44296-024-00024-w
5. Heinrich Böll Foundation (2020) PLASTIC ATLAS. Facts and figures about the world of synthetic polymers. Heinrich Böll Foundation, Berlin, Germany. ISBN 978-3-86928-211-4
6. Kumar R, Anurag Verma A, Rama Sinha S, Prakash Kumar Jha R, et al (2021) Impacts of plastic pollution on ecosystem services, Sustainable development goals, and need to focus on circular economy and policy interventions. Sustainability 13(17):9963. https://doi.org/10.3390/su13179963
7. Delterra (2023) Circular cities: the six dimensions of circular waste management. Delterra. https://delterra.org.
8. Solar Impulse Foundation (n.d.) How to shape the future of waste? Retrieved August 29, 2024, from https://solarimpulse.com
9. Plastic Odyssey (2024) The global network serving plastic emergency. Retrieved from https://plasticodyssey.org/en/
10. Plastic Odyssey (2024) The expedition of solutions against plastic pollution. PRESS KIT 2024. https://plasticodyssey.org/wp-content/uploads/2023/DP/EN-PRESS-KIT-2023-PLASTIC-ODYSSEY-CP.pdf
11. Arqlite (2022). 7 recycled plastic building materials revamping construction—Arqlite Smart Gravel—the light and sustainable alternative to gravel. Retrieved from https://smart.arqlite.com/recycled-plastic-building-materials/
12. Eyüboğlu H, Yanılmaz Z (2023) The use of recycled plastics in the production of building materials in the context of sustainability.
13. Geyer R, Jambeck JR, Law KL (2017) Production, use, disposal, and recycling of primary plastics in the United States. Sci Adv 3(7):e1700782

Cost-Effectiveness and Market Adoption of Sustainable Building Materials: An Analysis of Comparative Economic Viability and Barriers to Industry Integration

Ahmed Elyamany and Ahmed Osama Daoud

Abstract Construction is a sector that has come under heightened scrutiny owing to its environmental effects, and as a result, there is a growing movement towards the use of sustainable materials. This chapter explores the issues of cost-effectiveness and market adoption of these sustainable alternatives relative to their conventional counterparts. The economic viability for the sustainability of the materials was assessed throughout their life cycle in terms of extraction, production, utilization, and end-of-life management by a detailed analysis. The chapter will thus capitalize on such a multidisciplinary approach in comparing the performances of renewable resources and recycled materials on the one hand and traditional ones on the other. This was achieved in relation to factors such as raw material availability, manufacturing processes, and long-term durability. It also identifies and analyzes barriers that prevent or restrain the broad application of the sustainable building material concept in the construction industry. These barriers are economic, regulatory, and sociocultural, thus pointing to challenges in market penetration and mainstream acceptance. It is on this background that this study attempts to articulate, for policymakers, industry stakeholders, and researchers, ways of overcoming such barriers, and how transitions towards a sustainable built environment can be expedited.

Keywords Sustainable construction materials · Life cycle assessment (LCA) · Economic viability · Market adoption · Environmental impact · Cost-effectiveness

---

A. Elyamany (✉) · A. O. Daoud
Civil Engineering Department, The British University in Egypt (BUE), Cairo, Egypt
e-mail: Ahmed.elyamany@bue.edu.eg

A. O. Daoud
e-mail: Ahmed.daoud@bue.edu.eg

Z. ElZein and A. Negm (eds.), *Recent Approaches of Sustainable Architecture in Arid and Semi-arid Cities*, Urban Sustainability,
https://doi.org/10.1007/978-981-95-1645-2_7

1 Introduction

1.1 Background and Rationale

The construction industry is one of the major contributors to environmental impacts worldwide, causing a shift towards sustainable building material types. These were defined as building materials with a lower environmental load throughout their life cycle, from raw material extraction through production, use, and disposal. Sustainable building materials, on the other hand, are gaining the market, with products such as recycled content, renewed resources, and low-impact alternatives, making regular construction practices a viable solution to environmental challenges [1]. The transition to sustainability is motivated not only by environmental concerns but also by the growing recognition that the economic feasibility of such an approach will be necessary for wide-market adoption. Economic viability is one of the key factors driving the sustainable building material market. In the long run, these materials create operational and environmental value. Economic barriers are typically experienced at the initial cost level of such materials in the process of material production up to placement. In the case of products considered to be sustainable, cost-effectiveness must be derived from all life cycles of the construction right from extraction, manufacturing, utilization, and end-of-life management [2]. Understanding how sustainable materials compare their economic performance with conventional alternatives is essential to ensure their greater use in the construction industry.

1.2 Objectives of the Chapter

This chapter primarily intends to address issues pertaining to the cost-effectiveness and market adoption of sustainable building materials. It attempts to analyze the economic viability through the performance of such materials in different stages of the building life cycle. It also identifies and assesses barriers to the wide use of sustainable building materials from an economic, regulatory, and sociocultural perspective. This chapter covers a comprehensive analysis that was carried out to fill such information gaps and provide valuable input for policymakers, industry stakeholders, and researchers on overcoming these barriers to expedite a shift towards a sustainable built environment. The analysis compared various life cycle aspects of renewable and recycled materials with those of traditional construction materials. These considerations include the availability of raw materials, the manufacturing process, long-term durability, and life cycle costs. The broader implications of such findings in terms of broad-based adoption of sustainable building practices and the development of supportive policies and incentives are discussed later in this chapter.

2 Comparative Analysis of Life Cycle Costs

2.1 Initial Investment vs. Long-Term Benefits

The costs of building materials assist in the control of construction activities or work. Concrete and steel are materials that have always been planned and preferred owing to their low initial cost and availability. However, because of the recent focus on sustainability in the built environment, developers have been turned off although the initial cost of sustainable materials such as cross-laminated timber (CLT), bamboo, recycled materials, and the rest is relatively high. The initial cost of investment in sustainable materials can be significantly higher and might be 20% higher than that of traditional materials. This is due to the cost of raw materials, use of skilled labor, and scarcity [3]. It might be interesting to know that these initial costs are compensated for by the use of sustainable materials from a long-term perspective. These advantages mean that sustainable materials require low energy consumption and minimal maintenance, given that they are very durable and perform reasonably well. For example, Biolek and Hanák [4] demonstrated that in many cases, sustainable materials have a lower life-cycle cost (LCC) than traditional materials, despite their higher cost. They established that even conventional materials such as concrete may generally have lower first costs and higher life cycle costs owing to higher energy usage and maintenance costs [4]. In addition, sustainable materials can allow for high indoor environmental quality, which translates to the improvement of the health and productivity of occupants to justify the premiums of such materials. Some of the best materials used are natural, including wood and bamboo, making the structures have less embodied energy. In doing so, it will have a smaller carbon footprint and cater to the phenomenon of biophilic design, which has several psychological benefits [5]. Thus, sustainable materials incur high costs in the beginning and, thus, have low costs in the middle and at the end of the building's life cycle from economic, environmental, and social points of view.

This means that to obtain full economic consequences of materials in construction, life-cycle cost analysis (LCCA) should be impacted. Conventional materials include concrete and steel; they are often chosen based on the initial cost, which is not expensive, but the long-term expenses of the project in the long run. For instance, concrete structures may often require a refresher, and in terms of heating and cooling, they consume a lot of energy because they do not insulate well [6]. On the other hand, sustainable materials reduce expenses throughout the building life cycle. Studies have shown that some of the sustainable materials that are used in construction, including CLT, have negative impacts, whereby some of the new insulation systems and advanced systems significantly lower the energy costs. Some of the gaps include those discussed by Pučko et al. [7], who discussed the use of building information modeling (BIM) in performing multi-objective integrated, comprehensive energy, and life cycle cost analysis of building envelope components. According to the findings of their studies, it was realized that energy-efficient materials for use in employment, regardless of the fact that the initial costs may be high, cost less both in terms of operational energy use and maintenance costs. The use of sustainable materials means that such materials tend to be more durable, which means that repair and replacement are done less frequently than materials that are not sustainable. Examples of these systems include fiber cement siding and metal roofing. These have low maintenance expenses but a long cycle time with steep fixed costs. These materials do not cause problems such as rot, mold, or insect damage, which can be very expensive to mend. Additionally, sustainable materials are used at an advanced level of insulation materials to make them energy-efficient in nature, combining the two levels for lower utility bills, as the insulation material lowers heating and cooling loads and preserves energy consumption.

Recent studies highlight the potential for sustainable materials to reduce life-cycle costs while addressing environmental concerns. Hasan et al. [8] emphasize that incorporating bio-based and green composite materials into construction supply chains significantly reduces embodied carbon and operational energy use over time. This integration aligns with life-cycle cost analysis (LCCA) principles, providing both economic and environmental advantages in construction projects. Furthermore, Meinardi et al. [9] demonstrated the applicability of advanced luminescent solar concentrator glazing technologies in reducing energy consumption, particularly in energy-intensive climates.

2.2 Economic Evaluation Methods

As depicted in Fig. 1, LCCA is an overall system for the assessment of the full cost of a material in relation to construction for the life cycle of the material right from procurement to installation, use, maintenance, and disposal. LCCA is increasingly being used in sustainable construction to provide a better assessment of the long-term costs that may be created by using sustainable materials. This enabled Lakousha [10] to design an LCCA computerized tool for construction projects to ease the calculations depicted here. LCCA software also allows a comparison of materials to make choices concerned with life cycle costs rather than initial costs. It builds on factors such as pulling electricity bills, the cost of maintainability, and, at times, the cost of a material replacement in the future to provide a much broader picture of the lifetime cost of the building material.

Another aspect of LCCA, which is closely associated with the concept of the time value of money, is related to the effects of changes in discount and inflation rates on the current values of future costs. The rates selected are among the most significant results of LCCA, and as such, a sensitivity analysis is needed within the method. Sensitivity analysis enables the evaluation of the impact of the adjustment of key assumptions on the total life-cycle cost, with a range of potentials that assist in decision-making [11]. For instance, in areas with high inflation, the probability is that the future maintenance or energy costs are likely to be much higher than those in stable economies, so sustainable materials can have a completely different impact on cost-benefit analysis.

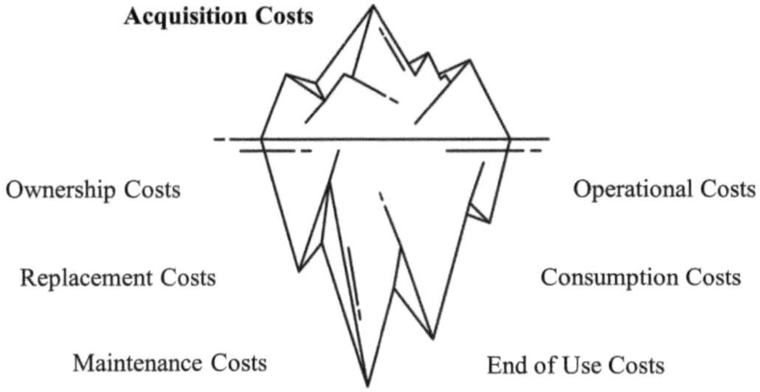

Fig. 1 Life Cycle Cost Iceberg (modified by the author(s) after [10])

Thus, for such comparisons, there are instruments such as Cost-Benefit Analysis (CBA) and Multi-Criteria Decision Analysis (MCDA). These frameworks involve integrating more than one economic, environmental, and social factor in a single decision-making process, thus offering a wider outlook on all the possible effects linked to the selection of a particular material. Another area where CBA is applicable is when trying to compare the costs of sustainable material to all the benefits that come from their use in the long run. Hence, CBA demonstrates logos of economic benefits accruing from using long-lasting, costly and 'green' inputs of construction and sustainable nature of energy-saving, lowered maintenance costs and durability longer than the service-life of the project. It also provides a method of pricing benefits in the environment, including, for example, the return of carbon emissions, which must be in line with the general ideas of cost and benefit analysis. However, MCDA can be a more permissive approach when it comes to the inclusion of qualitative factors, such as social and environmental factors, together with strict economic parameters. Abouhamad and Abu-Hamd [12] designed the concept of construction system selection using the MCDA, which involves LCCA and sustainability assessment. This has made it possible to conduct a comparative analysis of various construction systems in terms of economic implications and their correlation with the effects on the environment and social sectors. It was possible to establish which construction systems were most sustainable and efficient by applying the aforementioned residential and commercial building construction framework.

2.3 Implications for Market Adoption

To derive such findings based on life-cycle costs, one can only imagine the enormous significance of market affiliations in view of sustainable materials. However, perhaps the most massive observed concern among developers and construction experts is the entire life cost, not simply the construction cost. This changing paradigm is simply important for the broader application of more sustainable materials, many of which costs more initially but saves loads in the long run. It is clear that when LCCA is incorporated in the decision-making process, sustainability material will be the most likely to be selected. Xue et al. [13] also carried out an impact assessment of campus buildings in China and found that incorporating LCCA right from the beginning of design stages enhances greater value-based decisions and is sustainable. The work revealed that at other times, because of the rightful decision making involving the long-run cost advantage associated with the use of sustainable material, there is a likelihood that regardless of the high upfront costs of the material, the materials will be chosen. It is not only a way to use environmentally friendly materials but also a way to build the construction sector sustainably. In addition, incorporating of LCCA into certification systems, such as Leadership in Energy and Environmental Design (LEED), has led to using of sustainable materials [14–19]. The concept of considering life-cycle costs is embodied in the requirements of certification, as it decides that the user embraces materials that may cost more at the initial stages but are cheaper within the span of their "lifetime" or service. This has, in turn, boosted the market for sustainable materials in buildings, because developers need to achieve a higher certification that is likely to be in demand in the market for green buildings.

3 Barriers to Adoption: A Multifaceted Perspective

3.1 Market Perception and Consumer Awareness

Several major stakeholders, such as customers and builders in the construction sector, have strong perceptions that sustainable materials cost considerably more than their counterparts in the market. This is due to cost factors covering both the perceived costs of green building products and the real cost of utilizing these sustainable building materials. As highlighted by this perception, people remain ignorant of the fact that in the course of the building's life cycle, sustainable materials result in lower costs of energy consumption and maintenance charges [20]. The first limb of the problem is that sustainable materials cost more at the beginning, thereby resulting in reduced willingness to adopt them with an uncertain positive return in the future [21]. Awareness of the public and other industries is essential to enhance the use of sustainable building materials. This involves the use of public enlightenment programs and other means of informing the public of the relevant myth about the gains and pains of such materials. Finally, increased awareness can help change consumer behavior and the trends that dominate within the industry. However, the current level of awareness remains low, particularly in areas where sustainable construction practices are rarely implemented.

3.2 Regulatory and Policy Challenges

The other key impediment to sustainable building material manufacturing and use lies within the legal framework. Recycling systems have not benefited from regulatory frameworks as they should, and an inclination towards traditional, non-friendly products continues to be encouraged. This could be explained by the fact that building codes and standards may or may lack provisions that would permit the use of recycled or renewable materials, or by the fact that, where they permit them, they may set rigid standards for approval that serve sustainable materials becomes almost impossible. More importantly, policy gaps do not serve sustainable building needs, because there are no incentives for subsidies to developers who embrace sustainable building materials in construction [22]. Therefore, it is necessary to trace the changes in regulatory and policy frameworks to achieve a transition towards sustainable building practices. This evolution involved revising the building codes to incorporate information on new materials, thus providing special guidance on their application while also providing incentives to cater to the otherwise high costs of use. That is, the promotion of sustainable building materials remains a difficult endeavor, especially when the existing systems are not altered.

3.3 Industry Resistance and Institutional Barriers

The building materials industry does not have a disposition to accept new changes in production processes and profit models that have persisted for many years. This is particularly true of the traditional construction sector, where the adoption and acceptance of new materials on sustainability criteria is still highly embryonic, primarily as a result of question marks over performance, durability, and compatibility with well-established construction technologies [23]. This institutional inertia is compounded by the fact that sustainable materials are not encouraged as much as conventional ones, meaning that they are at a disadvantage. This is compounded by the fact that the construction industry is generally averse to risk bearing. No one will procure sustainable materials with good conscience if there is no guarantee that the use of the commodity will be available to the company in the long run or is demanded in the market. This resistance is especially significant when the area of construction companies is full of SMEs that might not be able to afford trial-and-error with new materials [24].

3.4 Financial and Economic Barriers

The key challenges that minimize the use of sustainable building materials are the financial and economic aspects. Despite a certain availability, such materials can hardly be called accessible to developers and builders as long as they require relatively high initial investments and, in addition, there is little money. This is true because many sustainable materials are relatively expensive to purchase and incorporate into a project, which may be a major drawback for projects with a low budget [25]. Financial constraints are further aggravated by the premise that savings through sustainable material usage, for instance, energy and maintenance, accrue over the long run, and are least appealing to short-term developers. In addition, high risk aversion and high uncertainty are related to the return on investment in relation to sustainable materials. New materials may not have the level of finishing and acceptance in the market among developers and investors, as well as the structural performance and potential for resale. This could deter creativity and ensure minimal change in the construction industry to more sustainable procedures [5].

Despite these advancements, barriers persist in widespread adoption. Shen et al. [26] highlight the limited scalability of innovative bio-based fillers derived from waste products, noting that market perception and regulatory hurdles hinder their full integration into mainstream construction practices. Additionally, Johnson and Liu [27] suggest that while energy-efficient insulation materials are technologically advanced, high initial costs and lack of financial incentives discourage adoption, particularly in low-income regions.

4 Regional Variations in Market Dynamics and Regulatory Environments

4.1 Market Dynamics Across Different Regions

Sustainable materials are widely adopted according to geographical location, new economic positions, legal requirements, and mentality. In Europe, the adoption of sustainable materials is a result of extensive environmental standards and high standards of cultural sustainability. This has been due to a series of actions the European Union took, such as reducing carbon emissions and enhancing energy efficiency, thus promoting the large-scale use of sustainable materials in the construction of houses and business entities. However, this uptake of sustainable materials is low when economic performance is poor. Another problem that has hampered the use of sustainable materials in developing countries is the high cost of capital required to acquire sustainable materials. Such people value immediate consumption in these regions and thus make un financially sustainable decisions.

In fact, following the awareness of the environmental impacts of the products that nations produce and use, Schneider-Marin and Lang [5] stated that the value information of both global warming potential (GWP) and resource depletion ensures that different regions adopt appropriate materials. As shown in Fig. 2, they also identified gain factors that point out that regulation and financial encouragement for using sustainable materials lead to its adoption. Where such support is unavailable,

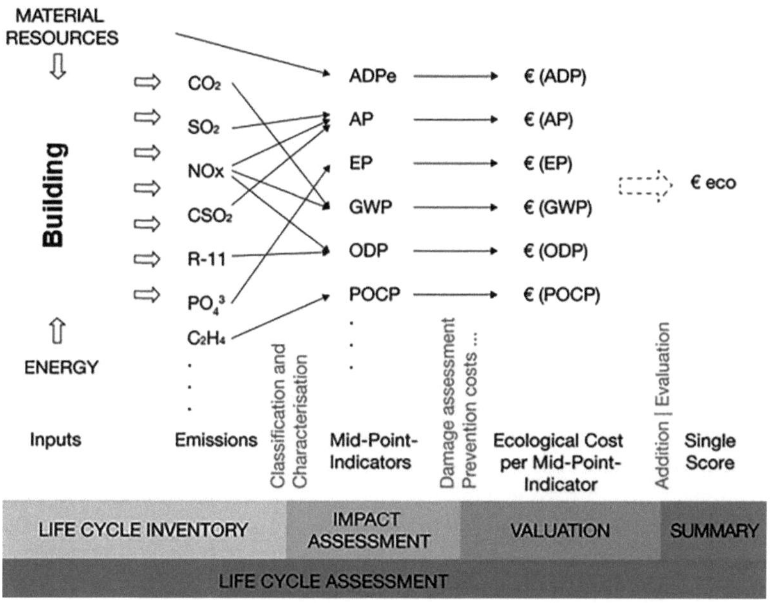

Fig. 2 Building LCA and monetary valuation of mid-point indicators to arrive at a single score (made by author(s) and adapted from [5])

conventional materials remain predominant in built landscapes worldwide. However, consumers are regulators of materials that can create or break sustainable material usage. High concerns about the environmental issues of construction projects are reflected more in countries that have already developed a pertinent culture, such as the Scandinavian countries. Thus, both regulation and awareness put pressure on these materials and such practices are in demand. On the other hand, in areas where the level of awareness of environmental issues is relatively low, the market for sustainable materials is not very large, therefore, the uptake is sluggish.

It has been established that economic factors are one of the main factors that define sustainability of the materials used. Other reasons are that regions characterized by higher disposable income of both consumers and developers are likely to register a higher incidence of sustainable materials because, while they are initially costly, more investment is channelled towards their procurement. These are areas where the long-term cost benefits from sustainable materials, although lower energy and maintenance costs, are appreciated [28]. However, buying sustainable materials is expensive, meaning that sustainability is usually costly, especially in less economically developed areas. Irrational immediacy is even more important here, immediate utility and the resulting satisfaction are the primary concerns with costs in the distant future. This is further aggravated by the inability to secure financing for sustainable materials so that they become cheaper in the market. For example, Schneider-Marin and Lang [5] proved that in regions with weak regulation and financial backing, traditional materials that may require less capital investment in the initial stages but more in the long run still dictate the market.

4.2 Regional Regulatory Environments

There is a reason that the environment specified by regulatory requirements is important in the selection of sustainable building materials. For instance, in areas of the European Union, where governments have set high environmental policies, this has triggered more utilization of sustainable materials in the market. Such regulations, which are commonly imposed, include energy efficiency, waste management, and the use of environmentally friendly materials, among other measures to ensure sustainability [6], [29–32].

For instance, the European Union's Energy Performance of Buildings Directive (EPBD) mandates that new buildings are nearly zero-energy buildings (nZEBs) by 2021. This directive has permitted the development of sustainable and energy-efficient materials in buildings, including superior insulation, energy-efficient windows, and even renewable energy sources. The directive also widens the application of life-cycle costing for assessing the economic sustainability of these materials to advance their development [11].

However, occurrences in areas with lax laws governing the rights of sustainable material adoption are not as fast as those in developed and developed developing countries. In Asia and Africa, in particular, the guidelines for construction are somewhat more liberal, and there is not much emphasis on the use of sustainable production. This further demonstrates the significance of proper laws and policies for promoting sustainable practices in the construction sector [29].

Areas with enabled environments have made great progress in matters concerning the sustainability of materials. For instance, Germany boasts of one of the most stringent green standards and has spent a lot of money subsidizing green-building projects, thereby making them revolutionaries in sustainable material consumption. The strategic focus on the efficiency of energy for buildings and sustainable energy sources has made the application of materials such as CLT and green concrete environmentally friendly and cost-effective in the country [6].

In the United States of America, places such as California have put strict measures on the codes of construction and even encouraged green buildings, hence causing the use of such green materials to be embraced. The Building Energy Efficiency Standards, known as Title 24, enacted in California are among the most stringent across the United States of America and demand that buildings be constructed with certain levels of efficiency. Voluntary standards, together with the provisions that offer incentives such as tax credit and rebates to use sustainable materials, have placed California as the forefront adopter [31].

While the developments that support such policies can easily incorporate sustainable materials and technologies, regions that do not have such policies in place continue to find it difficult to apply sustainable materials. Unfortunately, developers in many developing countries lack regulations and incentives to enable them to make business cases out of the increased cost of using sustainable materials. This highlights the importance of government subsidies as key drivers of sustainable development in the construction sector [28].

4.3 Impact of Local Culture and Practices

To a considerable extent, the use of sustainable materials is similar to local construction practices and cultural beliefs. A common problem is that people in certain areas are very conservative when it comes to both materials and techniques, and are rather protective when it comes to the paradigm shift that reflects more sustainable building solutions. For example, in many regions of Africa and Asia, it is impossible to conceive construction materials such as brick and concrete as alien to society's local circumstances. These are mostly used because of their ease of getting them, cheapness, and because most people are using conventional materials, it becomes difficult for new sustainable materials to capture the market [32].

On the other hand, global areas where the community embraces the need to protect the environment are more likely to consider sustainable materials. For instance, using environment-friendly materials is standard, and customers can become disappointed if they are not used. This has been facilitated by the region's understanding of the environmental conservation agenda and supportive regimes for the cause. In such regions, sustainable materials are viewed to reduce environmental impact and as a way to enhance the quality and value of buildings [33].

Bragadin et al. [32] discussed the perspectives and the obstacles of implementing sustainable materiality in various culture. In their study, they determined that where people value sustainability culturally, the reception towards the use of sustainable materials in construction will have better responses from consumers and the construction industry. On the other hand, in areas with heritage culture constraints, a lack of change enables firms to extend the use of sustainable materials.

Case studies provide a rich source of information on the lessons learned regarding the determinants of the use of sustainable products across geographical areas. For instance, in Scandinavian countries, the integration of sustainable material is promoted by government policies, cultural preferences, and consumer desire. The high concern for sustainability in the region has enhanced the formulation of constructive materials and practices that have turned out to be industrial norms [33].

However, the use of sustainable materials in Southern Europe has not been as successful as in Northern Europe countries. Economic considerations are also crucial because sustainable materials are usually expensive and therefore, unsustainable when conditions back home are not very good. Culturally, some factors favor local materials and traditional methods of construction. Zhang et al. [33] analyzed Spain, the Netherlands, and Sweden and discovered that sustainable materials were used most often, especially in Sweden, because of the government's support and people's willingness to embrace sustainable solutions. However, the adoption data rate was considerably lower in Spain, where economic and cultural barriers were major factors.

5 Strategies for Overcoming Barriers and Promoting Adoption

5.1 Policy and Regulatory Interventions

There is a need to establish a favorable regulatory framework that facilitates the use of green-building products. Governments could make a significant contribution by prescribing that such materials be used or recommending their use. For instance, providing rules that require green materials within construction and implementing floor standards on environmental performance may alter industrial practices. There is also a need to ease some of the rules and regulations that often act as barriers to implementing sustainable materials in construction projects [24].

Promotions, grants, and tax credits are some of the informative influences that the government can use to advance the use of sustainable building materials. However, the above materials have relatively high initial costs, and financial incentives can help bring the level of these costs to those of their conventional counterparts. The financial promotion of manufacturers of sustainable components along with the provision of tax incentives to the developers taking them into their projects will help develop the demand and use of sustainable materials [34]. Furthermore, policies such as carbon pricing mechanisms that are now being introduced can provide an economical means of minimizing the effects of construction materials on the environment [22].

5.2 Industry and Stakeholder Engagement

Stakeholders in the construction industry need to encourage the use of sustainable building materials. Owners, architects, engineers, contractors, and developers must be engaged in the implementation of sustainable best practices. The leaders in the industry can help in setting the pace by embarking on projects that will demonstrate the possibility of sustainability materials as well as the benefits that come with them [35].

The understanding that the government, industry, and academia need to come together to support the adoption of sustainable building materials is of equal importance. Partnerships are in building research collaborations and embarking on pilot projects and knowledge transfer activities that would create awareness of best practices. Academia can participate by investigating the current and future performance of sustainable materials in construction, whereas the government can assist with funding and policies [36]. Thus, the industry can contribute practical experience and advice to make scientific work corresponding to actual requirements.

5.3 *Financial Innovations and Incentives*

Financial constraints are a major hindrance to the use of sustainable building materials; therefore, innovative financing models should be employed. This is particularly so because conventional funding models cannot cater to the high initial costs incurred by these materials; hence, there is a need to develop new financial products for sustainable construction [6]. One such model is the performance payment model, in which payments are made contingent on results attained on environmental performance standards, such as energy efficiency and carbon emissions volume.

Sustainable building materials are best supported by green bonds, loans, and other products. For instance, green bonds are used to finance eligible projects with environmental benefits; thus, they can be used to finance sustainable materials for mass construction [29]. Similarly, green loans target borrowers who pledge to use green material and extend them easily, thereby lessening the financial load on developers while adopting green practices [33].

Innovative financial models and public-private partnerships could bridge the gap between high upfront costs and long-term benefits. For instance, Shen et al. [26] propose leveraging nature-inspired designs to overcome intrinsic weaknesses of waste-based materials, ensuring cost-effectiveness and scalability. Additionally, Hasan et al. [8] advocate for supply chain innovations that enhance the affordability and availability of bio-based materials, making them more competitive with traditional options.

5.4 *Education and Awareness Programs*

Exploding the market share of sustainable construction and educating potential consumers and contractors about available green building materials is another important prerequisite. These programs can help correct the wrong perception about the cost and effectiveness of eco-friendly materials, their sustainability, and advantages in the long run [37]. Segmented campaigns can be targeted at special interest groups of a couple of individual homeowners, high-rise developers, or any other interested party.

Mindful attention to these approaches was possible using case studies discussing various educational campaigns that were successful in implementation. For instance, helath literacy engagement through community engagement in a number of regions thrives through participatory workshops, seminars, and online portals and has enhanced composite utilization [38]. Moreover, training is important for architect engineers and builders, whereby professional development courses are helpful in providing knowledgeable and skilled professionals that can improve the industry [39].

6 Case Studies: Successful Implementation of Sustainable Building Materials in Arid and Semi-Arid Regions

6.1 Case Study in Saudi Arabia: Sustainable Use of Local Materials in Harsh Environments

The use of sustainable building materials in arid and semi-arid regions is crucial for addressing the unique environmental challenges of these areas. In Saudi Arabia, several case studies have demonstrated the successful application of local materials to create sustainable structures that align with the harsh climatic conditions. One notable example is the re-use of mining by-products for construction. A study conducted on mine tailings from a Saudi copper mine showed that these materials could be processed into cemented paste backfill, which, when reinjected into underground cavities, minimized environmental waste exposure while maintaining structural integrity [40]. This practice supports sustainable mining and reduces the environmental footprint of construction projects in arid regions.

Another successful implementation involves the use of traditional adobe bricks, which have been a staple in Saudi Arabian architecture for centuries. Research on these bricks highlights their effectiveness in providing thermal insulation and durability in hot, dry climates. By promoting the use of locally sourced materials like adobe, modern construction can achieve better thermal efficiency while reducing reliance on non-renewable resources [41]. This approach has proven beneficial in arid regions where temperature control and resource availability are critical considerations. Similarly, the traditional Bait Al-Sha'ar structures, which are made from goat hair and local plant fibers, continue to serve as models for sustainable building in desert environments due to their natural temperature regulation capabilities [42].

Additionally, frameworks for evaluating and selecting green materials using tools like Building Information Modeling (BIM) have been adapted specifically for the Saudi context, taking into account life cycle costs, environmental impact, and climatic conditions [43]. This approach enables the identification of optimal materials for use in arid regions, ensuring that construction projects meet sustainability goals. Another noteworthy initiative is the management of construction and demolition (C&D) waste, where a structured roadmap for recycling and reusing materials has been proposed to enhance sustainable practices in the construction industry [44]. These examples illustrate the diverse strategies employed to promote sustainability in building materials and construction in arid and semi-arid regions.

6.2 Case Study in Egypt: Revitalizing Historical Vernacular Techniques

The revitalization of historical vernacular building techniques in Egypt has become crucial to sustainable construction, particularly in arid and semi-arid regions. One example is the use of passive cooling techniques such as wind catchers (malqafs) and courtyards, which were widely used in ancient and medieval Egyptian architecture. These elements, originally found in traditional Islamic houses in Fatimid Cairo, have been adapted and incorporated into modern sustainable designs to reduce energy consumption while maintaining thermal comfort [45]. By integrating these low-tech solutions with contemporary technologies, architects have successfully created a sustainable framework for modern construction that is environmentally responsive and culturally appropriate.

Another prominent example is found in the Gharb Sohail village in Aswan, where traditional Nubian building techniques have been revitalized to promote regenerative sustainability. These vernacular structures, made from mud bricks, utilize local materials and passive solar design to achieve efficient thermal regulation and minimize energy consumption. A study on the village's architecture showed that these techniques significantly contribute to the creation of resilient communities by integrating environmental, social, and economic sustainability [46]. This regenerative approach conserves energy and preserves the cultural heritage and traditional knowledge embedded in local building practices.

The Siwa Oasis in Egypt is another noteworthy case study where traditional architectural practices have been leveraged for sustainable development. The vernacular buildings of Siwa, constructed using kershef—a mix of salt and mud—are renowned for their natural insulation properties and capacity to withstand harsh desert conditions. These structures have been studied for their potential to provide thermal comfort and reduce reliance on mechanical cooling systems. Lessons learned from Siwa have been used to inform sustainable construction practices for ecotourism projects in similar arid environments [47]. The successful adaptation of these techniques has led to the development of environmentally responsive, low-carbon buildings that align with both sustainability goals and local cultural values.

6.3 Case Study in Morocco: Green Building in a Semi-Arid Region

The development of green building initiatives in Morocco has gained momentum, particularly in semi-arid regions where sustainable construction techniques are critical to mitigate the effects of harsh climates. One prominent example is a Net Zero Energy Building (NZEB) constructed in Benguerir, which incorporates local materials and passive design strategies to optimize thermal performance. The building, made from earth-based materials, utilizes solar energy through a photovoltaic system and achieves energy self-sufficiency by producing 25% more energy than it consumes annually. This sustainable construction approach enhances the building's environmental performance and serves as a model for other semi-arid regions in Morocco [48].

In addition to energy efficiency, another successful implementation of green building techniques in Morocco involves the use of hemp concrete. A study conducted in Tangier evaluated the thermal performance of hemp-based materials compared to conventional alternatives. Results showed that hemp concrete significantly improved the internal thermal comfort and reduced energy consumption, making it a suitable option for passive low-energy buildings in Morocco's semi-arid regions [49]. Such sustainable materials lower the building's environmental impact and contribute to maintaining local construction traditions.

Furthermore, the implementation of green roofs and rooftop gardens has been explored as a strategy to enhance building performance in semi-arid areas. The administrative building of the Mohammed VI Polytechnic University in Benguerir serves as a case study where Integrated Rooftop Greenhouses (IRTGs) were used to improve energy efficiency. The greenhouses reduced energy demands by promoting natural ventilation and provided a CO_2-enriched environment for the building. This innovative approach exemplifies how green building practices can be adapted to semi-arid climates to achieve sustainable development goals while enhancing indoor environmental quality [50].

6.4 Case Study in the United Arab Emirates: Innovative Use of Recycled Materials

The United Arab Emirates (UAE) has made significant strides in utilizing recycled materials in construction projects to promote sustainable development. One notable example is the use of recycled aggregates sourced from demolished buildings to produce sustainable concrete. A study conducted in Abu Dhabi assessed concrete's mechanical and fresh properties which were made by replacing natural aggregates with recycled ones at varying percentages. The findings revealed that recycled aggregate concrete demonstrated similar compressive strength and slump retention as traditional concrete, making it suitable for use in various construction projects while reducing the environmental impact of construction waste [51].

Another innovative approach has been observed in the development of low-carbon concrete mixes using ground granulated blast-furnace slag (GGBFS) and fly ash. Masdar City, a flagship sustainable project in the UAE, utilized a blend of these supplementary cementitious materials to lower the carbon footprint of its construction activities significantly. By replacing up to 80% of ordinary Portland cement with GGBFS, the project achieved a notable reduction in CO_2 emissions and material costs while maintaining the structural integrity of the buildings. This sustainable concrete technology has set a new benchmark for eco-friendly construction in the region [52].

In addition, the Emirate of Sharjah has also played a leading role in advancing sustainable construction through the establishment of state-of-the-art recycling facilities, such as the Bee'ah facility, which processes construction and demolition (C&D) waste. This facility recovers a large proportion of usable materials, such as concrete, metals, and plastics, from the C&D waste stream. Despite these efforts, a study highlighted a gap between the current sustainable regulations and the practical adoption of recycling practices in construction projects, suggesting the need for improved policies and stakeholder collaboration to increase the use of recycled materials in the UAE [53].

7 Future Outlook

7.1 Emerging Trends in Sustainable Building Materials

It is conceivable that there will be bright advancements in sustainable building material space owing to increasing innovation. Therefore, there is evidence of an increasing trend in the utilization of bio-based materials generated from renewable biomass sources. Biomaterials in construction are becoming more popular because of the industry's goal of transitioning to low-carbon materials, such as engineered timber, hempcrete, and mycelium composites [32]. However, apart from the advantages of being sustainable in use, these materials also show prospective performance effectiveness, such as increased thermal conductivity and structural stability.

In addition, there has been enhanced research and technology on smart materials that are sensitive to changes in the environment or are generally efficient in energy usage. For instance, phase change materials (PCMs) that act as heat storages/releasers in buildings and self-healing concrete that can heal cracks on their own are considered great leaps in building technology [3]. It makes it possible for these innovations to go a notch higher in their efficiency in decreasing the negative impacts on the environment, especially through energy conservation and the durability of construction materials.

Both nanotechnology and 3D printing are planned to be incorporated into construction and are thought to change the construction industry. Using nanomaterials in construction can improve the general performance of the basic construction materials and hence the buildings they are utilized in because they are more durable, lighter, and incorporate energy-efficient features [5]. Simultaneously, 3D printing technology allows the creation of diverse shapes of structures without the removal of excess material and can be used for the production of environmentally friendly construction materials. All these technologies have significant potential to offer a gigantic impact on sustainable construction in the future.

7.2 Predicting Market Evolution

There is significant potential for the growth in the demand for sustainable building materials within the next few years owing to the general public's awareness, regulation, and awareness needs. Market trends suggest that this market will expand rapidly in the coming years, especially in areas with policies and favorable conditions for green building procedures [33]. Other developing markets in the Asia-Pacific and Latin America are also expected to grow significantly owing to ongoing industrialization and urbanization [54].

However, this growth presents some challenges. One cannot mention the trained assumption that the use of materials with a focus on sustainability has higher costs, while the latter is not always the case. To correct this perception, there is a need to sensitize key players in the market and enhance access to proper financing schemes that promote sustainability in construction [12]. Moreover, the industry seldom addresses material availability, and supply chain networks for sustainable materials are emerging [23].

Other areas of growth include the application of circular economy models in which products, materials, and parts are used, collected, and incorporated again into the market stream. Closed-loop systems for construction and design for reuse and recycling, where components are designed to be disassembled and used again, will be a focus of progression [28]. When the use of such practices becomes conventional, it will have a direct positive effect on the construction industry, as it will also foster new markets for recycled and upcycled products.

Therefore, it can be concluded that while the future of sustainable building materials has its bright sides as well as the problems it faces, the situation is more or less favorable. Sustained innovation of new materials, favorable and enabling policies, and changes in market perception are the key necessities required to enhance the performance of sustainable materials in the construction sector.

8 Conclusion: Fostering Sustainability in Arid and Semi-Arid Regions

8.1 Summary of Key Findings

This chapter has examined the unique challenges and opportunities for sustainable construction in arid and semi-arid regions. The harsh environmental conditions in these areas, characterized by extreme temperatures, limited water resources, and high solar exposure, necessitate specialized approaches to construction and material selection. Sustainable building practices are particularly critical, as traditional methods often fail to address the environmental stressors inherent to these regions. Findings from various case studies indicate that adopting locally sourced materials, such as adobe and mudbrick in Egypt or limestone and basalt in Saudi Arabia, can significantly reduce the carbon footprint and improve the thermal efficiency of buildings in arid climates. Furthermore, passive design strategies such as using courtyards and wind catchers have been shown to enhance natural ventilation and thermal comfort, reducing the reliance on mechanical cooling systems.

In addition, the economic benefits of sustainable materials become more evident over time in arid regions. While the initial costs of implementing sustainable materials and technologies may be higher, the long-term savings on energy costs, reduced maintenance, and increased building lifespan compensate for these expenses. This is particularly relevant in regions with high energy costs due to the intensive use of air conditioning and cooling systems. Several studies also highlight the social benefits of sustainable construction, such as improved indoor air quality and enhanced living conditions, which contribute to overall well-being. Therefore, the adoption of sustainable construction practices in arid and semi-arid regions not only supports environmental sustainability but also promotes economic resilience and social health in the long term.

8.2 Recommendations for Policymakers and Industry in Arid Regions

Given the environmental constraints in arid and semi-arid regions, targeted policies and industry practices are necessary to promote the use of sustainable materials and construction methods. Policymakers should prioritize the development of region-specific building codes that incorporate sustainability criteria tailored to local climatic conditions. Such building codes should mandate the use of energy-efficient materials, emphasize passive design elements, and include provisions for water conservation strategies, such as rainwater harvesting and the use of greywater systems. Additionally, green certification programs, such as LEED or BREEAM, should be adapted to reflect the unique sustainability needs of arid regions, with additional points awarded for innovations that address water scarcity, thermal comfort, and heat management.

Financial mechanisms are also crucial for driving the adoption of sustainable building practices. Policymakers can support the industry by providing financial incentives such as subsidies, low-interest loans, and tax credits for projects that meet sustainability standards. These incentives should be complemented by capacity-building programs aimed at educating developers, builders, and local communities about the benefits of sustainable construction. Governments can also play a role by investing in research and development for new materials and technologies that are specifically designed to withstand the harsh environmental conditions of arid regions. Public-private partnerships can be leveraged to pilot and scale up such innovations, helping to reduce costs and improve access to sustainable building solutions across the region.

Moreover, industry stakeholders should take proactive steps to incorporate sustainability into all stages of the construction process, from design to material selection and project execution. This includes sourcing local and renewable materials whenever possible, integrating smart building technologies to monitor energy and water use, and employing design strategies that take advantage of the natural environment to enhance building performance. For example, the use of reflective roofing materials and sun shading devices can significantly reduce indoor temperatures and lower cooling demands. Industry associations can also establish best practice guidelines and offer training programs to ensure that sustainable construction becomes the standard in arid and semi-arid regions.

8.3 Strategies for Overcoming Barriers in Arid Regions

Several barriers hinder the widespread adoption of sustainable building materials and practices in arid and semi-arid regions. High initial costs, lack of regulatory support, limited availability of suitable materials, and low consumer awareness are some of the most significant challenges. Addressing these barriers requires a multi-faceted approach that involves financial, regulatory, and social strategies.

To overcome financial barriers, governments and financial institutions should collaborate to establish funding mechanisms that specifically target sustainable construction projects in arid regions. Green bonds, micro-financing schemes, and revolving loan funds can be used to lower the upfront costs of sustainable construction, making it more accessible for smaller developers and individual homeowners. Additionally, incorporating long-term savings analysis into project planning can help stakeholders see the value of investing in sustainable materials, which often pay off through lower operating costs and reduced energy consumption.

Regulatory barriers can be addressed by revising existing building codes and standards to include sustainability requirements that are appropriate for arid climates. This may include mandating the use of reflective surfaces, minimum insulation values, and the incorporation of shading elements in all new constructions. Governments can also streamline the approval process for sustainable projects by establishing fast-track permitting procedures and reducing bureaucratic hurdles for developments that meet green building criteria. By creating a clear regulatory framework, policymakers can reduce uncertainty and encourage more developers to invest in sustainable construction.

Social and cultural barriers, such as limited awareness and resistance to change, can be addressed through targeted outreach and education campaigns. These initiatives should focus on demonstrating the long-term economic and health benefits of sustainable building practices, particularly for homeowners and small-scale developers. Community workshops, public information sessions, and partnerships with local influencers can effectively shift public perception and increase demand for sustainable materials. Additionally, involving communities in the design and construction process can ensure that sustainable buildings reflect local preferences and cultural values, thereby enhancing their acceptance and appeal.

Addressing these barriers requires collaboration between governments, the private sector, and civil society. Establishing multi-stakeholder platforms where policymakers, industry leaders, and community representatives can come together to discuss challenges and share best practices is essential for driving the adoption of sustainable construction practices in arid and semi-arid regions. Such platforms can also serve as a forum for exploring new solutions and fostering innovation in response to the unique challenges of building in these environments.

8.4 Relevance to SDGs

The findings and methodologies of the chapter align closely with several Sustainable Development Goals (SDGs) outlined by the United Nations. Below is a detailed analysis of how the themes and insights from the chapter contribute to specific SDGs:

SDG 7: Affordable and Clean Energy

Sustainable materials, particularly those that enhance energy efficiency, directly supports SDG 7. By highlighting the role of advanced insulation systems, energy-efficient building envelopes, and the integration of renewable energy sources in construction practices, the chapter provides actionable pathways to reduce energy consumption in buildings. For instance, the adoption of cross-laminated timber (CLT) and hempcrete not only reduces embodied energy but also contributes to creating structures with high thermal efficiency, minimizing heating and cooling requirements. Furthermore, the Life Cycle Cost Analysis (LCCA) methodology discussed in the chapter reinforces the long-term economic feasibility of sustainable materials, which is essential for promoting affordable energy solutions.

SDG 9: Industry, Innovation, and Infrastructure

The construction industry's transition to sustainable materials aligns with SDG 9 by fostering innovation in building practices and material sciences. The analysis of barriers to market adoption, including regulatory and financial challenges, identifies opportunities for reforming industry standards and practices. By advocating for innovative solutions such as Building Information Modeling (BIM) and smart materials like self-healing concrete, the chapter contributes to the development of resilient and sustainable infrastructure. These innovations not only enhance the durability and efficiency of buildings but also encourage the construction sector to adopt cutting-edge technologies that reduce environmental impact.

SDG 11: Sustainable Cities and Communities

SDG 11 aims to make cities inclusive, safe, resilient, and sustainable, and the insights from this chapter significantly contribute to this goal. By focusing on the economic, social, and environmental benefits of sustainable building materials, the chapter provides a blueprint for developing urban environments that prioritize sustainability. For example, the use of recycled materials and renewable resources in construction reduces urban waste and promotes circular economy principles. Additionally, the chapter's emphasis on creating structures with improved indoor environmental quality aligns with the goal of enhancing the health and well-being of urban residents.

SDG 12: Responsible Consumption and Production

The chapter's advocacy for sustainable material usage and waste reduction directly supports SDG 12. Through its focus on life-cycle assessments and the promotion of recycling construction and demolition waste, the chapter highlights strategies to optimize resource efficiency. By encouraging the adoption of renewable materials and technologies that minimize waste, such as fiber cement siding and metal roofing, the chapter advances responsible consumption and production practices. These practices ensure that construction processes are aligned with sustainable development principles, reducing the overall ecological footprint of the industry.

SDG 13: Climate Action

Construction activities are significant contributors to global greenhouse gas emissions. The chapter's discussion of sustainable building materials' potential to lower carbon footprints resonates strongly with SDG 13. By advocating for materials like green concrete and bamboo, which have lower embodied carbon compared to traditional materials, the chapter addresses the urgent need for climate action. Additionally, its emphasis on energy-efficient designs and practices, such as passive cooling techniques, demonstrates how the construction industry can contribute to mitigating climate change effects through reduced energy consumption and emissions.

SDG 15: Life on Land

The chapter indirectly supports SDG 15 by promoting sustainable practices that reduce environmental degradation associated with material extraction and land use. The analysis of renewable materials and recycled resources underscores the importance of conserving natural habitats and biodiversity. For instance, by advocating for the use of bamboo—a rapidly renewable resource—over traditional hardwoods, the chapter encourages practices that mitigate deforestation and promote ecosystem restoration. Furthermore, the emphasis on sustainable mining and the reuse of industrial by-products aligns with efforts to reduce land degradation and foster sustainable resource management.

SDG 17: Partnerships for the Goals

Achieving the widespread adoption of sustainable building materials requires collaborative efforts across sectors, a concept that aligns with SDG 17. The chapter's recommendations for engaging policymakers, industry stakeholders, and researchers highlight the need for partnerships to overcome barriers to sustainability. By promoting public-private collaborations and the integration of academia in developing innovative materials, the chapter exemplifies how multi-stakeholder efforts can drive progress toward achieving sustainability goals. These partnerships are essential for creating an enabling environment that supports the transition to sustainable construction practices globally.

9 Conclusion

This chapter makes a significant contribution to advancing multiple SDGs by addressing critical aspects of sustainability within the construction industry. By linking economic viability with environmental and social benefits, it provides a comprehensive framework for promoting sustainable development through innovative building materials and practices. Its alignment with SDGs 7, 9, 11, 12, 13, 15, and 17 underscores its relevance and impact on global sustainability efforts. Policymakers, industry leaders, and researchers can leverage the insights from this chapter to drive transformative change in the built environment, paving the way for a more sustainable future.

References

1. Attia S, De Herde A, Hensen J (2017) A critical review of methodologies to assess the life cycle costing of buildings. Energy Build 140:35–48
2. Barbosa L, Bergerson J, Birkved M (2018) Life cycle cost analysis of sustainable building materials: a review. J Clean Prod 177:491–501
3. Annibaldi V, Cucchiella F, Rotilio M (2021) Economic and environmental assessment of thermal insulation: a case study in the Italian context. Case Stud Constr Mater. https://doi.org/10.1016/j.cscm.2021.e00682
4. Biolek V, Hanák T (2019) LCC estimation model: a construction material perspective. Buildings 9(8):182. https://doi.org/10.3390/buildings9080182
5. Schneider-Marin P, Lang W (2020) Environmental costs of buildings: monetary valuation of ecological indicators for the building industry. Int J Life Cycle Assess 25(9):1637–1659. https://doi.org/10.1007/s11367-020-01784-y
6. Fantozzi F, Gargari C, Rovai M, Salvadori G (2019) Energy upgrading of residential building stock: use of life cycle cost analysis to assess interventions on social housing in Italy. Sustainability. https://doi.org/10.3390/SU11051452
7. Pučko Z, Maučec D, Šuman N (2020) Energy and cost analysis of building envelope components using BIM: a systematic approach. Energies 13(10):2643. https://doi.org/10.3390/en13102643
8. Hasan MR, Khan MA, Wuest T (2024) Towards industry 5.0: a systematic literature review on sustainable and green composite materials supply chains. J Clean Prod 421:129001. https://doi.org/10.1016/j.jclepro.2024.129001
9. Meinardi F, Bruni F, Castellan C, Meucci M, Umair AM, La Rosa M, Catani J, Brovelli S (2024) Certification-grade quantum dot luminescent solar concentrator glazing with optical communication capability for connected sustainable architecture. Sol Energy Mater Sol Cells 245:111832. https://doi.org/10.1016/j.solmat.2024.111832
10. Lakousha M (2023) Con-LCCA V1.0: a computerized tool for analyzing the life cycle cost of construction projects. SVU-Int J Eng Sci Appl 4(1):62–78. https://doi.org/10.21608/svusrc.2022.165183.1080

11. Fregonara E, Ferrando DG (2020) The stochastic annuity method for supporting maintenance costs planning and durability in the construction sector: a simulation on a building component. Sustainability 12(7):2909. https://doi.org/10.3390/su12072909
12. Abouhamad M, Abu-Hamd M (2019) Framework for construction system selection based on life cycle cost and sustainability assessment. J Clean Prod 241:118397. https://doi.org/10.1016/j.jclepro.2019.118397
13. Xue Z, Liu H, Zhang Q, Wang J, Fan J, Zhou X (2020) The impact assessment of campus buildings based on a life cycle assessment–life cycle cost integrated model. Sustainability 12(1):294. https://doi.org/10.3390/su12010294
14. Daoud AO, Othman AAE, Ebohon OJ, Bayyati A (2021) Quantifying materials waste in the Egyptian construction industry: a critical analysis of rates and factors. Ain Shams Eng J 12(4):4275–4289. https://doi.org/10.1016/j.asej.2021.02.039
15. Daoud AO, Othman AA, Robinson H, Bayyati A (2018) Exploring the relationship between materials procurement and waste minimization in the construction industry: the case of Egypt. The 4th NZAAR international event series on natural and built environment, cities, sustainability and advanced engineering, pp 76–85. https://static1.squarespace.com/static/565bcedee4b09e25856124af/t/5a6b801a0852293fdbf7ad07/1516994747704/NZAAR+Jan+2018+Proceedings.pdf
16. Daoud AO, Othman AAE, Robinson H, Bayyati A (2020) An investigation into solid waste problem in the Egyptian construction industry: a mini-review. Waste Manage Res 38(4):371–382. https://doi.org/10.1177/0734242X20901
17. Daoud AO, Othman AAE, Ebohon OJ, Bayyati A (2022) Overcoming the limitations of the green pyramid rating system in the Egyptian construction industry: a critical analysis. Architect Eng Des Manage 18(2):114–127. https://doi.org/10.1080/17452007.2020.1802218
18. Daoud AO, Othman AAE, Robinson H, Bayyati A (2018) Towards a green materials procurement: investigating the Egyptian green pyramid rating system. In Adel M, El Maghraby R, Fathi S (eds), Green heritage international conference (Chance—Change—Challenge), Elain Publishing House, pp 575–591. https://www.researchgate.net/publication/323588948_Towards_a_Green_Materials_Procurement_Investigating_the_Egyptian_Green_Pyramid_Rating_System
19. Daoud AO, Omar H, Othman AAE, Ebohon OJ (2023) Integrated framework towards construction waste reduction: the case of Egypt. Int J Civil Eng 21(5). https://doi.org/10.1007/s40999-022-00793-2
20. Ayodeji B, Oyediran L, Afolabi A (2018) Skills and expertise requirements for sustainable construction in developing countries: a case study of Nigeria. J Constr Eng Manag 144(7):04018063
21. Gibson L, Thring R (2018) Market transformation for sustainable buildings: the role of codes and standards. J Sustain Dev 25(3):112–128
22. Taher H, Tang P, Ballesteros M (2020) Policy instruments for promoting sustainable building materials: a comparative analysis. J Environ Plann Policy Manage 22(5):578–594
23. Rashed M, Gad A (2019) Economic barriers to adoption of sustainable building materials: evidence from the Middle East. J Build Environ 25(3):345–360
24. Goubran S, El-Hakim S, Abdelaty S (2020) Permitting and approval processes for sustainable building materials: a comparative analysis. J Sustain Dev 27(4):112–128
25. Mohamed M, Hameed Z, Azman M (2019) Cost-benefit analysis of sustainable building materials: a case study of green roof systems. J Build Performance 12(4):345–360
26. Shen SC, Spitzer B, Stefaniuk D, Zhou S, Masic A, Buehler MJ (2024) Valorizing sewage sludge: using nature-inspired architecture to overcome intrinsic weaknesses of waste-based materials. Mater des 225:111293. https://doi.org/10.1016/j.matdes.2024.111293
27. Johnson K, Liu Y (2024) Energy-efficient insulation materials: advances and market trends. J Clean Prod 420:128456. https://doi.org/10.1016/j.jclepro.2024.128456
28. Pomponi O, Rivela B, Neila J (2019) Life cycle thinking towards sustainable development policy-making: the case of energy retrofits. J Clean Prod. https://doi.org/10.1016/J.JCLEPRO.2018.09.173

29. Daoud AO, Othman AAE, Ebohon OJ, Bayyati A (2023) Analysis of factors affecting construction and demolition waste reduction in Egypt. Int J Constr Manag 23(8):1395–1404. https://doi.org/10.1080/15623599.2021.1974682
30. Zhang C, Hu M, Yang X, Amati A, Tukker A (2020) Life cycle greenhouse gas emission and cost analysis of prefabricated concrete building façade elements. J Ind Ecol 24(5):1016–1030. https://doi.org/10.1111/jiec.12991
31. California Energy Commission (2019) Building energy efficiency standards (Title 24, Part 6). Retrieved August 31, 2025, from https://www.energy.ca.gov/programs-and-topics/programs/building-energy-efficiency-standards
32. Bragadin M, Guardigli L, Calistri M, Ferrante A (2023) Demolishing or renovating? Life cycle analysis in the design process for building renovation: the ProGETonE case. Sustainability. https://doi.org/10.3390/su15118614
33. Zhang C, Hu M, Laclau B, Garnesson T, Yang X, Tukker A (2021) Energy-carbon-investment payback analysis of prefabricated envelope-cladding system for building energy renovation: cases in Spain, the Netherlands, and Sweden. Renew Sustain Energy Rev. https://doi.org/10.1016/J.RSER.2021.111077
34. Li X, Wang J, Cheng X (2021) Government policies and incentives for sustainable building materials: a comparative analysis. J Sustain Dev 28(1):89–104
35. Kim J, Park J (2020) Stakeholder perspectives on sustainable building materials: a qualitative study. J Sustain Constr 17(3):145–162
36. Vilches A, García A, Martín R (2019) Research and innovation in sustainable building materials: a review of recent developments. J Sustain Constr 16(4):112–128
37. Chen Y, Gallardo S (2023) A multi-objective optimization method for the design of a sustainable house in Ecuador by assessing LCC and LCEI. Sustainability. https://doi.org/10.3390/su16010168
38. Jiang H, Li Y, Sun Y (2020) Technical barriers to adoption of sustainable building materials: a case study of China. J Constr Build Mater 246:112378
39. Aliabadi M, Kermani A, Yazdani S (2021) A review of alternative building materials for sustainable construction towards sustainable development. J Mod Mater 7(1):68–78. https://doi.org/10.21467/jmm.7.1.68-78
40. Hefni MA, Ahmed HA, Omar ES, Ali MA (2021) The potential re-use of Saudi mine tailings in mine backfill: a path towards sustainable mining in Saudi Arabia. Sustainability 13:6204
41. Azouqah H, Alomar N, Kaadan L, Sonbul M, Abdulaziz H, Labib WA (2021) Sustainable local materials: a study of Adobe Bricks in Saudi Arabia. Mater Sci Forum 1047:163–173
42. Rafey ME, Mulhem RA, Wagdi M (2016) Sustainable construction materials in (Bait Al-Sha'r) and modern trends in Saudi Arabia. Fourth International Conference on Sustainable Construction Materials and Technologies, pp 1–10
43. Marzouk MM, Metawie MS, Hisham M, Al-Sulahi I, Kamal M, Al-Gahtani KS (2014) Modeling sustainable building materials in Saudi Arabia. Computing in civil and building engineering, pp 1546–1553
44. Blaisi NI (2019) Construction and demolition waste management in Saudi Arabia: current practice and roadmap for sustainable management. J Clean Prod 221:167–175
45. Attia AS (2019) Traditional sustainable architecture techniques and its applications in contemporary architecture: case studies of the Islamic House in Fatimid Cairo and Sana'a, cities in Egypt and Yemen. J Civ Environ Eng 13:266–271
46. Ahmed N (2023) Re-exploring Vernacular Architecture from the lens of regenerative thinking: a case study Gharb Sohail Village in Egypt. J Sustain Archit Civ Eng 32(1):58–76
47. El Haridi NMA, Ibrahim MA, Ayad HM, El Sayad ZT (2019) Comparative analysis of the desert and green Vernacular Architecture in the oases of Egypt. In Stankov U, Boemi SN, Attia S, Kostopoulou S, Mohareb N (eds) Cultural sustainable tourism. Advances in Science, Technology & Innovation, pp 37–46.
48. Kaitouni SI, Mghazli MO, Nait-Taour A, Es-sakali N, El Mankibi M, Ahachad M, Brigui J (2023) Empirical validation and analysis of the energy performance of an ecological Net Zero Energy Building (NZEB) in Benguerir-Morocco. E3S Web of Conferences.

49. Lidoh H, Ikken B, Kaitouni SI (2018) Thermal performance of a hemp concrete residential building envelope in Tangier-Morocco. 2018 6th International Renewable and Sustainable Energy Conference (IRSEC), pp 1–8.
50. El Aouni I, Mastouri H, Radoine H, El Asli A, Koubaa N (2021) The impact of integrated-rooftop greenhouses on buildings' efficiency: the case of the administrative building of UM6P. 2021 9th International Renewable and Sustainable Energy Conference (IRSEC), pp 1–6
51. Khartabil A, Al Martini S (2019) Fresh and mechanical properties of sustainable concrete using recycled aggregates. Key Eng Mater 803:239–245
52. Elchalakani M, Aly T, Abu-Aisheh E (2014) Sustainable concrete with high volume GGBFS to build Masdar City in the UAE. Case Stud Constr Mater 1:10–24
53. Białko M (2017) Bridging the gap between sustainability, the regulatory, and the recycling of construction and demolition waste in the UAE, with specific reference to the Emirate of Sharjah. MATEC Web of Conferences 120:05010
54. Suzuki T, Murakami S, Nakajima K (2019) Economic development and market dynamics for sustainable building materials: a comparative analysis of Asia-Pacific and Latin America. J Sustain Constr 17(2):112–128

Standards for the Use of Sustainable Green Finishing Material and Their Impact on Internal Spaces from Economic, Social, and Environmental Point of View

Sammer H. Salem

Abstract Environmentally Sustainable Interior Design (ESID) is a crucial issue in interior design, but many architects and engineers remain unaware of its importance. This is particularly evident when choosing sustainable building and finishing materials. This chapter aims to develop a comprehensive understanding of sustainable options for building materials and finishing, focusing on current green and sustainable products. The study also aims to improve access to information on sustainable residential spaces and provide updated information on sustainable building and finishing materials. The methodology proposed in this study focuses on evaluating sustainable materials used in residential buildings and their selection criteria to integrate environmental, economic, and social aspects. A descriptive and comparative analytical approach was used to analyze international and Arabic systems for auditing sustainable buildings and evaluating criteria related to standard materials and resources. The study also analyzes environmental, economic, and social aspects related to choosing suitable building and finishing materials. The proposed model measures the sustainability of building and fishing materials used in projects, leading to an analytical approach specialized in international and local experiments. This helps clarify how to use sustainable finishing materials in different buildings and measure the achievement of sustainability standards in both local and international model buildings. In conclusion, this chapter proposes strategies and policies for the selection of finishing materials in Egypt.

Keywords Sustainable design · Sustainable green material · Sustainable interior design

S. H. Salem (✉)
Associate Professor of Architecture, High Institute for Engineering and Technology, Giza, Egypt
e-mail: Marmarsaleh83@yahoo.com

Z. ElZein and A. Negm (eds.), *Recent Approaches of Sustainable Architecture in Arid and Semi-arid Cities*, Urban Sustainability,
https://doi.org/10.1007/978-981-95-1645-2_8

1 Introduction

The interior design profession is always known with a traditional meaning that means One-dimensional practice only aims to provide aesthetic improvements to interior spaces according to the client's requirements this resulted in narrowing of designer's job and make it limited only in focusing on the latest findings in the contemporary interior design field, and how to design small luxurious environments. This made it an approach that ignores the importance of saving energy and reducing harmful emissions, and lack of full awareness of the harmful effects on users' health either mentally or physically. Most of the pollutants suspended in the air inside the internal spaces are carcinogens, and irritants to the respiratory system and the percentage of these air pollutants found in the internal spaces reaches levels above 2 to 10 times their levels in the outdoor spaces [1, 2].

This Research paper will focus on studying the effects of the used finishing material in internal spaces from the point of view above mentioned various aspects to realize sustainability, the effects of these materials on environmental levels and resources are to be studied when evaluating each material at every stage of the manufacturing process, the resources it consumes, the resulting waste and the amount of energy consumed. Also, designing building materials that don't need a large quantity of water or power to be manufactured should be a priority. As presented in the research [3].

The present study will also focus on the benefits from renewable material, the importance of reducing wastes from manufacturing, reducing damaging of current energy resources and encouraging the recycling and reuse of these materials, as for the impact of materials on the social level, the importance of the evaluation of the companies responsible for selling and marketing construction and finishing products, comes first. This will be done in addition to taking account of the health of the users of those products and the suitability of the type of finishing material used for activities performed in the internal space [4].

2 Research Problem

Despite the continuous efforts of the state and international and local companies in manufacturing sustainable and environmentally friendly building and finishing materials, there are still deficiencies in the selection of these materials and how to use them in architectural spaces in general. This is in addition to not taking into account the consequences of choosing [5].

Traditional building and finishing materials and the extent of their impact on the health of occupants of the architectural space. This is because the process of choosing finishing material from the majority point of view is just choosing colors, raw materials, and harmonious materials according to a general sense, This, in addition to the unawareness of the importance and how dangerous is these materials that can directly affect our daily life.

Not this only but a large number of finishing materials is available at markets in communities of developing and developed countries, but despite these materials being used on the widest range inside or outside Egypt, all of these materials or most of it has negative the environmental, societal and economical negative effects. No attention is paid to these effects despite their severity, particularly on the environment and human health [6].

3 Research Aim

The research aims to check a set of foundations and requirements for the selection of building materials and sustainable finishing in the interior spaces of the buildings. This will allow the concept of sustainability of the building can be achieved and its impact on various environmental, economic and social dimensions.

This will lead to evaluating of the materials' effect on the economic level. Studying the evaluation of the life cost (LCC) of finishing material and the possibility of being recycled, the efficiency of using local materials and encouraging their consumption are important as shown previously [7].

4 Types of Interior Spaces

The internal spaces are of great importance, as it spends most of our time in designed architectural spaces. Some architectural schools divide architectural spaces into two main sections: residential, and non-residential spaces or commercial spaces. Despite the multiplicity of architectural spaces that designers deal with, and makes buildings find that all of these spaces with their multiplicity and different functions will fall according to the previous classification of spaces (whether residential or non-residential spaces). Residential spaces are all environments intended to be lived or stayed in for periods. But Non-residential spaces are all business-, enterprise or corporate environments, both for public and private purposes, and the (Fig. 1) is shown emphasizes the importance of the importance of internal spaces for the people [8].

Fig. 1 Shows user presence ratios within spaces, by Shows user presence ratios within spaces. *Source* Bonda P, Sosnowchik K (2020) Sustainable commercial interiors, 2nd edn, Wiley, NJ, p 203 (1)

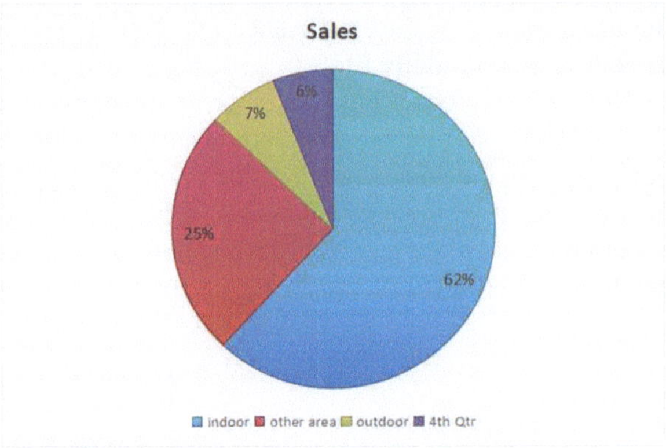

5 Common Materials in Different Internal Spaces and Their Seriousness

Interior finishing is the process of removing walls, floors, ceilings, and other surfaces within a space, which is determined by choosing the type of material applied and its treatment. Interior finishes are very important for buildings because they are the visible surface of all parts of the void. Also, the choice of the appropriate finishing materials for the vacuum is determined by several factors. The most important of which is the cost, the influence of architectural coordination, resistance to moisture, fire or sound, the lifetime of the material, and the resistance of the material to abrasion. In addition to that, the finishes are highly influenced by personal choice, professional experience and technical taste of the finishing materials and building type. With the multiplicity of those previous determinants in choosing interior finishing materials, the most important factors that affect the vacuum are neglected. The interior finishing materials must be chosen based on several additional factors, such as considering the health and environmental impact of these materials. Some previous studies research has confirmed that some of those materials used in finishing the various voids have a great relationship to the pollution of the internal space and the appearance of some health damages to the users of those spaces.

This is in addition to wasting natural resources. It is also clear that there is a lack of interest in studying the life cycle analysis of each subject [9].

6 Harmful Chemicals in Interior Finishing Materials

The most important harmful chemical compounds involved in some finishing materials are used in internal space as shown in Table 1. Such as paints—PVC flooring—flooring tiles—wallpaper—stones—wood—fire material—carpet.

Table 1 Shows the finishing materials that produce harmful emissions, by the researcher

Finishing material	Harmful compounds found in finishing materials	Pollutants				
		Asbestos	Rn	Pb	Formaldehyde	VOCs
Paints	Chemical compounds found in different types of paint			√	√	√
PVC flooring	PVC COMPOSITE				√	√
Floor tiles	Vinyl compounds	√	√		√	√
Wallpaper,	Harmful additives				√	√
Stones	Granite and marble	√	√		√	√
Wood	Harmful emissions				√	√
Fire-resistant materials	Harmful flame retardants	√			√	√
Carpet	Dyes and flame retardants				√	√

7 Sustainable Architecture

Architect William McDonough defined sustainable architecture as a set of principles, including respect for the relationship between the building and the surrounding environment, the creation of design solutions based on technology and the environment, as well as the recruitment of sources Natural energy and removal of construction waste [10]. Architect Charles Kebert has defined sustainable architecture as a set of principles to achieve its goals of establishing and operating healthy building buildings Healthy Built Environment, where sustainable architecture depends on the rationalization of energy, water, construction resources, and other resources [4]. Therefore, sustainable interior design is defined as the interior design in which all systems and materials are designed with a focus on integration in its entirety with the aim of minimizing the negative impacts on the environment and occupants and maximizing the positive impacts on environmental, economic and social systems over the course of the life cycle of the building. (Energy, water, construction resources, and other resources) [11].

7.1 Sustainable Interior Design Standards

The trend towards sustainable interior design refers to achieving quality standards, sustainability principles and strategies, high levels of efficiency in the use of energy and water sources, appropriate land use, site coordination and the use of compatible building materials to achieve the quality of the indoor environment and water efficiency and reduce the impact of the building during its life cycle and solid waste

Fig. 2 Shows sustainable interior design standards (by the author)

management. Sustainable interior design standards can be stated as shown in [8] (see Fig. 2).

7.2 Sustainable Building Rating Systems

A Methodology aimed at achieving a building assessment tool within the framework of the most important determinants affecting the performance of the building: energy, water, carbon emissions, transport and health of individuals. It is a good tool for achieving more comfortable buildings for individuals and more energy efficient by reducing negative environmental impact, relying on sustainable building materials, recycling materials, and relying on renewable energy [6]. Different major global systems have been selected and studied in terms of the method and determinants of evaluation and analysis of the material and resource component in each system, in addition to the study of 7systems from the Arab world, as shown in: [12] (see Fig. 3).

- **LEED** "Leadership in Energy and Environmental Design ranks"
- **PRS** "Performance Rating System"
- **BREEAM** "Building Research Establishment Environmental Assessment Method"
- **IGBC** "Indian Green Building Council"
- **CASBEE** "Comprehensive Assessment System for Built Environment Efficiency."
- **GSAS** "Global Sustainability Assessment System."
- **GPRS** "Green Public Rating System."

- **LEED** "Leadership in Energy and Environmental Design ranks" [13]
- **PRS** "Performance Rating System" [14]
- **BREEAM** "Building Research Establishment Environmental Assessment Method" [14]

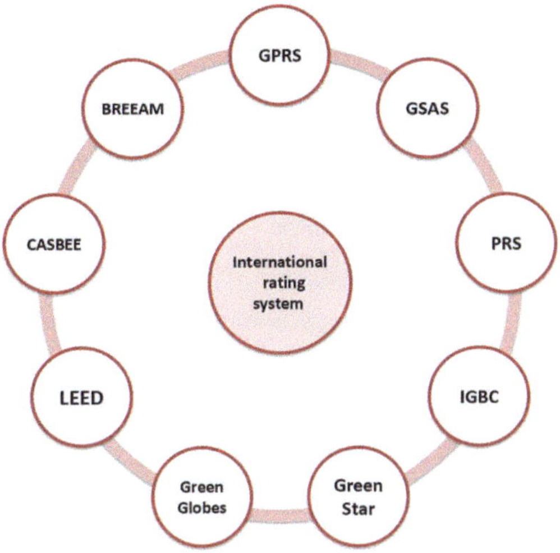

Fig. 3 Shows international sustainable building rating systems (by the author)

- **IGBC** "Indian Green Building Council" [15]
- **CASBEE** "Comprehensive Assessment System for Built Environment Efficiency". [16]
- **GSAS** "Global Sustainability Assessment System." [1]
- **GPRS** "Green Public Rating System." [17]

7.3 The Weight Ratio of the Material and Resource Standard in International and Arab Rating Systems

When comparing the different systems, it became clear that THE PRS "Performance Rating System" was the first sustainable building assessment system to have the highest weight ratio of materials and resources by 18%, followed by IGBC "Indian Green Building Council" by 16%. LEED "Leadership in Energy and Environmental Design ranks" fifth with a weight ratio of 13% materials and resources [18].

The Egyptian Green Pyramid Assessment System has only 10% of the material element. Some systems have allocated a separate complete evaluation system to assess sustainable internal spaces, including the internal material component. There is no system in the Arab world yet for assessing sustainable internal spaces, and these systems are: BREEAM "Building Research Establishment Environmental Assessment Method"—LEED "Leadership in Energy and Environmental Design ranks"—Green Star—Green Globes—IGBC "Indian Green Building Council" [19–23] (see Fig. 4).

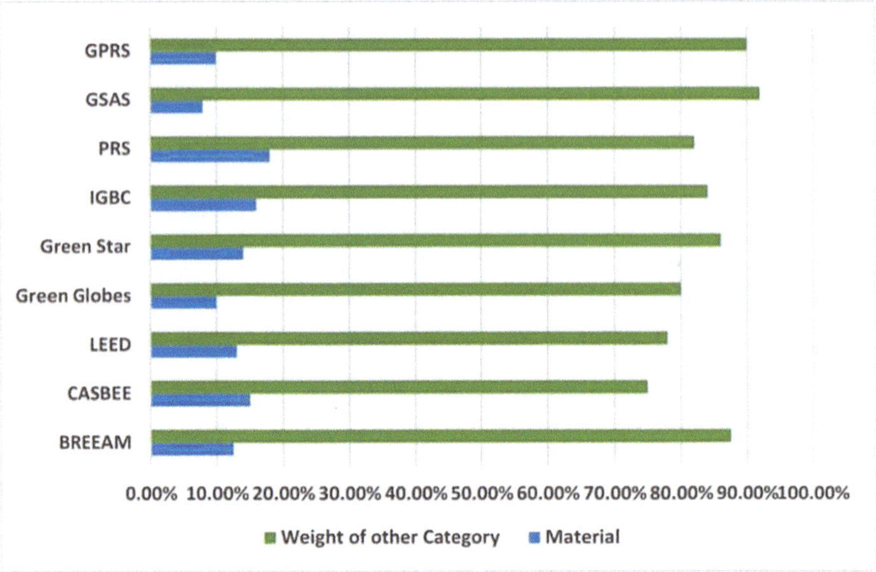

Fig. 4 Shows comparison between international rating system in material and resource component, by researcher source-28. Groves K, Valente M (2020) Sustainability in Interior Design, 2nd edn. Laurence King Publishing, London

8 Sustainable Finishing Materials and Evaluation Criteria

The concept of sustainable building and finishing materials is those that use natural resources from the earth but in a responsible environmental way while respecting the environment and ecosystems, and away from non-renewable resources such as coal and minerals. These are recycled materials, energy and water efficient, environmentally friendly and respectful of the resource life cycle [24] (Fig. 5).

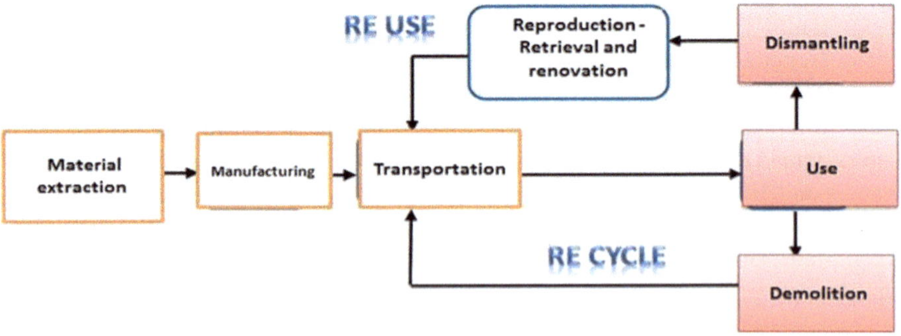

Fig. 5 Shows sustainable finishing material life cycle. *Source* Addis B (2019) Building with reclaimed components and material, Earth scan, UK, p 13

Fig. 6 Shows criteria for the selection of sustainable finishing. *Source* Green Building Materials: a guide to product selection and specifications, Wiley, New Jersey, 2012, p 27

8.1 *Criteria for the Selection of Sustainable Finishing Materials*

The attached table presents a methodology for the classification of sustainable materials (see Fig. 6).

8.2 *Analysis of the Economic, Environmental and Social Aspects of Finishing Material*

Sustainable green materials are employed in projects to achieve three main advantages, environmental, social and economic. As local and natural as possible, their use does not adversely affect health, is not toxic, renewable, durable, and recyclable and used, low- energy, low-cost, and cost-effective. The sustainability of materials is assessed in three key areas that must be adhered to to achieve sustainability: the environmental, social, and economic dimensions [25] (Fig. 7).

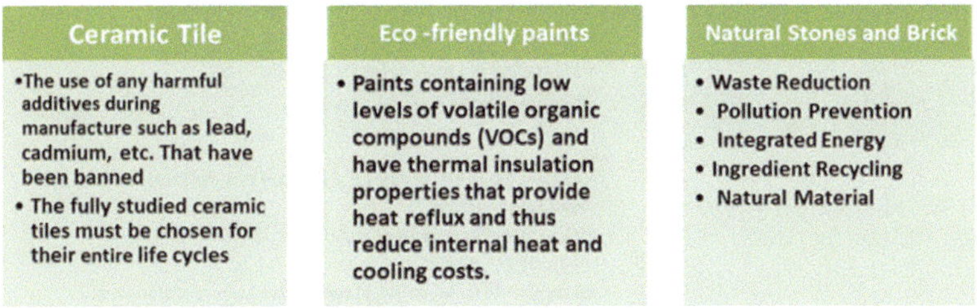

Fig. 7 Shows sustainable finishing treatments. *Source* Stelmack AK (2019) Sustainable residential interiors, 2nd edn, Wiley, Hoboken, NJ, p 327

9 Case Studies Methodology

This section reviews the components of the methodology used in the analytical study, and begins by identifying the objectives of the study by analyzing models of global and local projects that adopt a sustainable environmental trend, through these objectives [26], the descriptive methodology proposed in the analytical study is formulated.

This section outlines the methodology components utilized in the analytical study. It begins by clarifying the study's objectives through an analysis of global and local projects that embody sustainable environmental practices. These objectives inform the descriptive methodology proposed for the analytical research.

The process includes identifying the tools for data collection, defining the study's scope through specific criteria for selecting residential projects, and determining the relevant samples. Following this, the selected samples are analyzed, measuring the sustainability of the materials used by employing a proposed measurement table to derive the analytical research results [27] (see Fig. 8).

9.1 *Case Study*

The case study (4 buildings) was chosen by the previous criteria and what achieves the objectives of the study, and they are as shown per the (Tables 2 and 3).

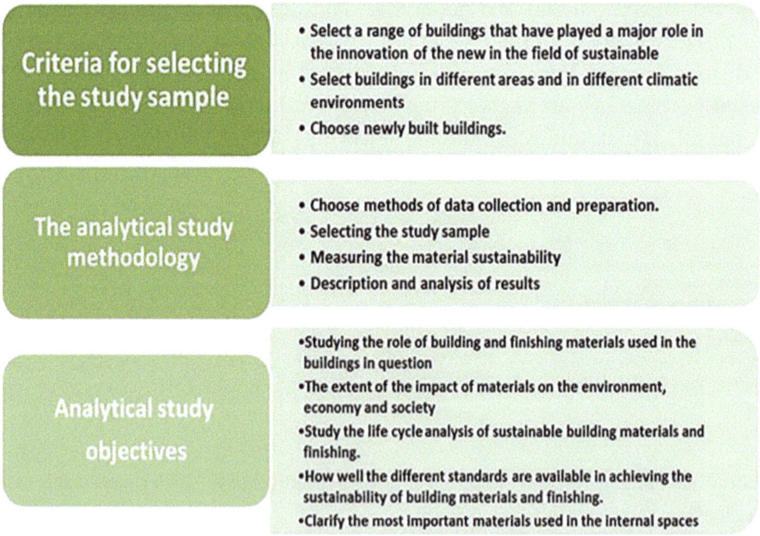

Fig. 8 Shows case studies methodology (reviews the components of the methodology used in the analytical study [by Author])

Table 2 Shows case study analysis by Researcher

Project Name	EcoVilla	Neil Road Home	Citriodora House	Hover House
Introduction	It is. a pilot project equipped with energy and water-saving technologies, and an Emirati family will move to test the experience of living in this typical sustainable housing	This home emphasizes preserving the heritage and identity of the ancestral homes while adding some modern changes to suit the current Inving requirements. It was revitalized by ONG * ONG Pte Limited in 2013	This house Designed for vacation, this home ts set amidst trees that smell lovely lemon and near the famous Great Ocean Road.	It is the third home in the Glen Irani Architects 'Hoover House' series. This series focuses on Maximizing outdoor living, making the most of daylight and building connectivity with nature and exterior location
A picture of the project				

Source Ching FDK, Binggeli C (2018) Interior Desig Illustrated, Wiley, Hoboken, New Jersey

Introducing the Building information [28]

Building 1:-('EcoVilla')
It is a pilot project equipped with energy and water-saving technologies, and an Emirati family will move to test the experience of living in this typical sustainable housing.

Building 2:-('Nile House')
This home emphasizes preserving the heritage and identity of the ancestral homes while adding some modern changes to suit the current living requirements.

Building 3:-('Citriodora House')
This house Designed for vacation, this home is set amidst trees that smell lovely lemon and near the famous Great Ocean Road.

Building 4:-('Hoover House')
It is the third home in the Glen Irani Architects 'Hoover House' series. This series focuses on maximizing outdoor living, making the most of daylight and building connectivity with nature and exterior location.

Table 3 Shows case study final concolution and analysis by researcher

Aspects of sustainability	Sustainability Standards		Furniture	Architectural vacuum determinants		
				Celling	Flooring	Walls
Aspects of sustainability	The environmental aspect	Hover House	Environment good			
			Social Very good			
			Economic good			
		Citriodora House	Environment very good			
			Social very good			
			Economic very good			
		Nile House	Environment good			
			Social veyr good			
			Economic very good			
		Eco villa	Environment good			
			Social very good			
			Economic good			

Source https://www.sustainablearchitecture.com

9.2 Analysis of the Case Study Sample

None of the projects in the analytical study sample received evaluation points regarding the criteria for selecting certified wood and choosing materials accompanied by environmental disclosure documents. This indicates a lack of awareness among interior design engineers, architects, and even users about the importance of adhering to these criteria and their significant impact on achieving sustainability in terms of the environmental dimension. The Citriodora House received the highest evaluation score for measuring the sustainability of the building and finishing materials used regarding the environmental dimension. All available materials and resources in the surrounding environment of the project site were optimally utilized to meet the interior requirements for building, finishing, furnishings, and furniture. This was due to various wood materials available in the local market within the project's country. Additionally, the woods used for flooring, walls, ceilings, or furniture production are free from any harmful chemical additives, have a lower embodied energy, and do not produce any harmful pollutants.

Both the Neil Road House, and Sitriodora House received the highest evaluation scores for measuring the sustainability of the building and finishing materials used in terms of the economic dimension. All these projects prioritized the use of locally source.

Finaly The analysis of the case study sample reveals significant insights into the sustainability performance of the selected projects, particularly in terms of environmental, economic, and social dimensions. Citriodora House emerged as the top-performing project, achieving the highest evaluation scores in both environmental and economic dimensions. This success can be attributed to its optimal use of locally sourced materials, which reduced the carbon footprint associated with transportation and supported the local economy. Additionally, the use of chemical-free, low-embodied-energy materials contributed to improved indoor air quality, enhancing the health and well-being of its occupants. In contrast, the other projects, including Neil Road House, demonstrated a lack of awareness regarding the importance of selecting certified wood and materials accompanied by environmental documentation. This gap highlights a critical need for increased education and awareness among architects, designers, and users about the long-term benefits of sustainable material choices. While Neil Road House performed well in the economic dimension by prioritizing locally sourced materials, it fell short in addressing environmental and social considerations, such as the use of non-toxic materials and ensuring occupant health. Overall, the findings underscore the importance of integrating sustainability principles into material selection processes, emphasizing the use of locally available, recyclable, and non-toxic materials to achieve comprehensive sustainability. To address these gaps, it is essential to promote awareness, provide training on sustainable practices, and encourage the adoption of green building standards across the industry. See Tables 2 and 3 [29].

It is clear from the result of the previous analytical study, both the Neil Road-house and the Citriodura house obtained the highest evaluation rate in measuring the sustainability of building and finishing materials used in terms of the economic dimension, as all of these projects took into account the use of locally sourced materials whenever possible [30].

10 Conclusion

Sustainable architecture is essential for promoting a healthier environment and improving community well-being. It is crucial to enhance social awareness of sustainability and the benefits of using eco-friendly building materials [7], especially in diverse settings like Egypt. By highlighting the importance of sustainable finishing materials, we can enhance the quality of interior spaces while respecting local environmental contexts [31].

Additionally, incorporating sustainable design principles into architecture education will equip future architects with the knowledge to prioritize environmentally responsible practices. This educational transformation can help dispel myths about the economic costs associated with sustainable materials, showcasing their advantages over the long term [32].

To overcome existing challenges such as limited awareness, misconceptions about high costs, and the scarcity of sustainable materials in the market, a collaborative effort from architects, educators, and policymakers is necessary. Together, we can foster a culture that values sustainability, ultimately creating a built environment that benefits both humanity and the planet [33].

11 Recommendation

- The field of sustainable architecture is one of the vital areas that concern everyone and in all environments. It is necessary to increase social awareness of the concept of sustainability and education with the advantages of using building materials and sustainable finishing in internal spaces.
- Increasing interest in the interior spaces of different buildings, and focusing on the use of sustainable and suitable building and finishing materials suitable for the local environment in Egypt
- Working to increase information and depth in analyzing the social and economic dimension of building and finishing materials used in any project
- Including the idea of sustainable spaces in the curricula of architecture education, to create a new generation of designers fully aware of the importance of designing internal spaces that are not harmful to humans or the environment.
- Encouraging the architects of all their customers to use and purchase sustainable building and finishing materials, changing the idea of the high economic cost of these materials, and raising awareness of the extent of the benefits of using those materials, whether in the short or long term.

- The lack of awareness among users of the importance of choosing sustainable building and finishing materials, the spread of the idea of the high economic cost of sustainable materials, and the lack of availability and spread in the local market.

References

1. Moxon S (2020) Sustainability in interior design, 2nd edn. Laurence King Publishing, London
2. Miller Boehr K, Knowles E (2019) The comprehensive guide for selecting interior finishes. Pearson, USA.
3. Bonda P, Sosnowchik K (2020) Sustainable commercial interiors, 2nd edn, New Jersey, p 203
4. Kibert C (2018) Sustainable construction: Gr. building design and delivery. Wiley, New Jersey
5. Moxon S (2021) Sustainability in interior design. Laurence King Publishing Ltd, China
6. Ching FDK, Binggeli C (2018) Interior desig illustrated. Wiley, Hoboken, New Jersey
7. Groves K, Valente M (2020) Sustainability in interior design, 2nd edn. Laurence King Publishing, London
8. Ashby M (2020) Materials and sustainable development. Elsevier Ltd, USA
9. Stelmack K (2019) Sustainable residential interiors, 2nd edn, Wiley, Hoboken, NJ
10. Bonda P, Sosnowchik K (2020) Sustainable commercial interiors, 3rd edn. Wiley, Hoboken, NJ
11. Binggeli C (2016) Building systems for interior designers. Wiley, Hoboken, NJ
12. Green Building Council (2019) LEED guide associate study, USA
13. https://inhabitat.com
14. https://www.worldgbc.org
15. https://www.usgbc.org/leed
16. https://www.greenpublicrating.org/
17. Leydecker S (2019) Nano materials in architecture, interior architecture and design, 2nd edn. Birkhäuser, Basel
18. Guerin D, Kang M (2019) The characteristics of interior designers who practice environmentally sustainable interior design. Environ Behav 41(2):170–184
19. Beckman SL, Barry M (2020) Design thinking and sustainable development goals. Emerald Publishing Limited, Bingley
20. https://www.ibec.or.jp/CASBEE/english/
21. https://www.sustainablearchitecture.com
22. https://www.breeam.com/
23. https://igbc.in/igbc/
24. Boyle C (2021) Sustainable design: circular economy in architecture and interior design. Laurence King Publishing, London
25. Butti L, Perini K (2020) Nature based strategies for urban and building sustainability. Butterworth-Heinemann
26. Chick A, Micklethwaite P (2020) Design for sustainable change: how design and designers can drive the sustainability agenda, 2nd edn. Routledge, London
27. Dixit MK (2019) Life cycle assessment of building materials and construction practices. Springer, Cham
28. https://www.gord.qa/gsas
29. Farren-Bradley J (2019) Sustainable design: a critical guide, 2nd edn. RIBA Publishing, London
30. Giebeler G et al (2020) Refurbishment manual: maintenance, conversions, extensions, 2nd edn. Walter de Gruyter, Berlin
31. Keeler M, Vaidya P (2021) Fundamentals of integrated design for sustainable building, 2nd edn. Wiley, Hoboken, NJ

32. Kibert CJ (2021) Sustainable construction: green building design and delivery, 5th edn. Wiley, Hoboken, NJ
33. Lawson B (2019) How designers think: the design process demystified, 5th edn. Routledge, New York

Sustainable Water Management in Urban Settings

Sustainable Water Management in Buildings: Innovative Strategies for an Ecological Future: The Case of Blida City

Faiza Hallouz and Mohamed Meddi

Abstract Strangely enough, it was not until the 1992 Dublin International Water Conference that the international community recognized water as an economic good. It is nevertheless obvious that, like any other scarce resource, water must be managed rationally to reconcile supply and demand. With climate change affecting weather patterns and their severity, population explosion, pollution, and inequitable access to drinking water, it is urgent to implement more sustainable water resource management. Due to its scarcity, water in Algeria, as in most countries, is a limiting factor for development and a source of social tensions. If this issue is raised today in Blida, it is rightly so because by placing this element in the historical context of this city, we better understand the concerns of officials wanting to raise awareness among the population and administrative, technical, or other structures about the severity of a problem whose permanence could jeopardize the future. The city of Blida is located at the foot of the northern slope of the Atlas Blidean and south of the Mitidja plain, at an altitude of 260 m. It is situated on a debris cone built by the Roumman-El Kebir river. The Tell Atlas protects the city from the dry winds of the south coming from the High Plateaus. Our chapter's objective is to minimize the environmental impact of water consumption in buildings, as is the case in the city of Blida, while ensuring the comfort and health of occupants. Finally, sustainable water management requires not only effective technologies and policies but also active community participation.

Keywords Drinking water · Pollution · Sustainable water · Blida · Water management · Technologies

F. Hallouz (✉)
FSTGAT, USTHB, Algiers, Algeria
e-mail: fhallouz@usthb.dz

F. Hallouz · M. Meddi
GEE Laboratory, ENSH, Blida, Algeria
e-mail: m.meddi@ensh.dz

M. Meddi
Higher National School of Hydraulics, ENSH, Blida, Algeria

Z. ElZein and A. Negm (eds.), *Recent Approaches of Sustainable Architecture in Arid and Semi-arid Cities*, Urban Sustainability,
https://doi.org/10.1007/978-981-95-1645-2_9

1 Introduction

Water plays a vital role in the continuity of life on Earth. Man has used this water as a source of life and socio-economic development. This resource is becoming more and more insufficient all over the world. This is due to its unequal distribution, high population growth, climate change, and pollution of water resources. Due to its scarcity, water in Algeria, as in most countries, is a limiting factor in development and a source of social tensions. During this century, our resources are becoming increasingly limited and difficult to exploit, especially in drinking water supply networks where large quantities are lost due to leaks in the network or by consumers. With less than 600 m^3 per inhabitant per year, Algeria (more than 46 million inhabitants in 2024) is in the category of countries poor in water resources, about the scarcity threshold set by the World Bank at 1000 m^3 per inhabitant per year [1]. Faced with the challenge of ensuring the coverage of the constantly increasing water needs (cities, industries, agriculture), an active policy of mobilization of water resources has been implemented, as well as new management instruments. Water is currently taking on the dimension of a national problem in our country because citizens and operators in various agricultural and industrial sectors sensibly feel the existing difficulties.

The preservation of water in the construction industry and within our society is of crucial importance, especially since many of our regions are facing increased drought conditions, exacerbated by the impacts of climate change in the residential sector, domestic water consumption represents 11% of the total global water consumption according to the IGB [2]. Indeed, the rapid increase in population requires the development of urban structures, which means that the water demand is greater, and the satisfaction of the requirements of the agglomerations becomes more difficult with the deterioration over time of the initial structure of the networks if they exist. With climate change exacerbating water shortages in many parts of the world, water management in buildings is becoming increasingly crucial. Cities such as Cape Town in South Africa and São Paulo in Brazil have experienced acute water crises, highlighting the need for sustainable water management strategies. The public water policy is, therefore, fully in line with sustainable development, aiming for balanced management of water resources. In the long term, it must enable the development of economic uses of water, and ensure access for all to quality water in the context of social issues, while guaranteeing the preservation of water resources. Limited access to water, in quantity and quality, is one of the main factors hindering sustainable development. The sustainable management of water resources must be based on an integrated approach linking the development and protection of the natural environment, the participation of all actors and parties concerned, and the recognition of the economic and social value of water. It is necessary to promote the integration of all aspects of the development, management, and protection of water resources by developing plans that aim to satisfy essential needs and promote an equitable and efficient distribution of water resources, the protection of ecosystems, and the maintenance of the water cycle [3].

According to Gro Harlem Brundtland, Prime Minister of Norway: ‘sustainable development is a development that meets the needs of the present without compromising the ability of future generations to meet their own needs’ United Nations [4, 5]. The sustainable development process aims to reconcile the ecological, the economic, and the social by establishing a kind of “virtuous” circle between these three pillars: it is an economically efficient, socially equitable, and ecologically sustainable development [5]. It respects natural resources and ecosystems (environment: preservation of natural resources), supports life on earth, and guarantees economic efficiency (economy: production methods that respect the planet and people), without losing sight of the social aims of development, which are the fight against poverty, inequality and exclusion, and the quest for equity (social: sharing and equity). Sustainable development, which is intended to be a new way of thinking about the world and “managing” it, which tries to provide answers to concerns about the planet, actually refers to many questions. Nevertheless, despite the conceptual and political vagueness, despite the doubts about the instruments to be adopted for its implementation, the difficulty of articulation and definition of the scales relevant to its implementation, despite the multitude of actors and the prevalence of conflicts that characterize their relations, sustainable development has acquired a global dimension whose interest it is necessary to emphasize as the contradictions [6].

Drinking water sources are threatened by various pollutions and global warming, with rainfall varying increasingly from one region of the world to another. Water is essential to life, and sustainable water management is becoming a crucial issue. There are also marked differences in domestic consumption. In wealthy countries, it includes toilet flushing (30 to 40% of household consumption, watering gardens, and even private swimming pools, and can exceed 5 m^3 per person per day) [7].

An American consumes about 2000 m^3 of water per year, while a Jordanian consumes only 100 m^3 and a Haitian 7 m^3. About 1.1 billion people are not connected to a piped water supply. In sub-Saharan Africa, only 58.5% of the population has access to it 27 [7]. In Algeria, there has never been a real debate on consumption, i.e. the role and weight of household consumption in investment and savings.

During the time of the administered economy, it had been decided to give priority to investment, and Algeria ended up with the investment rate—i.e. the ratio of gross fixed capital formation (GFCF) to gross domestic product (GDP)—reaching a peak of 48.60% in 1978 and gradually stabilizing at around 20% from the beginning of the decade to the present day [7]. However, the higher rate of investment compared with neighboring and developing countries has not ensured that Algeria has a higher growth rate than those countries.

Our chapter aims to minimize the environmental impact of water consumption in buildings, as is the case in the town of Blida while ensuring the comfort and health of the occupants. The city of Blida, in northern Algeria, is known for its hot, dry summers. The city faces major challenges in terms of water management, particularly due to the variability of rainfall and the increase in demand for drinking water. Given the pressing challenges involved in preserving water resources, it is important to invest in innovation to achieve effective water management focused on sustainable development. Technological advances play a crucial role in preserving the available water. Data science and artificial intelligence also offer the possibility of managing water in a predictive way by analyzing data to optimize its use and anticipate needs. Similarly, rainwater harvesting systems offer an ingenious solution by capturing rainfall for non-potable uses such as watering gardens.

At the same time, rainwater management represents an essential aspect of the water conservation strategy in buildings. In addition, ecological design, a fundamental principle of sustainable architecture, also offers innovative solutions for water management. The integration of green roofs, for example, not only promotes the thermal insulation of buildings, but also acts as a natural reservoir by absorbing and filtering rainwater. In addition, policies and regulations can stimulate innovation and promote sustainable water management practices.

2 Description of the Blida City

The town of Blida lies around 40 km southwest of Algiers, the capital of Algeria, in the central Tell region. It is located in the plain of Mitidja. A plain renowned for its fertile soils. It is an important city in the agricultural, commercial, industrial, academic, tourist and military fields. The city's population is constantly growing. Table 1 shows its evolution in comparison with other places. This allows us to understand how it has developed in relation to these areas.

The city of Blida which extends over an area of 150.15 km^2, composed of four municipalities, which are: Blida; Bouarfa, Ouled Yaich, and Beni Mered (see Fig. 1), is backed to the north by the chain of the Tellian Atlas and to the south by the plain of Mitidja of which is the capital.

The highest point of this town is 270 m above sea level at the place where the wadi El Kebir takes its source which sprinkles the city with its beneficial waters at the origin of its agricultural wealth [3]. A survey carried out in 2023 revealed that residential buildings in Blida consume an average of 150 L of water per person per day. The main sources of water consumption are toilets, showers, and cleaning activities [9].

Table 1 Population trends from 2017 to 2040 [8]

Population	2017	2020	2030	2040
Blida	183,751	191,010	217,346	247,313
Bouarfa	42,164	44,482	53,169	63,553
Ouled Yaich	131,729	151,189	239,325	378,840
Beni Mered	57,411	66,651	109,606	180,245

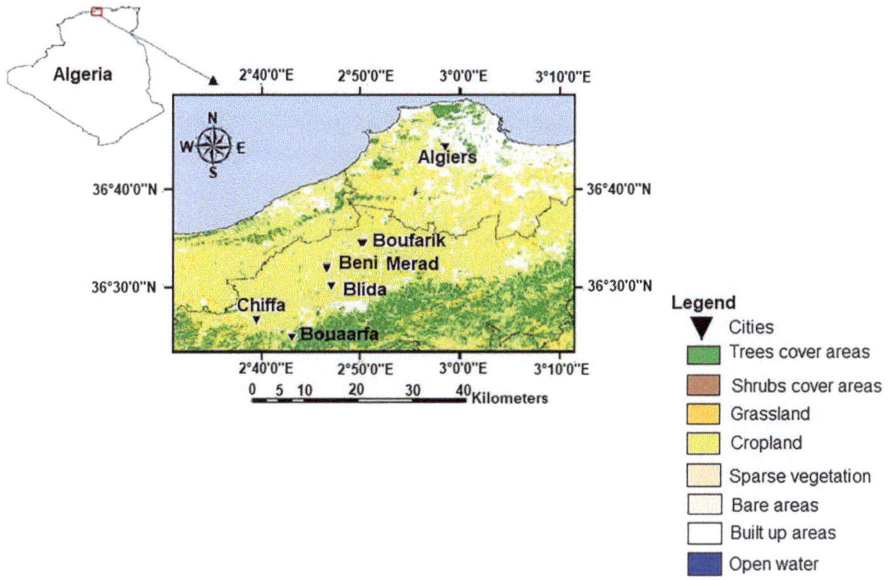

Fig. 1 Vegetal cover of the study area (by author(s))

3 Water Resources

The city of Blida is part of the hydrographic Region (Algerian Hodna Soummam). Covering the watersheds of the Algerian coastlines, Sebaou, Isser, Soummam, and Chott Hodna. It is part of the central region of Northern Algeria considered the most urbanized, the most industrialized, and also the richest in terms of agriculture.

This observation places water resources at the heart of a very difficult problem. Resources are becoming scarce and expensive to mobilize (see Fig. 2) [5].

a. **Surface water**

The Mitidja plain is crossed, from south to North, by several Wadis born in the Atlas that cross the Sahel. However, these wadis are not used to any great extent, due to the lack of water mobilization facilities in the Blida city [5].

b. **Underground water**

A large water table, mainly fed by water from the Atlas Mountains, known as the Mitidja water table, is located in the plain, estimated at 200 Mm^3, and whose real potentialities are of 180 Mm^3. This water table is over-utilized because of the weakness of the surface waters and the easy and well-estimated nature of a resource that must meet immediate and quickly satisfied needs [5].

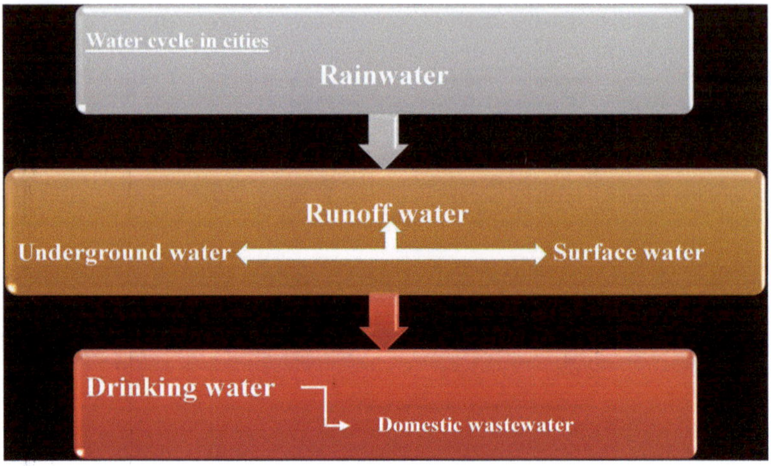

Fig. 2 Water cycle in cities from rainwater harvesting as input to domestic wastewater as output of the cycle (by author(s))

4 Climate Situation

a. **Rainfall**

The climate of the city of Blida is Mediterranean, mild and wet in winter, and hot and dry in summer. At a distance of 22 km from the sea, it is subject to marine influences. The climatic conditions are generally favorable. The rainfall is generally above 600 mm per year on average. It is particularly high in the Atlas Mountains. The rainfall peaks in December and February, which are months that account for about 30 to 40% of the annual rainfall. Conversely, the summer months (June, and August) are almost always dry [5].

b. **Temperature**

The temperatures are generally linked to the seasons, a cold season from October to January, and a wet season which can drop to below zero. A hot season when the temperature can reach more than 40 °C [5].

c. **Humidity**

Humidity levels vary according to the season, in summer, humidity can reach 57%, while in winter, from December to January, it can reach 82% [5].

5 Drinking Water Supply

The current drinking water supply system in Blida City is made up of two main systems.

a. **Drinking water supply system in Blida, Bouarfa, and Ouled Yaïch**

Samples are taken to supply drinking water to Blida, Bouarfa, and Ouled Yaïch [10].

b. **In the Mitidja aquifer (three wellfields)**

The pumping station of the Beni Tamou capture field delivers a flow rate of 15,300 m^3 / d in a pipe with a diameter of 800 mm then 600 mm and a length of 7240 m to the city of Blida, where it splits into two pipes [10]:

- A steel and HDPE pipe with a diameter of 350 mm and a length of 1325 m feeds the Touares reservoir (on circular ground with a volume of 5000 m^3 wedged to the 265 NGA coast);
- A cast iron pipe with a diameter of 600 mm and a length of 1150 m supplies the reservoir (on a rectangular floor with a volume of 4×5000 m^3 wedged to the 270 NGA coast) [10].

The Maramane wellfield pumping station delivers a flow rate of 16,500 m^3 / d through a steel pipe measuring 7512 m in length and 800 mm in diameter, then 700 mm, 600 mm, and 500 mm, to the reservoir (on the rectangular ground with a volume of $2 \times 5{,}000$ m^3 wedged to the coast 263.5 NGA), from the city of Bouarfa [10].

In the Chiffa wellfield, the pumping station delivers a discharge of 8400 m^3/ day through a 500 mm diameter concrete pipeline, 7300 m in length, to a storage reservoir with a capacity of 4000 m^3, located on a rectangular plot at an elevation of 281 m (NGA), supplying the city center of Blida [10].

The water intakes in the Wadis of Bab M'hamed, Hanous, and Tabarkachent convey captured water by gravity to a chlorination chamber located downstream of the Sidi El kebir capacity. From this chamber, a steel pipe with a diameter of 500 mm and a length of 2350 m carries a flow of 6100 m^3/d to the Downtown reservoir, which has a volume of 4000 m^3 and is located at the coast 281 NGA) [10].

Additionally, the Ben Achour intake currently conveys extracted water to the Touares reservoir by gravity, connected by a steel pipe with a diameter of 100 mm that conveys a flow rate of 1200 m^3/d over a length of 1450 m [10]. Lastly, The catchments of the Fresh Fountain and Plane Tree sources transport to the City Center reservoir by a cast iron pipe 300 mm in diameter and 800 m long which conveys a flow rate of 2000 m^3/d, on the supply pipe. The waters from the Plane Tree are directed to the distribution networks using a cast iron pipe with a diameter of 200 mm, a length of 2520 m, and a flow rate of 1300m^3/d [10].

c. **Beni Mered drinking water supply system**

The samples for the drinking water supply of Beni Mered are carried out at two production sites.

d. **Beni Mered wellfield**

This site includes eight boreholes, six of which discharge into the Beni Mered reservoir (with a volume of 1000 m^3 on a tower) via a 300 mm diameter steel pipe, and two boreholes which directly supply the new reservoir (with a volume of 1000 m^3 on a tower wedged to the 143.77 NGA coast) of Hai Fettel with a 160 mm PVC pipe, (the total flow rate is 4300 m^3/d) [10].

e. **Diar El Bahri wellfield**

It includes two boreholes that supply a recovery station, which supplies the tower reservoir with a capacity of 1000 m^3 through an HDPE pipe 315 mm in diameter and 4500 m long with a flow rate of 970 m^3/d in the Diar El Bahri sector. A third borehole feeds the 1000 m^3 reservoir directly [10] (Table 2).

Following the increase in agglomeration of the city of Blida, the water demand is growing rapidly, with more than 200,000 m^3/day expected by 2040.

Based on all the above information, we recognized that the city of Blida is facing major problems with its water supply system, which can be summarized as follows

- The issue of water will undoubtedly be a major preoccupation during this century, not to mention the fact that our water resources are becoming increasingly limited and difficult to exploit. All these surface and underground resources are conditioned by rainfall that is highly irregular in time and space;
- The problem of drought during the summer is clearly showing its harmful effects on water resources, with significant reductions in run-off, a drop in the piezometric levels of the Mitidja water table, leading to the drying up of shallow wells and the drying up of sources, particularly the Chiffa catchment;
- The city of Blida is characterized by a dense drinking water supply network that is diversified in terms of age, diameter, and materials, as it includes old pipes;
- In addition, the city of Blida has a large industrial base, but there has been no real survey of management because industries do not give their real needs, which makes it difficult to forecast future water requirements;
- However, the SP3 transfer operation does not concern the municipality of Beni Mered at the moment, which leads us to resort to this transfer to possibly counteract the deficit of this area;
- Lack of research into water resources.

Table 2 Current resources of Blida City [10]

Ressource	Daily production (m^3/d)
Sampling from the Beni Tamou wellfield	15,300
Sampling from the Maramene wellfield	16,500
Samples from the Chiffa wellfield	8400
Sampling from the Beni Mered wellfield	4300
Sampling from the Diar El Bahri wellfield	970
Intakes in wadis	7300
Total	62,300

6 Strategies for Sustainable Water Management

The combination of demographic growth, and the development of the industrial and agricultural sectors, has led to a spectacular increase in water requirements, a much greater increase than the quantities mobilized, to the point where the deficit between water supply and demand is likely to constitute, in the long term, an insurmountable handicap to development and a negative impact on socio-economic development.

Competition for freshwater resources is increasing due to growing demand, a greater variety of uses and users, the depletion of some water resources and the loss of other resources due to pollution. However, the view expressed in the World Water Vision: Making Water Everybody's Business [11] is that all uses could be met in the future, without depleting the planet's natural capital, if everyone committed to improving water management.

The first step is to create a management environment that allows new tools, institutions, and capacities to emerge. Among the strategies for sustainable water management in the city of Blida, we can propose certain solutions such as:

6.1 Techniques for Reducing Water Consumption

a. **Saving water indoors**

Reducing the amount of water you and your family use doesn't have to be complicated. It's as easy as making small changes to your consumption habits, repairing leaks, and, when the time comes, replacing old appliances and high-flow devices with water-saving models. Saving water has many advantages, not least of which is saving money.

Low-flow devices, such as toilets, shower heads, and faucets, can save water. For example, standard shower heads use about 2.5 gallons per minute (gpm), while low-flow ones use 2.0 gpm or less, which saves about 2700 gallons of water per year for an average family. Standard faucets consume about 2.2 gpm, while water-saving ones use only 1.5 gpm, which is a 30% reduction and can save up to 700 gallons per year. For toilets, high-efficiency toilets (HET) use about 1.28 gallons per flush compared to traditional models that can use up to 6 gallons. The choice of these devices and their installation is then crucial in the design of buildings whatever their use, to ensure sustainable management of water consumption. What's more, according to Project Drawdown, the increased adoption of low-flow devices between 2020 and 2050 could result in an average water savings of 15.8% [12].

b. **Finding leaks**

The first step to saving water at home is to check for leaks. These can lead to the waste of huge amounts of water, often without anyone noticing for long periods. First inspect all your appliances and devices using water, at a time when the house is quiet and there is no water flow Let your eyes and ears be on the lookout for leaks in bathrooms, kitchens, laundry rooms, and basement drains. If your house has a water meter, check regularly for hidden leaks by reading the meter (or taking a photo) at night after all the occupants have finished using the water. If your meter has a dial, note the position of the needle as well. The next morning, before anyone starts using the water, check the meter reading again. Compare the readings (and, if necessary, the position of the needle). If the morning meter reading is higher (or the needle has moved), someone or something has used the water in the house overnight (for example a water softener or a humidifier fitted to the furnace) or there is a leak in a pipe, appliance or device.

c. **Showers and baths**

Taking a shower usually requires less water than a bath. The relative efficiency of a shower compared with a bath depends on several factors, such as the bath's size, the showerhead's flow rate, and personal habits. While an average eight-minute shower with a showerhead flowing at 9.5 L per minute will use 76 L of water, a full bath may require 150 L or more. Of course, people who take longer showers or use an inefficient shower head could consume more than 76 L of water, and people who only partially fill the bath could use less than 150 L. Citing all these techniques, we can give the example of Singapore which introduced sensor faucets in public buildings which allowed a significant reduction in water consumption [13].

6.2 Water Reuse and Recycling

The reuse of purified wastewater is a widespread practice in the world and mainly in regions affected by water resource shortages (Fig. 3). Indeed, the recycling of gray water is a promising strategy for the sustainable management of water in urban areas. Greywater from sinks, showers, and washing machines accounts for 40 to 90% of domestic wastewater according to estimates. Well-designed greywater recovery systems can reuse up to 70% of the water used in a building. This reuse can significantly reduce the demand for drinking water, which is particularly valuable in regions facing water stress. Greywater recovery systems often use filters to clean the water coming from showers, sinks, and washing machines. Once treated, grey water can be reused for various non-potable applications, such as flushing toilets, garden irrigation, or car washing [13].

In Tokyo, reuse is strongly encouraged. Many buildings are equipped with a second distribution network to supply the toilet flushes. The purified effluent, treated by sand filtration and chlorination, is distributed to the reservoirs in each building concerned. The major disadvantage of such reuse is the financial cost of developing a second domestic distribution network with sufficient safety features to avoid any interconnections. In addition to these buildings supplied by wastewater treatment plants, it can be emphasized that other buildings recycle gray water for the supply of flushing toilets. This supply constitutes from 20 to 30% of the water used in a building [13].

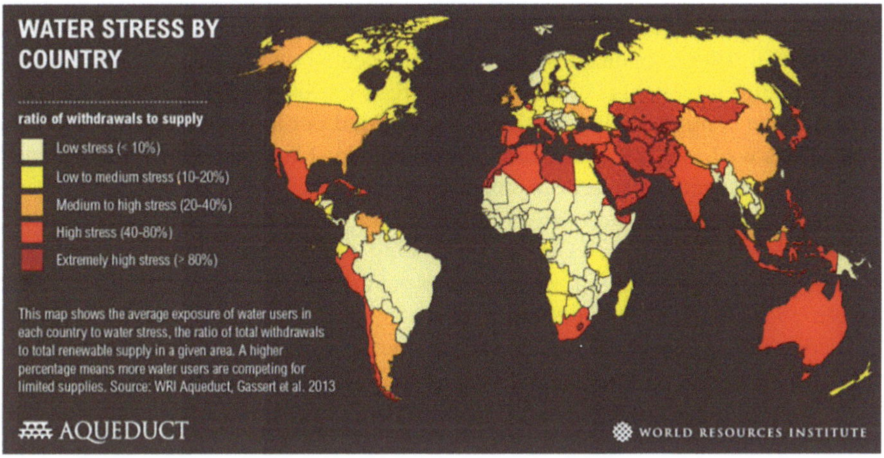

Fig. 3 Countries most affected by water stress and therefore concerned by the reuse of wastewater after treatment [14] (https://www.wri.org/data/water-stress-country)

This practice affects 33% of households in urban areas in Japan. Japanese law requires all new buildings in urban areas to have a greywater recycling system. In Tokyo, any building with a floor area of more than 30,000 m^2 or likely to reuse more than 100 m^3 of treated effluent per day must be equipped with a double distribution network. As another singular example of urban wastewater reuse, is the feeding of a dried-up urban river. The Nobidome stream in Tokyo, which had dried up when its source was diverted in 1976, was looking more and more like a landfill site. Australia is one of the driest continents. Rainfall intensity varies widely across the country, with 80% of rainfall concentrated in one-quarter of the continent. Since 1991, there has been a rapid increase in water recycling initiatives, particularly in the reuse of urban wastewater. These initiatives have reduced urban and industrial demands on water resources. Australia's population (18 million) is small compared to its vast territory (8 million km^2). A large proportion of the population lives in the metropolises located in the south of the continent, which benefit from high rainfall. On this continent, wastewater is reused to irrigate crops, grasslands, green spaces, or for industrial use. In the temperate zones of Australia, reuse mainly concerns the irrigation of crops such as sugar cane while in arid zones, it is the irrigation of tree plantations that predominates. This water is then locally demineralized before being used as cooling water. Initiatives to use wastewater to flush toilets began in 1996. Reuse for approval purposes has been practiced since 1955 in the United States. In addition to watering parks, golf courses, and public gardens, a singular example is the creation of artificial lakes fed in whole or in part by purified wastewater. For example, in 1961, Lake Santee in California was put into water and fed with purified and disinfected water. After several studies, bathing was authorized in 1965 [13]. In 1990, the supply of treated urban water to artificial lakes for recreational use accounted for 3% of reuse in California. To illustrate the technical advances made in the United States in the field of reuse, we can also cite the specific case of an urban wastewater treatment plant in Santa Monica. Located on the site of the reuse, the water from this treatment is used to irrigate green spaces and feed a leisure lake. On the American continent, this practice is also carried out in several South American countries as well as in Mexico. For example, Mexico City uses the effluents treated by its 16 wastewater treatment plants to irrigate parks, public gardens, and leisure facilities. It should be noted that only 6% of raw wastewater is purified. In 1996, raw wastewater from Mexico City was still being used for agricultural irrigation and in particular as part of the largest irrigation plan in the world (irrigation of 85,000 hectares of corn, barley, and tomatoes) [13].

There are many other examples, such as the town of Taif in Saudi Arabia. This city has a wastewater treatment plant that treats 70,000 m^3 of water a day. It is one of the largest plants in the world. The treated effluent is used to water parks and gardens and to clean streets, buses, and taxis [13].

In Singapore, sewage waste transformed into ultra-clean water: The small territory with limited water resources has long been dependent on supplies from neighboring Malaysia. To become more self-sufficient, the government has developed a sophisticated system for reprocessing wastewater with a network of pipelines and high-tech plants. Reprocessed water already provides 40% of Singapore's needs today, and the proportion is set to rise to 55% of consumption by 2060, according to the National Water Agency [15]. While most of this water is used for industrial purposes, some are also used for drinking water, replenishing the reservoirs on the island of 5.7 million inhabitants. This system helps to reduce marine pollution, as only a small proportion of wastewater is discharged into the sea. This contrasts with the situation in most other countries, where 80% of the world's wastewater is discharged into ecosystems without being treated or recycled, according to United Nations estimates [15].

6.3 Stormwater Management

Stormwater management practices have evolved rapidly over the past few decades. Traditionally, the focus was solely on controlling peak flow rates. Until the early 1980s, rainwater was considered a relatively negligible source of pollution. Since then, several studies, such as the Nationwide Urban Runoff Program (NURP) in the United States, and other studies conducted in Canada and Europe have demonstrated that runoff can be an important source of pollution [16].

The management of water in urban areas and discharges into the environments (rivers, soils, groundwater) remains a major concern for urbanized territories (cities, peri-urban areas). Simultaneously, over the past few decades, rainwater has increasingly been perceived as a valuable resource and an added value for the city (used for amenities, reused for various purposes, urban temperature regulation, biodiversity elements …). Water management, therefore, lies between a threat and an opportunity, presenting a complex yet stimulating challenge for water and urban stakeholders [17].

Located in the northwestern region of the United States, at the confluence of the Willamette and Columbia Rivers, the city of Portland is subject to frequent rainstorms. More than a third of the city has an old unitary sewer system that receives both sanitary wastewater and stormwater runoff. As a result, the city experienced several episodes of dumping billions of liters of wastewater into the Willamette River during the 1990s [18]. Subjected, in 1993, to a lawsuit for violation of the Clean Water Act, and to comply with the new federal drinking water regulations, the City of Portland implemented an integrated approach to rainwater management in the early 2000s. This approach is based, among other things, on the construction of gray infrastructure, the modernization of the sewer network, and the implementation of green infrastructure that promotes the natural management of rainwater. The Green Streets program, established in 2003, is part of this approach [18].

Add to this, sponge cities in urban areas mimic natural ecosystems (see Fig. 4), particularly in their approach to stormwater management. These are nature-based approaches that rely on blue and green natural infrastructures for stormwater management. Sponge cities aim to manage rainfall at the source rather than directing runoff to artificial (gray) stormwater management infrastructure systems. This makes it possible to manage stormwater on-site and reduce flooding. Since the water remains where it falls, it is also less likely to absorb pollutants harmful to the natural ecosystem. These systems also reduce the need to modernize and expand gray infrastructure over the years. Sponge city approaches are useful; it was found in the city of Portland (Oregon, United States) that after implementing sponge city-inspired strategies on NE Siskiyou Street, peak stormwater runoff was reduced by 81%.

Fig. 4 Small-scale sponge city application, green roof [19] (https://unsplash.com/fr/photos/maquette-de-batiment-nXJyZLBnyUQ)

Indeed, the first country to use the sponge city approach was China, where urbanization and a climate marked by typhoon seasons near its coasts led to water management issues. After years of research, in 2014, the Chinese government requested the national development of sponge cities in all new urban deelopments. By 2022, more than 64 Chinese cities had enacted legislation to implement the sponge city guidelines at the local level [20].

7 Innovative Solutions for Water Conservation

Smart technology has revolutionized various aspects of our lives, from our homes to our workplaces. It has become increasingly widespread in resource conservation and environmental protection efforts. When it comes to preventing leaks and minimizing water waste, smart technology offers innovative solutions that can help us address this urgent issue. By harnessing the power of innovation, we can not only save water but also reduce our ecological footprint. Innovation is essential in the field of water management to tackle sustainability challenges in the construction sector. Let's discover cutting-edge solutions that reinvent the use and management of water in modern structures:

7.1 Blue Roof Systems

The blue roofs, distinct from the green roofs, are designed to retain rainwater temporarily before gradually releasing it. This method helps to prevent flooding in urban areas and to reduce the pressure on municipal sanitation infrastructure. In Amsterdam, for example, the RESILIO project has transformed 12,683 m^2 of roofs into smart blue-green roofs, adding biodiversity and helping to retain excess rainwater, making the city more resilient to extreme weather conditions [12].

7.2 Smart Water Management Systems

The adoption of smart water management systems can also lead to significant water savings by quickly identifying leaks and optimizing water use according to real needs. Thames Water, in the United Kingdom, has implemented a smart water network using AI, which has resulted in a 13% reduction in water consumption. This technology has enabled the rapid identification of water consumption issues, such as undetected leaks or excessive use, thus allowing a rapid intervention to solve these problems. Additionally, the system has made it possible to provide fairer and more transparent billing for consumers, as they are now billed based on their actual water consumption.

7.3 New Water Absorbing Materials

Innovative materials, such as porous concretes and absorbent substrates, can retain large amounts of water, thereby helping to prevent urban runoff. These materials are already being used in various paving and coating projects in cities such as Philadelphia and Chicago [12].

7.4 Collection and Use of Condensation Water

In hot and humid regions, recovering water from the condensation of air conditioning systems represents a significant alternative source. For example, the condensate water recovery system of the Bahrain World Trade Center makes it possible to recover enough water to irrigate its green spaces throughout the year [12].

7.5 Modernization of Infrastructure

Investing in the modernization and maintenance of water supply infrastructure is crucial to solving the tap problem. By replacing old and leaking pipes, valves, and faucets with more efficient and sustainable alternatives, water losses can be significantly reduced. For example, the City of Los Angeles has implemented a comprehensive pipeline replacement program, resulting in a 40% decrease in water losses due to leaks [12].

7.6 Smart Water Meters

One of the most effective ways to prevent leaks is to use smart water meters. These devices provide real-time data on water consumption, allowing homeowners to monitor their consumption and quickly detect any anomaly or leak. By integrating with mobile applications, smart water meters allow users to track their consumption and receive alerts in case of excessive or unusual water flow. This technology allows individuals to take immediate action, preventing leaks before they become significant problems. For example, cities such as London have successfully implemented smart water meters, which has reduced water waste and raised awareness among residents [12].

7.7 Leak Detection Systems

Smart technology has also given rise to advanced leak detection systems that use sensors and artificial intelligence algorithms to identify leaks in real-time. These systems can be installed in various leak-prone areas, such as bathrooms, kitchens, and basements. When a leak is detected, the system sends an alert to the user's smartphone, allowing them to act quickly. This not only saves water but also prevents potential water damage on the property.

7.8 Automated Shut-Off Valves

Another smart solution to prevent leaks is to use automated shut-off valves. These valves can be installed on the main water supply line and are designed to automatically shut off the water flow in the event of a leak. Some automated shut-off valves are equipped with leak detection sensors, while others can be controlled remotely via mobile applications. By quickly stopping the water flow, these valves minimize water waste and mitigate the damage caused by leaks.

7.9 Political Reforms

Governments must implement effective water management policies that prioritize water conservation and address the tap problem. This includes stricter regulations on water use, incentives for water-saving technologies, and financial support for infrastructure modernization. For example, Singapore has implemented a comprehensive water management strategy, including strict regulations, educational campaigns, and investments in water infrastructure, making the city-state a world leader in water conservation [21].

By integrating these innovative technologies and methods, modern buildings can significantly reduce their water consumption and contribute to environmental sustainability and the resilience of cities to climate change and future challenges in terms of water resource management.

8 Implementation of Water Efficiency Measures

The effective implementation of water efficiency measures in the construction sector requires a structured and well-planned approach. Here are the key steps to ensure a successful transition to a more sustainable management of water in buildings:

- Start by carrying out a water audit to assess current consumption and identify areas of waste. This involves reviewing water bills, checking for leaks, and evaluating the effectiveness of existing appliances and systems. Based on this audit, establish a clear action plan, defining the specific objectives, the measures to be implemented, and the schedule. Based on this audit, establish a clear action plan, defining specific objectives, measures to be implemented, and the timeline.
- Select and install water-efficient devices such as low-flow toilets, showerheads, faucets, and gray water recycling systems. Ensure that these technologies are adapted to the building's specific needs and comply with local regulations.
- Invest in smart water management systems that can monitor water consumption in real-time and detect leaks. These systems can provide valuable data to adjust water use and improve efficiency.
- Organize training sessions and awareness campaigns for building occupants and management staff. Explain the importance of water conservation and show how everyone can contribute to reducing daily water consumption.
- Develop internal policies to promote the sustainable use of water in daily operations. This can include practices such as efficient irrigation of green spaces, regular maintenance to prevent leaks, and the use of eco-friendly cleaning products.
- Establish a monitoring system to measure the performance of the implemented strategies. Use the collected data to evaluate the effectiveness of water-saving measures and to identify areas in need of improvements or adjustments.
- Engage with local authorities, regulatory agencies, and other stakeholders to ensure that water efficiency measures align with community standards and objectives. Collaboration can also open up opportunities for grants, financial incentives, and sharing of best practices.
- When planning and implementing water efficiency measures, consider the requirements of environmental certifications such as LEED, BREEAM (Building Research Establishment Environmental Assessment Method), or HQE (High Environmental Quality) in France. These certifications evaluate buildings according to various sustainability criteria, including water management. By adopting recognized standards for water efficiency, such as those required for certifications, buildings can improve nont only their water management but also their overall sustainability performance. Certified buildings can benefit from a higher market value, a reduction in operational costs, and an improved brand image. Certification processes often require regular monitoring and reporting of water efficiency performance. This may include tracking water consumption, regularly checking systems for leaks, and maintaining water-saving devices.

Real estate developers, building managers, and occupants can contribute to more sustainable water management, reduce operational costs, and minimize their environmental impact by following these steps. Water efficiency in buildings is not only an ecological responsibility but also a long-term economic and social opportunity.

9 Conclusion

From all the above, it is clear that the city of Blida is moving towards a very hard future in terms of drinking water supply, which is why this chapter highlights the imperative need to adopt water conservation practices in the building sector to address current environmental challenges in Algeria in general and in the city of Blida in particular. The strategies presented, ranging from checking for leaky taps and installing low-flow systems to implementing green roofs, illustrate concrete ways to reduce the water footprint of buildings, while promoting sustainable resource management. Adopting these innovations is essential to ensure the resilience and sustainability of urban infrastructure in the face of future challenges. Effectively implementing water conservation measures in the building sector requires sustained commitment from all stakeholders involved, from planning to implementation. It is crucial for decision-makers, developers, architects, and building owners to integrate these smart and sustainable solutions into their projects.

10 Relevant SDGs

Our chapter aligns with SDG 6 (Ensure access to water and sanitation for all) and SDG 11 (Make cities and human settlements inclusive, safe, resilient and sustainable). Strengthening urban resilience is essential to preventing human, social, and economic losses. Enhancing the sustainability of urbanisation processes is vital to protecting the environment and mitigating disaster risks and climate change.

11 Recommendations

We recommend further research in the areas covered by our chapter, specifically the intelligent use of drinking water reducing the loss of mobilized water, and reusing water for irrigating green spaces as well as for cleaning streets and public roads. Expanding green spaces benefits public well-being. Additionally, children will be considered as part of developments designed to accommodate cultural and sporting activities. Added to this, awareness-raising campaigns and educational programs need to be implemented to inform the population about the importance of sustainable management of water resources and green spaces, involving schools and various local associations. Ultimately, it would be necessary to develop appropriate policies and incentives to promote water reuse, the use of water-efficient technologies and the financing of environmentally-friendly projects, while ensuring a precise and coherent regulatory framework to oversee these actions. These integrated measures will promote the sustainable use of natural resources, while addressing environmental challenges and raising living standards.

References

1. Fanack Water (2019) Water of the Middle East and North Africa. Algeria Water Report. Algeria's per capita water availability: renewable water resources and water crisis. Published on July 30, 2019. Retrieved from https://water.fanack.com/algeria/waterresources
2. Euzen A, Jeandel C, Mosseri R (eds) (2015) Water uncovered (L'eau à découvert). CNRS Éditions. https://doi.org/10.4000/books.editionscnrs.9746
3. The 4th International Symposium on Water Resources and Sustainable Development (CIREDD4, 2011). Organized by the National Higher School of Hydraulics on February 22–23, 2011, in Algiers. Retrieved from http://www.semide-dz.org/FR/Events/news_item.asp?NewsID=11022200
4. Brundtland Report (Our Common Future. Report of the World Commission on Environment and Development, 1987). Sustainable development is "development that meets the needs of the present without compromising the ability of future generations to meet their own needs", quote from Mrs Gro Harlem Brundtland, Norwegian Prime Minister (1987). https://www.are.admin.ch/are/fr/home/media-et-publications/publications/developpement-durable/brundtland-report.html
5. Abdelli R (2008) Strategies for sustainable water management in Remchi (Des stratégies pour une gestion durable de l'eau à Remchi). Thesis for State Engineer in Urban Hydraulics, Abou Bekr Belkaid University, Tlemcen.
6. Granier G, Veyret Y (2006) Sustainable development: what are the geographical challenges? (Développement durable: Quels enjeux géographiques?). Documentation photographique, No. 8053. Retrieved April 23, 2009, from http://www.ladocumentationfrancaise.fr/dossiers/developpement-durable/developpement-durable.shtml
7. IFEN (2005) Water abstraction in France and Europe (Les prélèvements d'eau en France et en Europe). Series "4 pages," July
8. NOS (2017) (National Office of Statistics). Algerian demographics 2017. Algiers:-Directorate of Publications and Dissemination.
9. Algerian Waters, Ministry of Hydraulics (2024) (Algérienne Des Eaux, Ministére de l'Hydraulique), ADE, Blida, Algeria. https://www.ade.dz. Accessed 4 January 2025
10. Bemrah H (2013) Strategies for sustainable management of drinking water (Des stratégies de la gestion durable de l'eau potable). Master's Thesis in Hydraulics, Abou Bekr Belkaid University, Tlemcen, 101 p.
11. Cosgrove WJ, Rijsberman FR (2000) World water vision: making water everybody's business. Earthscan Publications Ltd., London, UK
12. Green Building Consulting & Engineerin (2023) Sustainable buildings: challenges and solutions for water management (Bâtiments durables: Enjeux et solutions pour la gestion de l'eau). https://fr.greendesignconsulting.com/single-post/b%C3%A2timents-durables-enjeux-et-solutions-pour-la-gestion-de-l-eau
13. Puil C (1998) Reuse of urban wastewater after treatment (La réutilisation des eaux usées urbaines après épuration). Mém. D.U.E.S.S. "Eau et Environnement," D.E.P., Université de Picardie, Amiens, 62 p. Retrieved from https://www.u-picardie.fr/beauchamp/duee/puil.htm
14. Reig P, Gassert F, Luo T, Maddocks A (2013) Water stress by country. World Resources Institute. Retrieved December 12, 2013, from https://www.wri.org/data/water-stress-country.
15. AFP (2021) In Singapore, sewage discharge transformed into ultra-clean water (À Singapour, les rejets des égouts transformés en eau ultra propre). A Singapour, les rejets des égouts transformés en eau ultra propre. https://www.ouest-france.fr/economie/a-singapour-les-rejets-des-egouts-transformes-en-eau-ultra-propre-7378829#:~:text=Le%20petit%20territoire%20aux%20ressources,et%20des%20usines%20high%2Dtech
16. Ministry of the Environment, Climate Change, Wildlife, and Park, Québec (2024) Stormwater management (Gestion des eaux pluviales). https://www.environnement.gouv.qc.ca/eau/pluviales/contexte.htm

17. Bacot L, Barraud S, Lipeme Kouyi G (2022) Stormwater management in cities: 20 years of research in the service of action (Gestion des eaux pluviales en ville: 20 ans de recherche au service de l'action). OTHU-GRAIE. Retrieved from www.graie.org
18. Everett G, Lamond JE, Morzillo AT, Matsler AM, Chan FKS (2015) Delivering green streets: an exploration of changing perceptions and behaviours over time around bioswales in Portland, Oregon. J Flood Risk Management.
19. Chuttersnap (2018) Greener Cities (Published on August 31, 2018, Free to use under the Unsplash License). https://unsplash.com/fr/photos/maquette-de-batiment-nXJyZLBnyUQ
20. Szujo J (2023) Sponge cities (Villes éponges). CLIMAtlantic, novembre 9, 2023. Retrieved from https://climatlantic.ca/fr/blog/villes-eponges/
21. Faster Capital (2024) Stop the leaks: ecological solutions to faucet problems (Arrêtez les fuites: Des solutions écologiques au problème du robinet). (https://fastercapital.com/fr/contenu/Arretez-les-fuites---des-solutions-ecologiques-au-probleme-du-robinet.html)

Integrating Indigenous Knowledge for Sustainable Water Management in Buildings: Towards Liveable Communities

Sayam Singh

Abstract Globally, the issue of water scarcity has received widespread attention. Water shortage is felt most severely in arid and semi-arid regions, which cover a significant portion of the Earth. Historically, these desert regions have developed creative ways of managing water sustainably. This chapter proposes that architects and urban planners can adapt indigenous ways of water management while designing buildings. Instead of arguing for wholesale adoption of these practices, it is suggested that adopting a transdisciplinary approach would augment benefits that both offer and widen the scope of implementation. Despite the challenges inherent in synergising these two forms of knowledge, it contends that fostering of trust and genuine engagement can overcome these hurdles. The case of Egypt demonstrates the potential of indigenous knowledge in sustainable architecture, on the one hand, and argues against its wholesale adoption in construction of modern architecture. Further, the chapter explores the potential of integrating indigenous knowledge with modern technology by bringing to the fore the example of Seville, Spain where a novel strategy of implementing qanats in a modern way has been applied. Consequently, the innovative application of this ancient water system lowered ambient temperatures, reduced dependence on energy, and provided adequate water supply, thereby promoting sustainability and liveability. Through these case studies, this chapter contends that bridging the gap between indigenous knowledge and modern technology would produce sustainable water management practices that preserve culture and enhance the quality of life of residents, in even the most arid and semi-arid regions.

Keywords Indigenous knowledge · Qanats · Transdisciplinary approach · Water management

S. Singh (✉)
AGS Buildcon Pvt. Ltd., Delhi, India
e-mail: sayam.singh@agsbuildcon.com

Z. ElZein and A. Negm (eds.), *Recent Approaches of Sustainable Architecture in Arid and Semi-arid Cities*, Urban Sustainability,
https://doi.org/10.1007/978-981-95-1645-2_10

1 Introduction

Water scarcity is an increasingly acute crisis worldwide. The cause for water scarcity, that is, the condition wherein the demand for water exceeds its availability, is attributed to urbanisation and climate change [1]. It is estimated that people living in urban areas around the globe who are currently facing water shortages would rise to 2.373 billion by the year 2050, as compared to 933 million in 2016 [2]. This indicates that this is a serious issue that requires urgent attention.

Due to the lower rate of mean precipitation, the problem of water scarcity is severely aggravated in semi-arid regions, affecting not only the ecosystem, but also food production and people's health [3, 4]. Thus, water scarcity affects food security and public health. Out of the various regions in the world, the Middle East and North Africa (MENA), owing to their large population, rapid urbanisation and desert topography constitute the region with the most water scarcity in the world [5]. Therefore, it is this region that requires immediate solutions.

Water shortage in the region is further propelled by rapid urbanisation. A study of five cities within the MENA region, including Doha, Cairo, Casablanca, Dubai and Riyadh revealed that urban development caused loss of natural resources within the city and its surrounding areas [6]. Within this region, urbanisation was not accompanied by an increase in the quality of life as it led to a severe increase in carbon emissions and deterioration of social activity [7]. Buildings due to materials used in their construction and high demand for electricity and heat are considered to be a leading contributor to increase in carbon emission within urban areas [8]. Thus, urbanisation is intimately linked to pollution and water scarcity.

Owing to its geopolitical position, rapidly depleting groundwater, and large water consumption to meet the needs of its rapidly growing population, the Middle East is at a critical juncture [9]. It is predicted that Lake Urmia in Iran will soon dry up, as will other water bodies in the Middle East as climate change causes a rise in temperature and persistent droughts, along with overuse of groundwater through dams to meet food requirements, would cause water scarcity to such an extent that areas may become inhabitable and lead to conflict within the region [10]. Poorer countries in the Middle East may not withstand the crisis as they lack the technology and infrastructure to adapt to the changing conditions, leading to increased chances of conflict as inequalities may magnify. Economic inequalities within this region further compound the issue. Countries in the richer part of the Middle East may rely on technology like desalination to navigate the crisis but the fiscal strain from global transition to green energy makes them unsuitable for long-term use [11].

There is a need to develop sustainable strategies for water conservation within the region that promote equity.

1.1 Water Shortage in Buildings

Due to their high consumption, globally buildings are considered as the leading contributor to depletion of water resources [12].

Socio-economic inequalities such as education, income and household size as well as building characteristics, such as, water fixtures and age of the building influence water consumption patterns in households [13].

Rapid development, increase in population and vulnerability to climate change has worsened the crisis of water shortage, and is likely to continue doing so in the future, threatening the prospect of sustainable development in these regions [4, 14].

Therefore, there is a need to develop alternate solutions that can address the issue of water scarcity, while simultaneously mitigating the impact of climate change and ensuring continued development and economic progress. For this purpose, it is suggested the integration of traditional knowledge for water management in buildings in arid and semi-arid regions. This chapter contends that these solutions can also enhance liveability for urban residents.

Residential buildings, in Saudi Arabia, on average, consume more than 50% of the total consumed energy, of which, a substantial share is utilised in the air-conditioning of buildings [15], This problem is not limited to Saudi Arabia alone. Cooling of buildings during summer months is a major environmental problem in many other Middle Eastern countries as well, especially since the electricity in this region is highly dependent on fossil fuels [15].

In addition to emitting high quantities of carbon emissions and causing severe air pollution, use of fossil fuels also leads to dwindling of water supply and increased water pollution in dry regions that are prone to droughts [16].

Thus, increased dependence on fossil fuels for meeting energy requirements so as to keep buildings cool further intensifies the problem of water scarcity. Reliance on sophisticated technology and abundant fuel for maximising human comfort has led to architects ignoring the influence of climate on architecture, which was practised in the past, though recently the strategy of designing buildings in accordance with climate of a region, is being re-adopted, leading to sustainable architecture [17].

Urban areas in this region are prone to urban heat island effect where heat from the sun and vehicular emissions get trapped by heat-absorbing infrastructure and materials, like asphalt used to construct roads [18]. This means that buildings in urban areas are more vulnerable to heat. Compared to suburban and rural areas, cities have a higher positive thermal balance as the heat emanating from anthropogenic activities, lack of green cover and improper ventilation, get stored in buildings and infrastructure which aren't designed to withstand heat, leading to further rise in temperature and increased usage of energy [19]. Thus, buildings in the city aren't designed to absorb heat, leading to intensification of the problem, and underlining the need to develop an architecture that can bring down temperature levels, and consequently reduce energy consumption, thereby promoting sustainability. Later sections in the chapter will be enumerating upon the traditional strategies that can be integrated in modern buildings so as to act as thermal insulation leading to a decrease in dependence on fuels for energy consumption.

Massive infrastructural investments and other strategies to mitigate water scarcity, such as increasing desalination, are neither fiscally feasible nor environmentally sustainable, paving the need for reforms for improved management of the allocation of water across competing requirements [20]. Thus, technological approach by itself is not a long-term solution for sustainable water management. Therefore, this chapter proposes the adoption of traditional water conservation strategies in a manner that can address current needs of people in urban areas.

1.2 Water Scarcity and Liveability

As a concept, liveability can be understood as the quality of life, encompassing various mutually intersecting dimensions including environmental and cultural preservation, social cohesion, built environment and community participation [21].

According to the Global Liveability Index of 2024, Abu Dhabi ranks the highest amongst cities in the MENA region, however globally it was placed only on the 76th position [22].

In cities, where infrastructure influences lifestyles, water management plays an important role in liveability as supply, drainage, sewage system and potability affect people's daily life [23]. Therefore, water in urban areas is an intrinsic component of liveability.

Liveability and availability of water in urban areas is intertwined. Urban development without accounting for this relationship as insufficient water will hamper the liveability of the people [23]. This means that in case of water scarcity, liveability would not be achieved, and therefore one cannot be improved without another.

This study contributes to the emerging body of literature calling for the integration of indigenous strategies of water conservation in the mainstream by looking at how they can be utilised in the water management of buildings within arid and semi-arid regions. While considerable research has been done on the merits of traditional water management practices, a gap exists in exploring how can these be integrated with modern technologies in urban areas to enhance sustainability and liveability. Employing a transdisciplinary approach, this study highlights the usefulness of integrating indigenous knowledge with technology to overcome the limitations of both and provide cost-effective, energy-efficient, equitable and environmentally friendly strategies of water management in buildings in arid and semi-arid regions. It aims to contribute to the development of a water management model that can be replicated in other areas within the region. Furthermore, it hopes to broaden the understanding of the relationship between urbanisation, sustainability and liveability.

This chapter will first present the merits of using indigenous knowledge for sustainable water management. It suggests adopting a transdisciplinary approach, wherein both indigenous and modern forms of knowledge can find common ground and be utilised to enhance sustainability and liveability within arid and semi-arid regions. It will showcase the advantage of construction in the Siwa Oasis, Egypt with local material (Kashrif particles) and reveal its gaps, highlighting the need to incorporate modern technique to create resilient and eco-friendly enclosures. This would also showcase that wholesale adoption of indigenous knowledge may not be completely suitable for modern climate and lifestyles, and a transdisciplinary approach is required to actualise the potential of both. It would be followed by showcasing the case study of Qanats, shedding light on their advantages, disadvantages and scope for modern buildings, such as their use in Spain to cool buildings and thereby reduce energy consumption. In addition to it being a practical application of adapting indigenous knowledge for water conservation in urban buildings, through the case of Qanats, it is important to highlight the potential of integrating indigenous and modern knowledge systems to develop a sustainable method of water management in buildings that is environmentally viable and promotes liveability. The purpose is to present the usefulness of adapting indigenous methods as a mitigation measure of managing water resources for contemporary urban needs.

For this purpose, this chapter presents an extensive literature review, relying on qualitative data sourced from a variety of secondary data sources including peer-reviewed journals, books, reports published by international organisations, and reputed websites. More than half of the data collected and presented is taken from sources published between the years 2020–2024, so as to ensure that the study is in congruence with current research.

By presenting sustainable water management in buildings in arid and semi-arid regions, this study aims to facilitate mutual co-operation and dialogue between indigenous knowledge embedded in traditional strategies of conservation and Western science rooted in technological methods. It contends that adopting a transdisciplinary approach would bridge the gap between the two forms of knowledge, and lead to development of solutions that promote sustainability and liveability, thereby preserving the environment, culture and enhancing the quality of life for people.

2 Indigenous Knowledge

Knowledge held by a group of people which is specific to a particular space and/or culture, is holistic and adaptive is known as Indigenous Knowledge [24]. This knowledge is also referred to as traditional knowledge, though traditional here should be understood as a dynamic concept, as this knowledge includes constant adjustment to local situations [25, 26]. In this chapter, the two terms though used interchangeably, do not undermine the vast repository of knowledge held by members of indigenous communities.

2.1 Traditional Water Management Systems

Historically, water has always been at the centre of human activity, addressing relevant needs of people and fostering social communities, unlike modern technological interventions that disregarded this balance [27]. Thus, water was a cornerstone for indigenous communities as various aspects of their life revolved around it.

Historically, local communities had developed sophisticated water management systems. The earliest evidence of hydraulic knowledge emerges from the Indian subcontinent's remains from the Indus Valley Civilisation, and the first true hydraulic civilisation with knowledge of measurement of rainfall, water balance, and had dams with spillways, reservoirs and channels containing spillways is considered to be the Mauryan Empire [28]. Thus, since ancient times, local communities had developed sophisticated ways of water management.

Unlike Western science that perceives water as a commodity, indigenous knowledge views it as a precious resource that needs to be valued [29].

Due to their dependence on the Nile for their sustenance, ancient Egyptians worshipped the river, as they believed that it would lead to bestowal of favours upon them [30]. Pater et al. [31] suggests that spiritual connection with natural resources are essential components in people's motivation to manage them sustainably. Thus, reverence towards water enabled them to treat it as a valuable resource.

Even today, water is an intrinsic part of socio-economic development of a country as it sustains ecosystems, industrial growth, and boosts food productivity [32]. Thus, water continues to remain at the cornerstone for development.

Though globalisation fostered economic growth worldwide, it also led to furthering of inequalities for indigenous communities, both amongst themselves and with other non-indigenous populations [33]. Additionally, modern systems of water management, involving the use of pumps and deep-pumped wells have led to the establishment of a technocratic culture that overshadowed indigenous knowledge and hindered development at the local level [34]. Thus, indigenous knowledge was pushed to the background.

Since time immemorial, local communities have developed intricate networks of water management systems that were efficient, with distinctive design that fostered socio-economic growth and positively impacted culture in the region in which they were housed [35]. Thus, traditional knowledge systems are instrumental in providing sustenance, economic prosperity and cultural development of a society. Even today several people across the world are reliant on their traditional sources of water [32, 36]. They are not obsolete forms of knowledge as they continue to be used by people across the globe.

Recently, researchers have called for the integration of indigenous knowledge in sustainable development in various fields from mitigating the impact of climate change to architecture to conservation and even to improving liveability [37, 38]. At present, the immense wealth of knowledge held by local people is gaining recognition worldwide and the advantages of the strategies used by these indigenous communities to manage and conserve water are being brought to light in different parts of the world [27, 34, 35].

There is a resurgence of the value of knowledge possessed by indigenous communities. For instance, in the Atacama Desert, characterised by poor soil quality and extremely arid climate, traditional strategy of water conservation, involving the use of collecting and harvesting fog was studied to analyse the usefulness of hydroponics to assist food production in greenhouse crops [39].

Additionally, stepwells in Rajasthan are venerated for their environmental, functional and aesthetic appeal, and it was suggested that modern engineering take a leaf from their notebook to design equitable and sustainable water infrastructure [40]. Thus, the built environment around the water network was developed in a way that promoted holistic growth encompassing sustainability and liveability. However, digging of deep bore wells to produce cash crops that require large quantities of water has led to over exploitation of water resources and rapid depletion of groundwater [41]. This indicates that modern methods aren't sustainable for long-term as they cannot adequately maintain the level of groundwater.

Urban planning used to be designed around meeting the needs of the people, and understood their dependence upon water. Most of the houses in Sanaa's city, Yemen had a water well, and those in villages had a basin to collect and store rainwater, as to help in agriculture, and the sewer system protected the water from pollution [42]. Therefore, traditional architecture in cities and villages was conducive for water conservation.

2.2 Indigenous Knowledge and Sustainable Architecture

Globalisation has led to homogenisation of architecture separating modern from traditional knowledge systems in architecture [43]. Elites in developing societies following Western design methods, while disregarding the value of local building materials, construction techniques and even climatic conditions [37]. This led to a divorce of traditional strategies from modern methods.

Using local materials makes the process of construction cost-effective and quicker [37]. In Riyadh, Saudi Arabia it was found that buildings made with traditional design principles and local material were easier to maintain as clay is readily available and people possess knowledge on reaping maximum benefit from it [44].

A study in Yemen showcased that traditional architecture built according to the topography of the area with local materials, and following design principles rooted in culture, led to spaces that provided natural lighting and ventilation while protecting privacy, and also ensured decrease in energy consumption due to the construction system of walls protecting the building from responding heavily to hot and cold climate [42]. Thus, in addition to preserving the environment, traditional architecture enhanced privacy of the residents. A study in low-income households of Dutse, Jigawa state and Nigeria found a positive correlation between privacy and life satisfaction of residents, determining privacy as an important component of liveability

[45]. Therefore, buildings constructed with traditional design strategies can effectively improve sustainability and liveability, through its provision of cheaper, efficient and environmentally friendly alternatives.

2.3 Traditional and Modern Knowledge

Unlike modern knowledge which is developed in laboratories, traditional knowledge is empirically validated through its application by a particular community in a specific environment and is developed through constant experience [26]. Thus, indigenous knowledge continually evolves through constant engagement with the environment and evolves through practical application in the specific setting for which it is developed.

There are certain ideas and guiding principles that distinguish traditional knowledge from modern knowledge. The purpose of modern knowledge is to maximise profit while that of traditional knowledge is to minimise chances of risk [26]. Due to this, strategies formulated by modern knowledge may not always lead to holistic and sustainable solutions. For instance, Singh [38] sheds light on the negative effects of dams and how mobilisation by displaced local communities was effective against long-term harm to the environment. This does not mean that indigenous knowledge hinders economic progress. Ebhuoma [46] has documented the positive impact on economic progress by involving indigenous communities in preserving natural infrastructure and cultural heritage. Thus, in addition to sustainable development, knowledge possessed by indigenous communities has the potential to drive economic growth.

However, in order to conserve resources adequately, particularly water conservation, indigenous knowledge must be balanced with modern systems of knowledge to cultivate maximum potential by using expertise of both, as modern knowledge with its attention to specifics such as conserving an endangered species over others as compared to indigenous knowledge, which is broader, would be more suitable in certain circumstances [47]. While indigenous knowledge is embedded within a local and culturally specific network, Western science has a broader scope for practical application [48]. Thus, integrating the two would allow each to transcend its limitations, and produce adaptable, holistic solutions that can be applied in a broader setting (Table 1).

2.4 Egypt: Underlining the Scope for Integration

This section enumerates upon the need for integration of Indigenous Knowledge with modern methods.

Table 1 Comparison of traditional and modern approaches of water management (by Sayam Singh, Author)

Feature	Traditional/Indigenous Knowledge	Modern Approach
Perception	Precious resource	Commodity
Knowledge	Holistic Intricate hydraulic network suitable for local needs	Focus solely on economic growth, Technological can be applied in diverse settings
Management	Equal participation; community led Limited scope of application	Inequitable; Top-down approach Broader scope of application
Examples	Stepwells in Rjasthan, qanats in Middle-East, fog harvesting in Atacama desert	Large dams, deep borewells, desalination plants
Environmental Impact	Generally low impact, integrated with local systems Long-term sustainability	High level of pollution and groundwater depletion Unsustainable for long period of time
Social Impact	Fostered community cohesion	Furthered inequalities
Current Status	Gaining recognition for its holistic potential	Dominant paradigm where scope is being widened to address issues of inequality and short-term sustainability

Due to its location and increasing population, Egypt is facing a rising demand for an increase in per capita water share. However, it is continuously declining, making it a region with one of the lowest per capita water shares in the world [49].

Climate change, increasing population and a need to lower dependency on water from the Nile River due to dams being constructed in Ethiopia [50] has led researchers to suggest alternate methods for energy generation in Egypt. Ghoneim et al. 51 suggests that due to high temperature even underground, presence of hot springs and favourable geothermal gradients, geothermal energy should be considered to generate energy in the basins of the Siwa Oasis, as it'll utilise the high temperature and Thus, researchers are exploring sustainable solutions that effectively harness the region's existing climatic conditions. This focus on providing sustainable solutions that make use of the region's natural resources and environmental conditions has been applied to the field of architecture as well.

Research from Egypt contends that when compared to modern buildings, buildings constructed with locally available materials extracted from the surrounding environment, such as using Karshif particles for walls, wood from palm trees for opening of windows, special kind of green silt found only in Siwa for plaster of facade and traditional style of architecture, were more economical and eco-friendly [52]. Locally available material as it is already available nearby would save on transport costs and be more economical.

Karshif particles are used to make bricks which are then stacked in a way that optimal temperature of the building is maintained during varying weather conditions, as the material prevents carbon emission and acts as a thermal insulator, keeping inside of the building cool even during warmer months [53]. In addition to protection from solar radiation, cross ventilation was also ensured with the strategic positioning of windows [54]. Thus, traditional strategies promoted cross-ventilation, protected the buildings from solar radiation and maintained optimal temperature inside the buildings, particularly during summer months. Maintenance of temperature reduces the dependence on energy for keeping cool. Thermal insulation reduces energy consumption, thereby conserving water [55].

However, though these traditional buildings styles were socio-economically viable and promoted sustainability, site investigations revealed that buildings constructed with just vernacular techniques of construction were fragile and prone to decay [52].

This chapter proposes against the wholesale adoption of traditional techniques, rather we need to integrate local knowledge with modern methods of construction in a way that enhances the longevity of buildings. The literature presented above showcases the potential of integrating indigenous knowledge in architecture, particularly buildings.

This is supported by researchers who've called for harmonising development with cultural heritage and called for construction of resilient buildings that were economical, eco-friendly and promoted socio-cultural heritage [56].

The case of construction in Siwa Oasis reveals the gaps that exist between traditional and modern techniques of construction, undergirding the need for an integrated approach that provides holistic benefits to the people.

3 Transdisciplinary Approach

In light of resolving global water challenges and water scarcity, a transdisciplinary approach allows for conflicting frameworks of knowledge to hold a space and understand each other, collectively working towards sustainable development [57].

Transdisciplinary approach bridges the gap between knowledge and collective action through synthesis of non-scientific knowledge, complex everyday problems and a systematic perspective, leading to solutions that are acceptable to different stakeholders [58]. Incorporation of their knowledge, values and partnership with indigenous communities is vital for the realisation of the potential of transdisciplinary approach [59].

However, transdisciplinary approach is not consistent in delivering desired results [60]. This lack of consistency in achieving desired results would discourage policymakers and urban architects to rely on it.

However, this study contends that the potential of this approach transcends its limitations, and constant research can refine this approach and improve its scope in applied settings.

Inefficiency of transdisciplinary approach can be overcome by treating it as a way of being, rather than a form of knowledge [60]. This means that instead of viewing the transdisciplinary approach as a method, it should be seen as the fundamental form of worldview, thus calling for a radical change in perspective.

Adoption of a transdisciplinary approach to provide nature-based solutions within the realm of architecture revealed its great potential in sustainable development, and suggested that greater collaboration was an important tool for its further improvement [61].

Research in Anchorage, Alaska revealed that challenges, such as power imbalance and lack of trust hinder long-term collaboration with indigenous knowledge holders [62]. The example of Kubu Raya District in West Kalimantan, revealed that mutual dialogue and a genuine engagement with local communities fostered trust and led to their effective participation in local sustainability initiatives [63].

Transdisciplinary approach is successful when multiple stakeholders participate together in decision-making with a collaborative intent to produce a new set of insights and solutions, that involve mutual exchange of information [64].

Through the case of qanats, the potential for a collaborative approach between indigenous and modern form of knowledge will be explored. It will shed light on how the two can inform one another, leading to greater sustainability and liveability in buildings in arid and semi-arid regions (Fig. 1).

Fig. 1 Depiction of advantages of transdisciplinary approach for sustainable development. (By Sayam Singh, Author)

4 Qanats: An Ancient System Designed to Meet Modern Needs

Due to its ancient use in irrigation, Qanats have primarily been studied for their usefulness in agriculture [65]. This sub-section analyses the potential of qanats in urban buildings in arid and semi-arid regions. I'll be shedding light on the usefulness of this system, its limitations and how integration of this traditional system of water management with modern technology can overcome a few of these limitations.

Within desert regions, the Qanat is a multi-layered system of water distribution that can be seen as a type of horizontal well that extracts water from aquifers, and through gravity, conveys it from hill to surface [66]. Thus, there is no need for electricity as gravity is being utilised to transport water.

These water channels are found in different parts of the globe, be it China, Africa or the Middle East and are known by various names: karez in Afghanistan, gallery in Spain and so on [65]. Presently, in Iran, qanats are concentrated in villages and towns, rather than urban areas [67]. However, in other countries, such as Afghanistan, a large section of the population continues to depend on karez as a source of water [36].

4.1 *Construction of Qanats*

See Fig. 2.

Generally, a qanat consists of these main parts- an underground tunnel also known as a gallery, responsible for collecting and transporting water; the mother well, which is the primary and the deepest well; the shaft wells; the outlet; and the surface channels responsible for distribution of water [69].

Construction of Qanats is not a simplistic process as it requires great knowledge about the area and engineering skills [70]. Since qanats are constructed with some

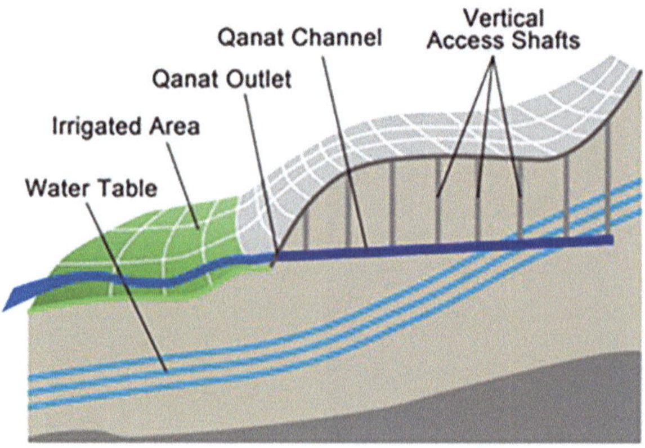

Fig. 2 Depicting the different layers of a Qanat [68]

extensive criteria of groundwater such as topography, kind of aquifer, depth, length and so on, it is environmentally sustainable as it ensures that groundwater is not depleted [71]. It can manage the level of groundwater.

A site is carefully chosen based on its proximity to a source of water that is underground [70].

Then, a diverse team of labourers came together to use their expertise to construct the tunnel [66]. Vertical shafts lined with water-resistant material like clay and cement were constructed to connect the horizontal channels, followed by an irrigation system on the surface level that connects to the channel below [70].

4.2 Advantages of Qanats

Since, Qanats were primarily constructed in regions with limited water availability, these are useful for mitigating challenges pertaining to water supply in arid and semi-arid regions [71]. Therefore, qanats have been studied to assess their usefulness, when integrated in urban areas within these regions.

In Iran, where it is believed to have originated, qanats were an important part of culture as they influenced territory organisation and social life [67]. They acted as social mechanisms to mitigate class disparities and improved social bonds and cohesion [72]. Thus, it has the potential to increase cooperation amomgst the public.

Within Iran, even in areas, where it wasn't in use anymore, qanats produced a culture of cooperation amongst the local population [73].

Altrock [74] contends that increasing social integration within urban areas, particularly in poorer neighbourhoods increases liveability of its residents. Thus, if used in urban areas, qanats have the potential to increase the quality of life of people living in these regions.

The advantages of qanats transcend beyond socio-cultural domain, offering scope for ecological and economic growth as well.

From hemp rope makers, oilers, stone cutters, to carpenters and so on, the process of building qanats requires gathering a diverse group of labour, each with its own area of specialisation [66]. Thus, it generates employment for a large and varied section of skilled labour.

Furthermore, its lack of dependence on electricity or fossil fuels for its smooth operation ensures that there is minimal cost in its maintenance, which is calculated at being less than 1% of its construction cost [66].

Qanats rely on local materials, cause no pollution, prevent desertification as it keeps the soil moisture balanced in the mountains and plains, only cause minute discharge of underground water, and balances water level even during drought [66]. Therefore, it's an ecologically sustainable and resilient system of water transportation.

It was postulated that the frequency of flooding is likely to increase in arid and semi-arid regions and countries, such as Saudi Arabia and Riyadh, who had faced it in the past were unaware of mitigation measures [75]. Qanats prevent flooding as their vertical shafts absorb the excess surface water [76]. Conversely, it also maintains an adequate level of water in times of drought [77]. Therefore, they are resilient to natural disasters on both ends of the hydraulic spectrum.

Experimental investigation to gauge the performance of a qanat source heat pump, revealed that it could maintain stable temperature under changing environmental conditions [78]. Studies have documented that a heat pump system using qanat water for heating and cooling performs better than an air heat pump system, and this system's efficiency and ability to conserve energy, can be further improved through interventions such as, increasing the working fluid's flow rate [79]. Hence, this ancient water system is compatible with modern technology, acting as an energy-efficient source for heating and cooling of buildings.

4.3 Limitations of Qanats and Scope for Integration

Thus, this indigenous system of water management offers multivalent benefits in terms of liveability and sustainability.

Despite the various benefits it offers, this water management system has its limitations.

Firstly, they can only be built in areas with suitable geographic formulations that have an adequate level of groundwater [80]. Scouring a suitable site may become tedious and time-consuming.

Using statistical analysis and geographical information system (GIS) method, researchers have simplified the process of identifying the best location for construction of a qanat [69].

Anthropogenic activities render groundwater vulnerable to pollution and increase the quantity of heavy metals and trace elements, which on racing an excessive level can become toxic for humans [81].

Contamination of water poses a major risk to potability, and hence would be a cause of concern for using Qanats in residential buildings, as it can transport diseases [76]. The water quality of qanat needs to be maintained.

This can be overcome by using modern technologies, such as using geotextile pipes and geomembrane materials and covering the inlet of shafts that are vertical with permeable materials to protect it from erosion, facilitate maintenance and filter sediments to reduce contaminants [82]. This would ensure rudimentary cleaning of water before its use. It can be subsisted further through tanks within households. Another strategy to maintain quality of water within residential buildings could be the construction of a reservoir tank, separate from the main channel that can be used specifically for domestic use [76].

Secondly, as it requires a slope it can only be constructed in a mountain or valley [80]. Hence, it can only be constructed in certain areas and may not be suitable for all kinds of buildings.

Finally, over exploitation of the water resource can cause qanats to dry up [80]. Increasing urbanisation of desert regions leads to consumption of large quantities of water, even in domestic supply [83]. Therefore, in order for its incorporation in buildings to be beneficial, qanats would have to meet the requirements of people. This can be achieved through technological means. Incorporation of artificial tools to recharge the water table, such as, an underground dam being used in Jandaq Qanat in Iran, can be used to support its replenishment [76].

4.4 Integration with Modern Technology

The scope of using qanats is not limited to the Middle East alone. Researchers around the globe are exploring the value of qanats and discussing their usefulness within urban areas.

In Spain, researchers looking for an energy-saving alternative to air conditioners, are using qanats to keep buildings cool; using its natural nocturnal cooling to transport water through strategically placed vertical vents in an effort to keep buildings cool as temperatures soar [84]. This is aided by modern technological intervention. A thin film of water flows over the solar panels, which is, created by pumping water to the roof of the amphitheatre in the building, which is then circulated through fans to cool the lower levels within the building, while nozzles placed outdoors use evaporative misting in a bid to expedite the cooling process [84]. Thus, the cooling process is enhanced and quicker. This technology has the potential to naturally mitigate urban heat. The Cartuja Project in Seville, Spain used qanats and integrated it with water tanks, canals and air-cooling tubes, to lower ambient temperature to 27 degree Celsius, and further maximised water cooling through nocturnal pulverisation and utilised solar power to broaden its potential application to other parts of Andalusia [85].

Therefore, through technological interventions, this sustainable method of water transportation can be harnessed to suit modern lifestyles and needs.

Thus, the integration of traditional knowledge within urban architecture, by relying on a combined approach that picks the best of both, has immense potential for sustainable development and liveability within arid and semi-arid regions. In the future, different types of traditional water management systems, such as Johads or Khadins of Rajasthan, owing to their ability to recharge groundwater [32] can be used with qanats and modern technology to develop a sophisticated system of water management in buildings.

5 Limitations of the Study

Due to limited availability of research on this topic, the data gathered and presented on effective harnessing of qanats for sky-rise buildings within the region was insufficient. Furthermore, since these water management systems require large areas, the construction and maintenance of traditional water management systems may not be feasible for urbanised areas, though it may prove to be effective for areas that are in the process of being urbanised. This study does not delve adequately into how this approach can be translated into wastewater management and drainage.

6 Conclusion

This study highlighted the importance of incorporating indigenous knowledge into contemporary water management practices in buildings in arid and semi-arid regions. It suggested that in order to bridge the gap between the two a transdisciplinary approach must be adopted. This approach sought to synthesise indigenous knowledge, rooted in local understanding with modern technological advancements to develop resilient and sustainable solutions for water management in buildings that are not only able to meet contemporary needs, such as by increasing energy efficiency, but also enhance quality of life and well-being of residents. Though integrating the two knowledge systems is fraught with challenges, mutual dialogue, genuine engagement and fostering of trust can overcome these hurdles.

The chapter shed light on the drawbacks of modern technological solutions, and underlined the gaps that need to be filled. The case of Egypt exemplified that despite the various advantages provided by indigenous techniques, for their full potential to be realised, they need to be subsisted by modern forms of knowledge.

Qanats, ancient subterranean channels showcase the sophisticated form of knowledge and technique that indigenous communities possessed. Its reliance on gravity instead of energy, minimal cost of maintenance and facilitating socio-cultural co-operation highlight its ability to provide holistic solutions for water management in a way that promotes liveability. Furthermore, the chapter suggests gaps within this system that can be fulfilled by modern knowledge, and demonstrates, through the example of buildings in Spain, the practical application of a transdisciplinary approach, wherein researchers leveraged a traditional system with technological innovation to sustainably manage water to harness energy and keep buildings cool. Thus, combining strategies from both forms of knowledge led to optimal benefits for the building and people. Researchers from Spain have paved the way for other countries to follow suit and create sustainable strategies of water management that prioritises people's needs. Future research can focus on developing creative ways to scale these solutions and implement them in high-rise buildings within the area. Successful integration in different settings might inspire other rapidly urbanising areas to implement holistic growth initiatives that understand that humans and the environment are intimately linked, and therefore ecological and anthropocentric development must go hand in hand. By bridging the gap between traditional and modern forms of knowledge, we can develop sustainable water management solutions that can address contemporary needs while simultaneously preserving cultural heritage and enhance the scope of liveability in urban areas. A sustainable future is inclusive not just of environmental resilience, but also of, social equity, health and economic viability. This holds true not only for arid and semi-arid regions, but different parts across the globe.

7 Relevant SDGs

Sustainable Development Goals (SDGs) are a set of 17 inter-related goals launched by the United Nations to promote social justice and environmental conservation, ensuring equitable growth for various countries. This study supports the achievement of these goals by addressing the critical issue of water scarcity and energy efficiency.

It is estimated that due to increasing desertification, by 2050, 1 in 4 people are likely to experience recurrent water shortage. Since it addresses the critical issue of water scarcity in arid and semi-arid regions, this study is linked to SDG 6, that is, clean water and sanitation. SDG6 calls for improved water quality and efficiency of usage of water, in addition to increased participation of local communities in water management [86]. Through its focus on utilising community knowledge of water management and fostering collaboration between indigenous and modern knowledge using a transdisciplinary approach, within arid and semi-arid regions, this study supports the components of SDG6. Future research can broaden the scope of this study by exploring ways to adapt not only qanats but other traditional systems of water conservation in a manner that it remains potable and accessible.

Integration of traditional water management strategies, such as qanats in modern architecture such as urban buildings aligns with SDG7 that calls for access to reliable, affordable, sustainable and modern energy [87]. The chapter highlighted the capacity for traditional water management systems in providing sustainable and cheaper alternatives for providing energy. The example of Seville, Spain which demonstrates how qanats can be adapted to reduce energy consumption in keeping buildings cool, effectively showcases how adaptation of traditional water management systems reduces reliance on non-renewable energy resources.

The study relates to SDG11, Sustainable Cities and Communities, that advocates for making human settlements particularly cities safe, sustainable, inclusive and resilient [88]. By shedding light on the advantages and gaps of aligning traditional strategies in urban design with local environmental conditions, and advocating for an integrated approach that combines indigenous knowledge with modern approach in architectural design, this research foaters collaboration between different stakeholders and underscores the importance of creating an urban design that promotes sustainability through innovative solutions that are inclusive and resilient.

Exploring the technological and socio-economic dimensions of scaling indigenous solutions to suit urban environment, lifestyle, and infrastructure is crucial for enhancing sustainability and liveability within urban areas and in meeting the different SDGs outlined by the UN.

8 Recommendation

By integrating traditional means of water conservation such as Qanats with modern strategies and technology, policymakers, urban planners and architects can devise sustainable water management practices that are efficient and tailored to address the specific challenges in the local region. Further, this can be facilitated through policy-driven initiatives, such as tax benefits for sustainable design of buildings and regulations to preserve the integrity of traditional systems.

Adopting a transdisciplinary approach would provide a framework for co-ordination between different stakeholders including policymakers, planners, architects and local communities, easing the process of collaboration, and provide effective and integrated sustainable solutions that fulfil diverse needs. Additionally, this approach can widen the scope of designing and implementing sustainable solutions beyond the field of water management.

Urban planners and architects must conceive of designs that are energy efficient and resonates with local culture and communities. Researchers must delve deeply in identifying and highlighting the gaps that exist in integrated solutions in urban environments to fully realise the potential of these solutions in enhancing sustainability and liveability.

References

1. He C, Liu Z, Wu J et al (2021) Future global urban water scarcity and potential solutions. Nat Commun 12:4667. https://doi.org/10.1038/s41467-021-25026-3
2. Sahu P, Debsarma C (2023) Urban water scarcity: a global challenge and impending solutions. https://doi.org/10.1007/978-3-031-21618-3_7
3. Garcia-Franco N et al (2018) Climate-smart soil management in semiarid regions. Soil Management and Climate Change, pp 349–368. https://doi.org/10.1016/b978-0-12-812128-3.00023-9
4. Morante-Carballo F et al (2022) What do we know about water scarcity in semi-arid zones? A global analysis and research trends. Water 14(17):2685. https://doi.org/10.3390/w14172685
5. Roudi-Fahimi F, Creel L, De Souza RM (2002) Finding the balance: population and water scarcity in the Middle East, pp 1–8. https://u.demog.berkeley.edu/~jrw/Biblio/Eprints/PRB/files/FindingTheBalance_Eng.pdf
6. Jaad A, Abdelghany K (2021) The story of five mena cities: Urban growth prediction modeling using remote sensing and video analytics. Cities 118:103393. https://doi.org/10.1016/j.cities.2021.103393
7. Elgendy K, Abaza N (2020) Urbanization in the MENA region: A benefit or a curse?, fes_international_de. https://mena.fes.de/press/e/urbanization-in-the-mena-region-a-benefit-or-a-curse.html
8. United Nations Environment Programme (2022) 2022 global status report for buildings and construction: towards a zero-emission, efficient and resilient buildings and construction sector. UNEP, Nairobi
9. Odhiambo GO (2017) Water scarcity in the Arabian Peninsula and socio-economic implications. Appl Water Sci 7:2479–2492. https://doi.org/10.1007/s13201-016-0440-1

10. Pleitgen F et al (2021) The Middle East is running out of water. CNN. Available at: https://edition.cnn.com/2021/08/22/middleeast/middle-east-climate-water-shortage-iran-urmia-intl/index.html
11. Wehrey F (2023) Climate change and vulnerability in the Middle East—Carnegie Endowment for international peace, Carnegie Endowment for international peace. https://carnegieendowment.org/2023/07/06/climate-change-and-vulnerability-in-middle-east-pub-90089
12. Mannan M, Al-Ghamdi SG (2020) Environmental impact of water-use in buildings: latest developments from a life-cycle assessment perspective. J Environ Manage 261:110198. https://doi.org/10.1016/j.jenvman.2020.110198
13. Zeroual M, Hani A, Boustila A (2021) Assessing domestic factors determining water consumption in a semi-arid area (Sedrata City) using artificial neural networks and principal component analysis. J Water Land Dev 49(IV–VI):219–228. https://doi.org/10.24425/jwld.2021.137115
14. Karimi M, Tabiee M, Karami S, Karimi V, Karamidehkordi E (2024) Climate change and water scarcity impacts on sustainability in semi-arid areas: lessons from the South of Iran. Groundwater for Sustainable Dev 24:101075. https://doi.org/10.1016/j.gsd.2023.101075
15. Wahl, E (2017). Buildings in arid desert climate, Luleå tekniska universitet. https://ltu.diva-portal.org/smash/get/diva2:1084934/FULLTEXT01.pdf
16. Melillo G (2023) In drought-stricken states, fossil fuel production jeopardizes limited water supplies. The Hill. https://thehill.com/changing-america/sustainability/energy/3847883-in-drought-stricken-states-fossil-fuel-production-jeopardizes-limited-water-supplies/
17. Varzaneh EH, Amini M, Bemanian MR (2014) Impact of hot and arid climate on architecture (case study: Varzaneh Jame Mosque). Procedia Eng 94:25–32. https://doi.org/10.1016/j.proeng.2013.10.010
18. Mahmoud M (2024) The looming climate and water crisis in the Middle East and North Africa, Carnegie Endowment for International Peace. https://carnegieendowment.org/research/2024/04/the-looming-climate-and-water-crisis-in-the-middle-east-and-north-africa?lang=en
19. Santamouris M, Vasilakopoulou K (2021) Present and future energy consumption of buildings: challenges and opportunities towards decarbonisation', e-prime—advances in electrical engineering. Electron Energy 1:100002. https://doi.org/10.1016/j.prime.2021.100002
20. de Waal D, Khemani S, Barone A, Borgomeo E (2023) The economics of water scarcity in the Middle East and North Africa: institutional solutions. World Bank, Washington, DC. https://doi.org/10.1596/978-1-4648-1739-7
21. Tennakoon TMMP, Kulatunga U (2019) 'understanding liveability: related concepts and definitions', Proceedings of the 8th World Construction Symposium [Preprint]. https://doi.org/10.31705/wcs.2019.57
22. The Peninsula (2024) Doha fourth most liveable city in mena: EIU report, ZAWYA. https://www.zawya.com/en/world/middle-east/doha-fourth-most-liveable-city-in-mena-eiu-report-cbqcthi1
23. Sochacka BA, Kenway SJ, Renouf MA (2021) Liveability and its interpretation in urban water management: systematic literature review. Cities 113:103154. https://doi.org/10.1016/j.cities.2021.103154
24. Mistry J (2009) Indigenous knowledges. In: Kitchin R, Thrift N (eds) International encyclopedia of human geography. Elsevier, pp 371–376. https://doi.org/10.1016/B978-008044910-4.00101-2
25. Bruchac, MM (2014). Indigenous knowledge and traditional knowledge. In: Smith C (ed) Encyclopedia of global archaeology. Springer, New York, NY. https://doi.org/10.1007/978-1-4419-0465-2_10
26. Ghosh PK, Sahoo B (2011) Indigenous traditional knowledge. Orissa Review, January, pp 65–70. https://magazines.odisha.gov.in/Orissareview/2011/Jan/engpdf/66-71.pdf
27. Hein C, van Schaik H, Six D, Mager T, Kolen JCA, Ertsen M, Nijhuis S, Verschuure-Stuip G (2020) Introduction: connecting water and heritage for the future. In: Hein C (ed) Adaptive strategies for water heritage: past, present and future. Springer Open, pp 1–17
28. Singh PK, Dey P, Jain S, Mujumdar P (2020) Hydrology and water resources management in ancient India. Hydrol Earth Syst Sci 24:4691–4707. https://doi.org/10.5194/hess-24-4691-2020

29. Martinez-Cruz TE et al (2024) Water is more than a resource: Indigenous peoples and the right to water. Social Sci Humanities Open 10:100978. https://doi.org/10.1016/j.ssaho.2024.100978
30. Mays L (2014) Integrated urban water management: arid and semi-arid regions: UNESCO-ihp. CRC Press, Boca Raton, FL, Florida
31. Pater C et al (2023) Spiritual values in forest management plans in British Columbia and the Netherlands. Forest Policy Econ 151:102955. https://doi.org/10.1016/j.forpol.2023.102955
32. Kumawat S et al (2023) Study of traditional water resource management in Rajasthan. J ReAttach Therapy Dev Diversities [Preprint]. https://doi.org/10.53555/jrtdd.v6i1.2609
33. Toti F (2024) The effects of economic globalization on indigenous communities. Int J Humanity Soc Sci 2(3):1–14. https://doi.org/10.47941/ijhss.1782
34. Boullosa N (2023) Qanats: ancient water management system guaranteed survival, faircompanies. https://faircompanies.com/articles/qanats-ancient-water-management-system-guaranteed-survival/
35. Murthy RNS, Kasinath IV, Srikonda R (2022) Traditional water management systems of India. J Int Soc Study Vernacular Settlement 9(2):61–77. https://isvshome.com/pdf/ISVS_9-2/ISVS9.2.5Satyanarayana.pdf.
36. Ghai R (2021) Will Afghanistan's centuries-old 'karez' system of irrigation survive the Taliban, Down to earth. https://www.downtoearth.org.in/water/will-afghanistan-s-centuries-old-karez-system-of-irrigation-survive-the-taliban-78451
37. Akbar N, Abubakar IR, Bouregh AS (2020) Fostering urban sustainability through the ecological wisdom of traditional settlements. Sustainability 12(23):10033. https://doi.org/10.3390/su122310033
38. Singh S (2024) The sustainable tapestry: fusing mainstream and indigenous insights for healthy and liveable cities. In: Singh P, Yadav N (eds) The climate-health-sustainability nexus. Springer, Cham. https://doi.org/10.1007/978-3-031-56564-9_11
39. Albornoz F, del Río C, Carter V, Escobar R, Vásquez L (2023) Fog water collection for local greenhouse vegetable. Production in the Atacama desert. Sustainability 15(22):15720. https://doi.org/10.3390/su152215720
40. Selvaraj T, Devadas P, Perumal JL, Zabaniotou A, Ganesapillai MA (2022) Comprehensive review of the potential of stepwells as sustainable water management structures. Water 14(17):2665. https://doi.org/10.3390/w14172665
41. Khan S (2024) Rajasthan groundwater overexploitation report 2023: Jaipur News—Times of India, The Times of India. https://timesofindia.indiatimes.com/city/jaipur/rajasthan-groundwater-overexploitation-report-2023/articleshow/110130463.cms
42. Attia AS (2022) Learned lessons from traditional architecture in Yemen -towards sustainable architecture. Int J Sustain Dev Plan 17(4):1197–1204. https://doi.org/10.18280/ijsdp.170418
43. Bhaskaran AS, Kumarankutty AP, Kurukkanari C (2024) Traditional architectural knowledge systems and the Sustainable Development Goals. The Elgar companion to the built environment and the Sustainable Development Goals, pp 420–438. https://doi.org/10.4337/9781035300037.00035
44. Dhingra M, Chattopadhyay S (2016) Advancing smartness of traditional settlements-case analysis of Indian and Arab old cities. Int J Sustain Built Environ 5(2):549–563. https://doi.org/10.1016/j.ijsbe.2016.08.004
45. Bashari S et al (2021) The moderating effect of privacy in the relationships between residential livability and residents life satisfaction. J Constr Dev Countries 26(1):45–62. https://doi.org/10.21315/jcdc2021.26.1.3
46. Ebhuoma EE (2024) Indigenous knowledge and natural infrastructure resilience to climate change in developing countries: a bibliometric analysis. Front Environ Econ 3. https://doi.org/10.3389/frevc.2024.1295690
47. Borthakur A, Singh P (2020) Indigenous knowledge systems in sustainable water conservation and management. Water conservation and wastewater treatment in BRICS Nations, pp 321–328. https://doi.org/10.1016/b978-0-12-818339-7.00016-3
48. Bohensky EL, Maru Y (2011) Indigenous knowledge, science, and resilience: what have we learned from a decade of international literature on "integration"? Ecol Soc 16:4. https://doi.org/10.5751/es-04342-160406

49. Abd- Ellah RG (2020) Water resources in Egypt and their challenges: Lake Nasser case study. Egypt J Aquat Res 46:1–12. https://doi.org/10.1016/j.ejar.2020.03.001
50. Cohen Y (2021) Egypt has a water problem—and no, it's not only the GERD. Atlantic Council. https://www.atlanticcouncil.org/blogs/menasource/egypt-has-a-water-problem-and-no-its-not-only-the-gerd/
51. Ghoneim E, Healey C, Hemida M, Shebl A, Fahil A (2023) Integration of geophysical and geospatial techniques to evaluate geothermal energy at Siwa Oasis, Western Desert, Egypt. Remote Sens 15(21):5094. https://doi.org/10.3390/rs15215094
52. Mohamed A (2019) Comparative study of traditional and modern building techniques in Siwa Oasis, Egypt. Case studies in Constr Mater 12:e00311. https://doi.org/10.1016/j.cscm.2019.e00311
53. Altawansy A, Ezzat A (2024) Assessment of the environmental impact of building envelope in Egyptian oasis villages using Design Builder simulation program. J Al-Azhar Univ Eng Sect 19:111–129. https://doi.org/10.21608/auej.2023.226284.1389.
54. Mokhtar-Elalfy FN, Mahmoud A (2016) Harmonization between architectural development and heritage. In Siwa Oasis—Egypt. Arpn J Eng Appl Sci 11:3. http://www.arpnjournals.org/jeas/research_papers/rp_2016/jeas_0216_3595.pdf
55. Paraschiv S, Paraschiv LS, Serban A (2021) Increasing the energy efficiency of a building by thermal insulation to reduce the thermal load of the micro-combined cooling, heating and power system. Energy Rep 7(5):286–298. https://doi.org/10.1016/j.egyr.2021.07.122
56. Elalfy A, Farrag N (2016) Harmonization between architectural development and heritage in Siwa Oasis—Egypt. J Eng Appl Sci 11:2005–2015
57. Brennan M et al (2021) Conceptualising global water challenges: a transdisciplinary approach for understanding different discourses in sustainable development. J Environ Manage 298:113361. https://doi.org/10.1016/j.jenvman.2021.113361
58. Renn O (2021) Transdisciplinarity: synthesis towards a modular approach. Futures 130:102744. https://doi.org/10.1016/j.futures.2021.102744
59. Baker V et al (2023) Transdisciplinary science and the importance of indigenous knowledge. Integr Environ Assess Manag 20(3):805–816. https://doi.org/10.1002/ieam.4847
60. Rigolot C (2020) Transdisciplinarity as a discipline and a way of being: complementarities and creative tensions. Human Soc Sci Commun 7:100. https://doi.org/10.1057/s41599-020-00598-5
61. Butt AN, Dimitrijević B (2022) Multidisciplinary and transdisciplinary collaboration in nature-based design of sustainable architecture and urbanism. Sustainability 14(16):10339. https://doi.org/10.3390/su141610339
62. Lauter O (2023) Challenges in combining indigenous and scientific knowledge in the arctic. Polar Geogr 46(1):62–74. https://doi.org/10.1080/1088937X.2023.2233578
63. Moallemi EA et al (2020) Achieving the sustainable development goals requires transdisciplinary innovation at the local scale. One Earth 3(3):300–313. https://doi.org/10.1016/j.oneear.2020.08.006
64. Kassam KA (2021) Transdisciplinary research, indigenous knowledge, and wicked problems. Rangelands 43(4):133–141. https://doi.org/10.1016/j.rala.2021.04.002
65. Sultanbekova AK et al (2021) Karizes and current prospects for their use in Kazakhstan. Central Asian J Water Res, pp 181–198. https://doi.org/10.29258/cajwr/2021-r1.v7-2/181-198.eng.
66. Abedi S, Ansari M, Haghighatbin M et al (2023) Comprehensive classification and categorization of Qanat features: an interdisciplinary exploration using landscape infrastructure concept and semi-systematic review. Environ Syst Res 12–35. https://doi.org/10.1186/s40068-023-00318-3
67. Bensi SN (2020) The Qanat system: a reflection on the heritage of the extraction of hidden waters. In Hein C (ed) Adaptive strategies for water heritage. Springer, Cham. https://doi.org/10.1007/978-3-030-00268-8_3
68. Hejazi S (2023) Private property and the commons: the case study of water distribution in Persian Qanats: from abundance to drought: cultural patterns of water management and knowledge before private property went mainstream. Global Jurist 23(3):305–316. https://doi.org/10.1515/gj-2023-0125

69. Mansouri-Daneshvar MR, Ebrahimi M, Ahmadi FS (2023) Efficiency assessment of the environmental variables in the Qanat systems based on major hydrological basins in Iran. Appl Water Sci 13:99. https://doi.org/10.1007/s13201-023-01905-6
70. Izadipour S (2023) Iran underground qanats facts and features, Eavar. Available at: https://www.eavartravel.com/blog/2023/5/8/130631/qanat/
71. Mohajerani M et al (2024) Geospatial distribution of qanats in Middle Eastern countries: potential for sustainable groundwater system. J Arid Environ 222:105170. https://doi.org/10.1016/j.jaridenv.2024.105170
72. Khaneiki ML (2022) Hydro-social cohesion in Iranian local communities. Indigenous Water Drought Manage Chang World, pp 243–266. https://doi.org/10.1016/b978-0-12-824538-5.00013-3
73. Naghavi A, Shaeyan M (2024) Qanats. Centro Studi Luca d'Agliano Development Studies Working Paper No. 486, Available at SSRN: https://ssrn.com/abstract=4696527 or https://doi.org/10.2139/ssrn.4696527
74. Altrock U (2022) Urban livability in socially disadvantaged neighborhoods: the experience of the German program "socially integrative city." Front Archit Res 11(5):783–794. https://doi.org/10.1016/j.foar.2021.12.006
75. Nabinejad S, Schüttrumpf H (2023) Flood risk management in arid and semi-arid areas: a comprehensive review of challenges, needs, and opportunities. Water 15(17):3113. https://doi.org/10.3390/w15173113
76. Taghavi-Jeloudar M et al (2013) Review of ancient wisdom of Qanat, and suggestions for future water management. Environ Eng Res 18(2):57–63. https://doi.org/10.4491/eer.2013.18.2.057
77. Haeri MR (2003) Kariz(Qanat); an eternal friendly system for harvesting..., UNFCC. https://unfccc.int/sites/default/files/121103_iran.pdf
78. Delfani S, Esmaeelian J, Karami M (2021) Experimental investigation on the thermal performance of a qanat-source heat pump. Geothermics 89:101933. https://doi.org/10.1016/j.geothermics.2020.101933
79. Karami M, Abdshahi H (2023) Energy and exergy analysis of the transient performance of a qanat-source heat pump using TRNSYS-MATLAB co-simulator. Energy Environ 34(3):560–585. https://doi.org/10.1177/0958305X211068152
80. Mahed G (2024) An ancient system that could bring water to dry areas, Down to Earth. https://www.downtoearth.org.in/water/an-ancient-system-that-could-bring-water-to-dry-areas-93926.
81. Kazemi A et al (2022) Health risk assessment of total chromium in the qanat as historical drinking water supplying system. Sci Total Environ 807:150795. https://doi.org/10.1016/j.scitotenv.2021.150795
82. Nasiri F, Mafakheri MS (2015) Qanat water supply systems: a revisit of sustainability perspectives. Environ Syst Res 4:13. https://doi.org/10.1186/s40068-015-0039-9
83. Gober P (2010) Desert urbanization and the challenges of water sustainability. Current Opinion Environ Sustain 2(3):144–150. https://doi.org/10.1016/j.cosust.2010.06.006
84. Chase-Lubitz J (2024) Can ancient tunnels cool 21st-century heat in summer? Al Jazeera. https://www.aljazeera.com/features/2024/4/28/how-an-ancient-design-is-cooling-21st-century-streets
85. López-Pérez ME, Reyes-García ME, López-Sanz ME (2023) Smart mobility and smart climate: an illustrative case in Seville, Spain. Int J Environ Res Public Health 20(2):1404. https://doi.org/10.3390/ijerph20021404
86. UNDP (n.d.) Goal 6: Clean Water and Sanitation. United Nations Development Programme. https://www.undp.org/sustainable-development-goals/clean-water-and-sanitation
87. UNEP (n.d.) Goal 7: affordable and clean energy. United Nations Environment Programme. https://www.unep.org/topics/sustainable-development-goals/why-do-sustainable-development-goals-matter/goal-7-affordable
88. UNEP (n.d.) Goal 11: sustainable cities and communities. United Nations Environment Programme. https://www.unep.org/topics/sustainable-development-goals/why-do-sustainable-development-goals-matter/goal-11

Proposing Water Sensitive Urban Design Solutions in Residential Areas, Case Study: Dar Misr, Helwan, Egypt

Yara Walid Omara, Sarah Rashed, Malak Mohamed, Abdelazim Negm, and Zeina ElZein

Abstract One of the most important challenges facing Egypt is water scarcity and its per capita ratio. From around 1893 cubic meters per capita per year in 1959 to approximately 900 cubic meters in 2000 and finally, to 700 cubic meters in 2012, Egypt's average annual freshwater availability has been declining. Egypt's rapid pace of population increase means that, by 2030, the proportion of water per person should drop to 534 cubic meters, or almost half of the international standard for water poverty, As Egypt suffers from an unequal distribution of population in its governorates where Cairo and Alexandria are the most densely populated and increasing demands on water. Egypt actually has two problems with water: first, there is not enough pure water to drink or use to irrigate crops; second, a major portion of farmers' irrigation practices are still inefficient, which damages the soil and wastes water through evaporation and over-irrigation. However, one of the most important water management methods is Water Sensitive Urban Design (WSUD) in the context of wastewater management in urban areas that helps improve public health and increased resilience to climate change impacts. This chapter proposes urban design interventions for Dar Misr Residential compound in

Y. W. Omara · S. Rashed · M. Mohamed · Z. ElZein (✉)
Architecture Department, Faculty of Fine Arts, Helwan University, Cairo, Egypt
e-mail: Zeina.ElZein@f-arts.helwan.edu.eg; ZeinaBaher@gmail.com

Y. W. Omara
e-mail: yara.walid.2021-01203@f-arts.helwan.edu.eg

S. Rashed
e-mail: sarah.rashed.2021-07968@f-arts.helwan.edu.eg

M. Mohamed
e-mail: malak.mohamed-0789@f-arts.helwan.edu.eg

A. Negm
Water and Water Structures Engineering Department, Faculty of Engineering, Zagazig University, Zagazig, Egypt
e-mail: amnegm@zu.edu.eg amnegm85@gmail.com

Z. ElZein and A. Negm (eds.), *Recent Approaches of Sustainable Architecture in Arid and Semi-arid Cities*, Urban Sustainability,
https://doi.org/10.1007/978-981-95-1645-2_11

Cairo, Egypt, to provide insights into the potential of WSUD as a sustainable solution that includes rainwater harvesting systems, permeable pavements, green roofs, constructed wetlands, and efficient irrigation techniques.

Keywords Water sensitive urban design · Sustainable urban development · Wastewater management · Dar Misr compounds · Water quality enhancement · Nature-based solutions

1 Introduction

Water is a crucial element for life and development. The majority of the Earth's water is stored in its oceans, while permanent snow covers and groundwater account for approximately 3.44%, serving as vital sources of freshwater. However, uneven distribution of water resources across regions poses challenges to both local and regional development efforts [1, 2]. One of humanity's greatest challenges is how to use scarce resources in an equitable and sustainable way. Water scarcity is now the single greatest threat to human health, the environment and the global food supply. Besides the qualitative aspect of water (e.g., polluted water affecting human health). The available quantity of water has a direct impact on a region's potential to produce food. Planning is central in water resources management, as it is needed to develop strategies to reduce water use and use it more efficiently, to optimize the supply and to maintain or improve water quality (pollution control). Water scarcity is an increasingly pressing issue faced by urban areas around the globe, particularly in regions where rapid population growth and urbanization strain existing water resources.

The global water crisis is less about the availability of water and more about how it is managed. Inefficient water management exacerbates issues such as pollution, flooding, and shortages. Rapid urban growth further strains freshwater resources due to rising demand, increased wastewater production, and reduced groundwater recharge. By 2025, an estimated 1.8 billion people will live in regions facing extreme water scarcity, with two-thirds of the global population experiencing water stress. Additionally, between 24 and 700 million individuals may be displaced due to water shortages in arid and semi-arid regions.

Water scarcity has profound implications for public health. For instance, during the COVID-19 pandemic, people in areas with limited water access struggled to follow hygiene protocols, highlighting the critical role of water in disease prevention. Moreover, measures like border closures, movement restrictions, and the economic hardships caused by the pandemic have intensified food and water shortages, raising concerns that these crises may surpass the virus itself in terms of long-term impact [3].

As cities expand, the water demand intensifies, leading to water supply, quality, and management challenges. In this context, Water Sensitive Urban Design (WSUD) emerges as a vital strategy that seeks to harmonize urban development with the sustainable management of water resources. In Egypt, the population has been increasing steadily, with the urban population growing at a rate of 3.2% per year (2020 estimates). The rapid urbanization has led to the emergence of new urban communities, these communities often lack adequate infrastructure, housing, and services, leading to social and environmental issues.

Community design and land use significantly influence water consumption, costs, and management. Low-density developments with large lots tend to use more water for outdoor activities like irrigation and car washing, while also requiring extensive infrastructure, which increases costs and the risk of water loss through leaks. Efficient landscape design, such as incorporating native plants, can significantly reduce irrigation needs and conserve freshwater.

In response to growing populations and ecological concerns, water resource management has shifted toward strategies that prioritize sustainability. Many cities and countries now integrate water-efficient land use practices to address issues like contamination, stormwater runoff, and water supply reliability. This growing focus on water-conscious community design reflects the increasing importance of sustainable development in urban planning [4].

WSUD is an approach that combines urban environment planning and design with the management of the urban water cycle [5]. This methodology seeks to create sustainable, resilient, and livable urban spaces by efficiently managing water resources, enhancing biodiversity, and improving the general quality of life for inhabitants.

The chapter attempts to offer practical suggestions for incorporating WSUD principles into residential developments ultimately contributing to more resilient and sustainable urban environments in the face of water scarcity by identifying best practices and potential implementation barriers.

In this chapter, we selected Dar Misr residential compound as a case study to illustrate how successful water management techniques can be incorporated into residential design. In particular the use of permeable surfaces rainwater harvesting systems and green infrastructure are just a few of the WSUD practices that will be examined in this report as they are applied in Dar Misr. It will also explore how these practices affect the neighborhood's sustainability as a whole local environment and community well-being. This chapter aims to offer important insights and recommendations for future urban developments in Egypt and similar contexts by analyzing the best practices and challenges encountered in integrating WSUD in Dar Misr. It demonstrates the efficacy of WSUD in promoting community engagement, enhancing water security, and improving the general quality of life for residents.

2 Egypt Water Status

Egypt is currently using more water than its internal renewable resources supply and is expected to use even more water in the near future. Egypt is projected to face severe water stress by 2030 due to its limited water resources. With approximately 95% of the population residing in urban areas within the Nile Delta and Valley, the nation has already reached a critical point of water scarcity [1, 2, 6]. According to the Egyptian statistics agency CAPMAS, the country's per capita water resources have fallen from 2526 m^3/year in 1947 to less than 700 m^3/year in 2013, This leaves the country at a water deficit as it annually needs at least 90 m^3 billion to satisfy national needs. Egypt's annual per capita share of water declined to 570 m^3 in 2017, which is below the international standards at 1000 m^3/year *Egypt's updated nationally determined contributions, (2023).* The threshold deemed necessary by the UN to provide enough water for drinking, agriculture and nutrition. This trend is expected to continue, leading to a possible figure of less than 350 m^3/year by 2050 [7].

Similarly, aggregate water consumption is expected to increase by more than 20% in the next few years [8], while upstream development projects on the Nile risk reducing the amount of water flowing down to Egypt. As a result of growing water scarcity, Egypt could face rising food insecurity and unemployment, which, in turn, could revive anti-state grievances or even lead to political instability in the Nile basin region [9].

In Egypt, rapid population growth increases water stress by augmenting water requirements for domestic consumption and increasing irrigation water use to meet higher food demands (Dakkak 2016).

Egypt's' dependence on the Nile River is significant since it supplies 95% of the freshwater resources in the nation. For industrial drinking water and agricultural purposes the Nile is essential as shown in Fig. 1. Egypt is categorized as a water-scarce country because of its growing population and rising demand which has resulted in a decline in the amount of water available per person [10]. Effective management of the nation's limited water resources is a challenge. Egypt's water consumption is primarily attributed to agriculture which uses a large amount of the nation's water resources. To increase water efficiency in agriculture, the government has launched a number of irrigation projects. The Egyptian government has constructed water treatment plants and dams among the strategies to improve water management; like Aswan High Dam: This is one of

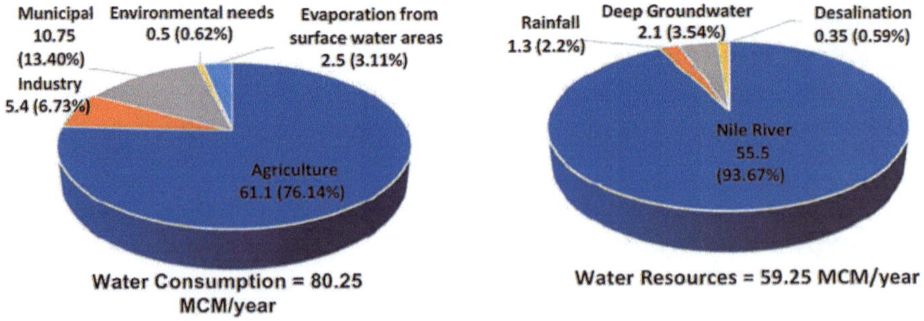

Fig. 1 Water resources consumption (demand) in Egypt (left) water resources supply in Egypt in 2020 (right) [11]

the major infrastructural projects in Egypt, completed in the year 1970. The dam helps control the flow of the Nile for preventing floods and droughts, storing water for agricultural use, and providing hydroelectric power.

and Aswan Low Dam: Built in the early years of the twentieth century and later raised, it was the first major dam on the Nile and a forerunner to the Aswan High Dam. It also helps with water storage and flood control It has also promoted the adoption of water-saving methods. The administration of the Niles water resources has led to disputes in the area, particularly with countries upstream like Ethiopia, which constructed the Grand Ethiopian Renaissance Dam (GERD) on the Blue Nile [12]. Although the process of desalination is expensive and energy-intensive, it promotes access to safe, clean drinking water. Wave power and other forms of energy capture are being used creatively to lessen the need for fossil fuels and cut down on emissions from desalination [13].

2.1 The Average Water Consumption Per Capita

The average water consumption per capita in Egyptian households is around 380 L per day (L/d) [14].

Residential Buildings:

- Bathing and hygiene: around 40–50 L/d per capita [15]
- Laundry: around 10–20 L/d per capita [16]
- Other uses (e.g., cleaning, toilet flushing): around 10–20 L/d per [17]

Commercial and industrial buildings have significant water demands, particularly for cooling systems and process water [18].

2.2 Increasing Urban Agglomerations in Egypt

Egypt's current annual urban growth rate of 2 per cent means that Egyptian cities have to accommodate almost 1 million new citizens—every year. Cairo alone saw 104 million new inhabitants in 2017 [14]. Making it the fastest growing city in the world Nearly two thirds of people in the Greater Cairo Region reside in informal settlements which are currently the nation's most common form of urban development. Egypt is developing residential areas in response to the country's rising housing needs which are being fueled by urbanization economic growth and population expansion as shown in Fig. 2 [19].

Due to poor water and sewage management and increasing droughts caused by climate change, Egypt is expected to face a critical shortage of water by 2025. This makes Cairo one of the most threatened cities worldwide that could soon be running out of drinking water. Egypt has only 13 cities with a population of more

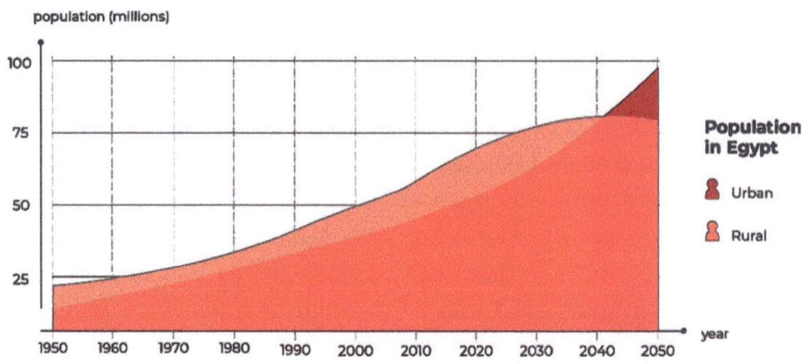

Fig. 2 Urban and rural population growth in Egypt. Egypt Infographics © GIZ [19]

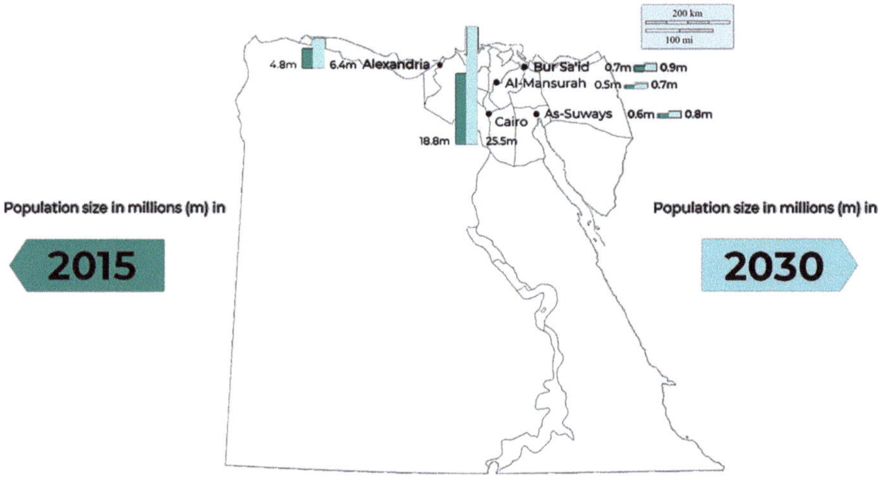

Fig. 3 Population size of Egypt's six largest cities in 2015 and 2030. Egypt Infographics © Modified after GIZ [19]

than 300.000, a rather small number considering its area. While 11 of these cities have a population of less than 500,000, the two largest cities have populations many times that of the next smaller ones: Alexandria with more than 5 million inhabitants and, far ahead, the capital Cairo with more than 20 million as shown in Fig. 3 [19].

2.3 Water Scarcity in Egypt

Water scarcity can be measured by the physical availability of water in a country or by the amount of water that is withdrawn compared to the available amount of internal renewable and inflowing water from other countries. Egypt is facing

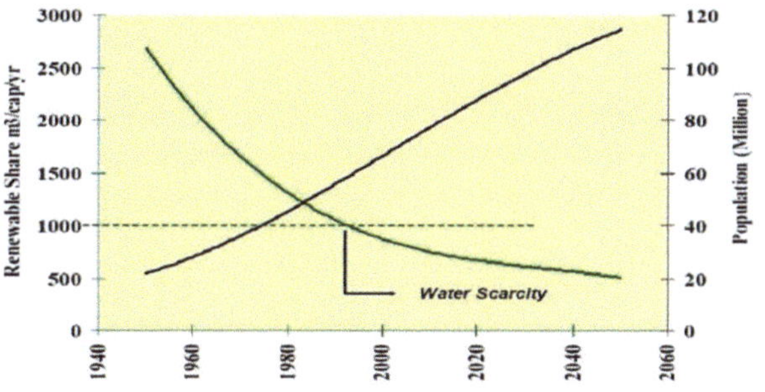

Fig. 4 Population growth in Egypt graph [20]

worsening water shortages as climate change brings more extreme heat, shifts in precipitation that causes increasing in the frequency of flooding and drought. Due to the progressive increase of population and the continuous expansion of urbanized areas, pollution dilemma has increased too. Egypt's population growth rate remains high compared to growth rates of less than 1% in Europe. Fresh water availability in Egypt dropped from 1893 cubic meters in 1959 to 900–950 cubic meters in 2000 and tends to decline further to the values of 570 cubic meter by 2017 and 500 by 2025 [14]. The main reason behind this rapid fall is the fixed water resources and the rising pressure from population growth as shown in Fig. 4.

Although the main challenges regarding freshwater in Egypt include increasing urban demand, land-use changes and environmental requirements [21], the climate change is happening with severe and diverse impacts, [22] and the Ethiopian Renaissance Dam represents the most prominent challenge [23].

Water quality is a problem around urban centers and in the Egyptian delta. Pollution increases downstream, especially downstream of Cairo. The river water quality is better than that of many other water bodies especially in high flood years, because the river is flushed periodically. High concentrations of fertilizers and salts in irrigation drainage water are likely to be the main challenges in the future.

Many residential buildings lack efficient water management systems, leading to wastage and overconsumption. Poor infrastructure and outdated plumbing can result in leaks and water loss [24].

3 Water Sensitive Urban Design

An all-encompassing framework for managing the urban water cycle is called WSUD. It integrates stormwater drainage sewage systems and potable water supplies into urban design. With an emphasis on social environmental and economic factors WSUD is consistent with integrated urban water management and sustainable urban water management. It encourages methods that are regenerative, equitable and resilient. In contrast to traditional urban water management, WSUD adopts a systems-based approach that changes the way water and related resources are taken into account at all scales during city planning density [25].

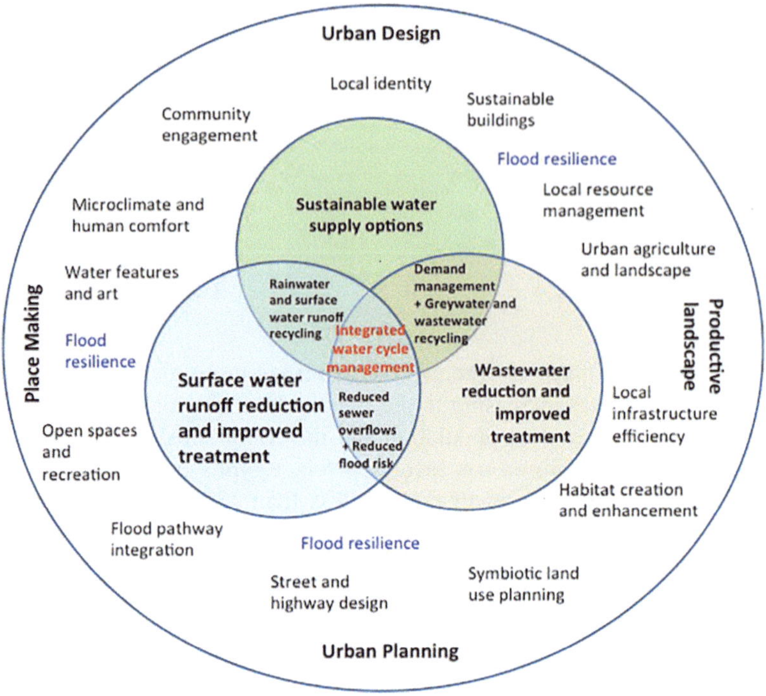

Fig. 5 Components of water-sensitive urban design and their interactions highlighting the place of flood resilience and other aspects of WSUD [27]

With its comprehensive approach and eco-friendly nature, WSUD encourages the wise use and administration of water in urban settings [26]. The concept of WSUD emerged as a response to the growing concerns over urbanization's adverse effects on water quality, flooding, and ecosystem health. Traditional urban development practices often lead to increased runoff, pollution, and strain on existing water infrastructure as shown in Fig. 5. WSUD offers a holistic and environmentally friendly alternative that promotes the sustainable use and management of water within urban environments [26].

3.1 Applications of WSUD

An assortment of applications targeted at incorporating water management into urban planning and design are collectively referred to as Water Sensitive Urban Design (WSUD). These include green infrastructure that makes use of vegetation soils and other natural systems in order to control stormwater runoff enhance air quality and lessen the impact of the urban heat island effect as shown in Fig. 6. These practices provide shade and evaporative cooling to mitigate the urban heat

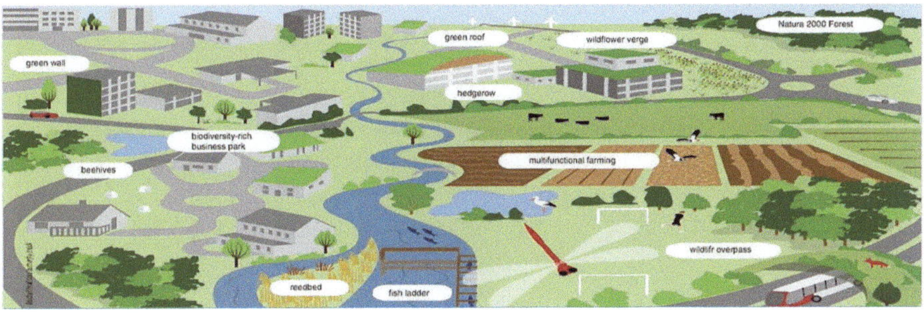

Fig. 6 Green infrastructure examples in urban scale [28]

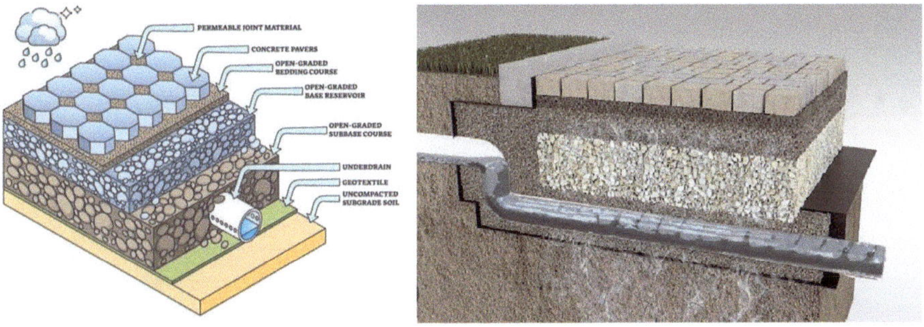

Fig. 7 Represents axonometric section for permeable pavements [28]

island effect green infrastructure improves air quality by reducing particulate matter and ozone and it also improves biodiversity and aesthetics.

Another practice is the use of permeable pavements in parking lots, and walkways to allow rainwater to seep into the ground as shown in Fig. 7 As a result, there is less surface runoff and more groundwater recharge. Enhancing urban wetlands and natural areas is imperative for enhancing the resilience and sustainability of cities. Incorporating urban wetlands and natural areas into city planning and development can help address some of the problems associated with urbanisation, such as stormwater management, heat island effects, and habitat loss [28].

Bioretention systems are designed to filter and treat large amounts of stormwater runoff as part of a treatment train as shown in Fig. 8. After being filtered the stormwater is collected in pipes under the systems and sent to retention basins natural streams or stormwater systems. The primary objective of bioretention systems is to filter out excess nutrients and contaminants. These additional nutrients which are primarily composed of potassium phosphorus and nitrogen are typically acquired from fertilizers. Additional advantages of bioretention systems include water quality control conservation of water and enhanced amenities. Although they are capable of controlling flooding their small size usually limits this as shown in Fig. 9 [29].

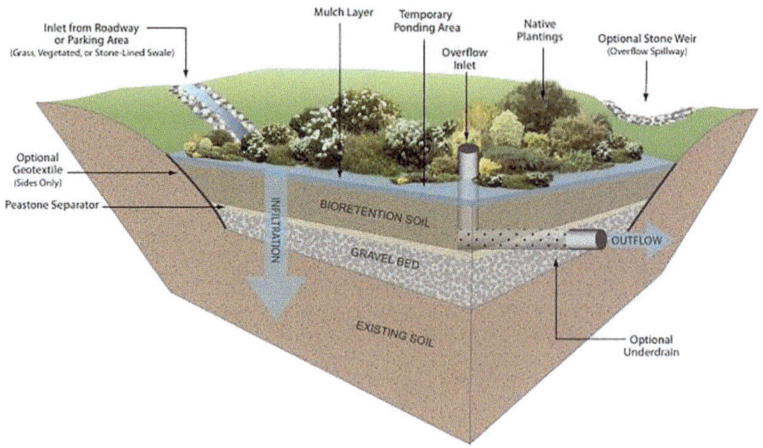

Fig. 8 Represents bioretention systems [30]

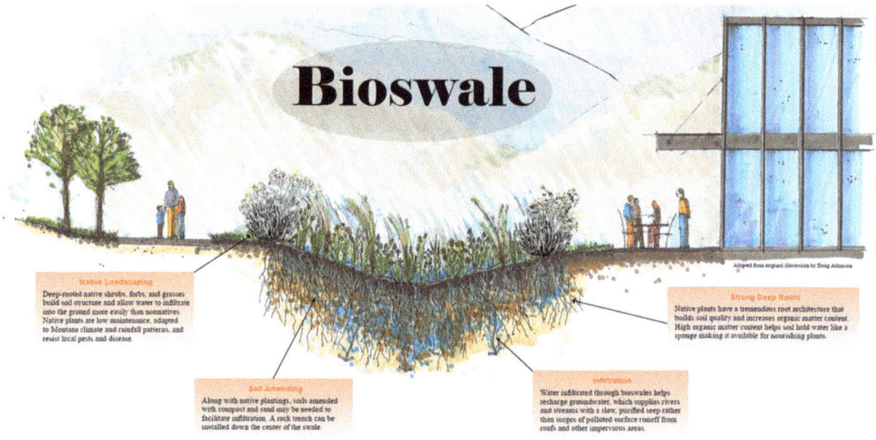

Fig. 9 Represents a typical bioswale [31]

Numerous concerns must be evaluated as part of a feasibility study prior to the creation of a wetland. They consist of, but are not restricted to: Levels and topography, Soils, Hydrogeology and Geology, Protection of groundwater, Drainage and hydrology, Risk of flooding, Sites and species that are protected, Use of land, Possession, Heritage and archaeology.

The way a treatment is designed Wetland development should be an iterative process that takes into account the goals while also addressing the limitations given by a particular location as determined by the feasibility study [32].

A technical specification and environment management plan that includes all the information required for a contractor to construct the Treatment Wetland would typically be created using the design data. Usually, this would include information about the environment, waste management, incident management, and safety and health risks specific to the site. Along with pertinent method statements, it should include details about infrastructure, planting, earthworks, and site clearance.

3.2 Constructed Wetlands

Constructed wetlands (CWs) are specially designed systems that use natural elements like plants, soil, and microorganisms to treat wastewater. These systems replicate the functions of natural wetlands but operate in a controlled environment to improve efficiency. This study explores the use of CWs as a localized solution for managing wastewater in communities, emphasizing their added advantages. By analyzing real-world examples, the research offers recommendations for implementing CWs in Egypt and addresses potential challenges they may face [33].

3.2.1 Surface Flow Wetlands

Surface-flow constructed wetlands are similar to natural swamp areas in which plants are rooted in a submerged layer of sand or gravel. The unique capacity of helophyte plants which serve as oxygen pumps and give dissolved oxygen to a variety of microorganisms via their roots permits the sediment to be aerated. Surface flow constructed wetlands are generally employed in scenarios involving highly variable flow rates like runoff from roadways and where anaerobic pretreatment in a septic tank or biodigester is deemed unnecessary due to the consequent odour nuisance. The following variables largely affect the design: matrix characteristics organic and hydraulic load surrounding temperatures and spatial restrictions [33, 34].

3.2.2 Horizontal Subsurface Flow Constructed Wetlands

This type of constructed wetland is most commonly used for aerobic post-treatment of household wastewater because it can withstand a higher hydraulic load than a surface flow constructed wetland. To dissolve solid organic matter anaerobic pretreatment in a biodigester or septic tank is required. Static air is retained above the aquifer by a thick layer of gravel which keeps the surrounding area free of offensive odors. Aeration happens similarly to wetlands created by surface flow as shown in Fig. 10. Still the wastewater must pass through the matrix in order to ensure that it gets in close proximity to the bacteria in the rhizosphere or plant root zone. This method of treatment is applied to all wastewater since there is no possibility of a short circuit [33, 34].

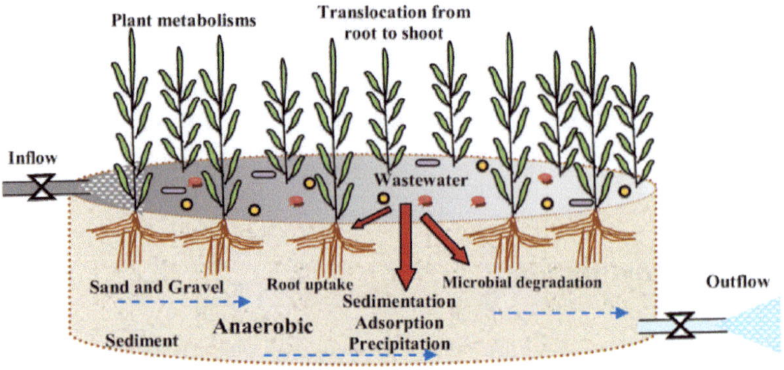

Fig. 10 Represents horizontal subsurface flow constructed wetland and associated removal mechanisms of PhCs and TPs [35]

3.2.3 Vertical Flow Constructed Wetlands

A further reduction in the size of constructed wetlands led to the development of vertical flow constructed wetlands. The produced wetland is topped with intermittent pumps of anaerobic pre-treated wastewater from a septic tank or biodigester. The wastewater effectively draws air into the newly formed wetland when the pump is turned off, forcing the rhizosphere to aerate. Compared to constructed wetlands with horizontal subsurface flow this can increase the aeration capacity by up to twenty times. Apart from that, no short circuit flows are possible and lower oxygen levels in the matrix under anoxic conditions lead to the elimination of nitrate. The depth and the level of the aquifer can be changed according to the design for instance [33, 34] (Fig. 11).

A stormwater harvest and reuse system is a constructed system that captures and retains stormwater for beneficial use at a different time or place than when or

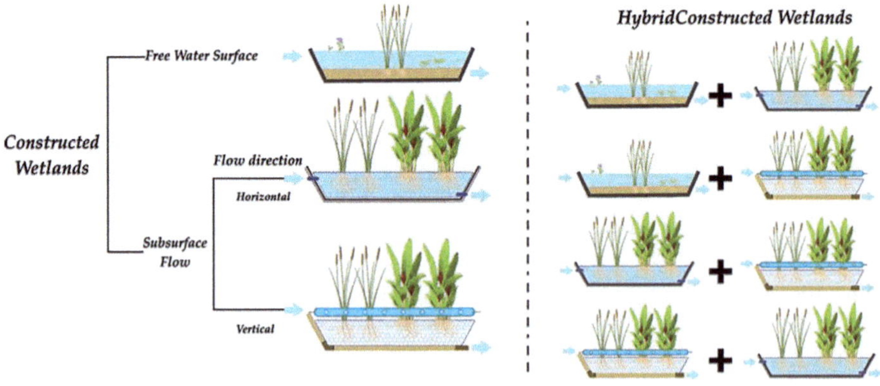

Fig. 11 Represents scheme of constructed wetland types including constructed wetlands both free water surface and subsurface flow (left) and hybrid constructed wetlands (right) [36]

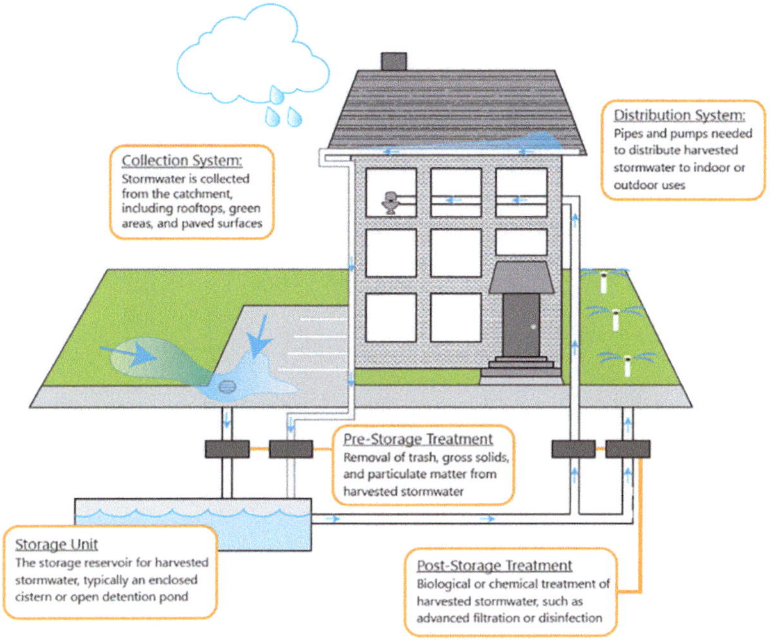

Fig. 12 Represents stormwater harvesting and use system schematic [38]

where the stormwater was generated. If required, the distribution system (which consists of pumps, pipes, and control systems) and the collection system (which might comprise the catchment area and stormwater infrastructure like curb, gutters, and stormsewers) as shown in Fig. 12, storage unit (like a pond or cistern), and treatment system (that removes solids, pollutants, and microorganisms, including any necessary control systems). The quality of the stormwater supply in a stormwater harvest and reuse system is mostly determined by the source location of the captured stormwater [37].

4 Water Sensitive Urban Design in Residential Buildings:

Exploring the application of WSUD in residential buildings can help to better understand how water management concepts are applied in urban development. The challenges and limitations faced by residential developments utilizing WSUD concepts can be identified by looking at case studies involving residential projects that incorporate the concepts of WSUD. By lowering stormwater runoff conserving water and establishing green areas enhance environmental sustainability helps to manage water resources effectively cut down on waste and guarantee a sustainable supply of water for locals. Reduced negative effects of buildings on the water

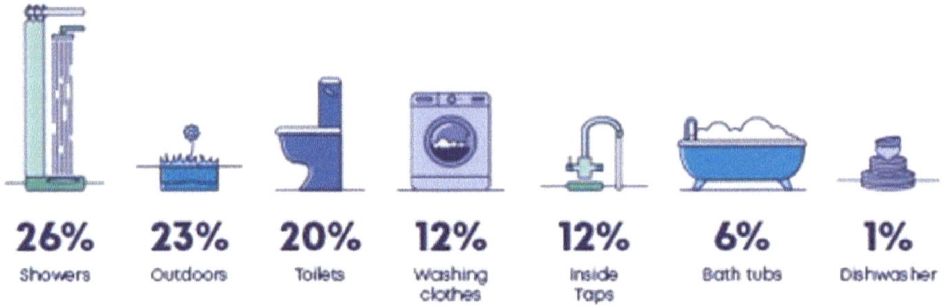

Fig. 13 Represents average daily water usage at home [40]

balance, preserved or improved water quality, promotion of water conservation, and preservation of water-related recreational and environmental opportunities are the goals of WSUD [39] (Fig. 13).

4.1 Application Case Study: Dar Misr

Helwan is a suburban district in the Southern Cairo in Egypt. There has been activity in the Helwan region (which includes 15th May City) since the prehistoric ancient Egyptian Roman and Muslim periods. Umayyad Egypt's capital Fustat was temporarily replaced by the city of Helwan which was established in 689 CE. Helwan changed dramatically during the 1960s becoming an industrial suburb of Cairo and a vast steelworks zone [41].

In 1978 a presidential decree established the city of 15th May. The city which has steelworks and auto factories was created as part of President Gamal Abdel Nassers attempts to industrialize Egypt. The city is strategically located between Greater Cairo and Upper Egypt about 35 km southeast of Cairo as shown in Fig. 14.

4.1.1 Overview

The Egyptian government launched Dar Misr, a well-known residential compound in the 15th of May City, as part of a larger initiative to provide affordable housing solutions. Dar Misr is a planned urban area located on the outskirts of Cairo, Egypt as shown in Fig. 15. Developed as part of the Egyptian government's initiative to provide affordable housing solutions, Dar Misr aims to cater to the growing demand for quality residential spaces in a rapidly urbanizing environment [41]. Numerous facilities are included in the compound with the intention of improving the living conditions of its occupants. These comprise vital services like shopping malls, medical facilities, and educational institutions, in addition to leisure amenities like parks, playgrounds, and sports field.

Fig. 14 Represents Helwan district: **a** Helwan boundaries in the nineteenth century and **b** South Helwan steelworks [42], Credit Sofinter

4.1.2 Climate Analysis

Egypt has a subtropical desert/low-latitude arid hot climate (Köppen-Geiger classification: BWh) subtropical hot desert climate (BWh). Helwan is situated in or near the subtropical desert biome, The mean temperature is 21.3 degrees Celsius (70.3 degrees Fahrenheit), Average monthly temperatures vary by 14.7 °C (26.5°F). This indicates that the continentality type is oceanic, subtype truly oceanic and the Total annual precipitation averages 18 mm (0.7 inches) which is equivalent to 18 Litres/m^2 (0.44 Gallons/ft^2) [43].

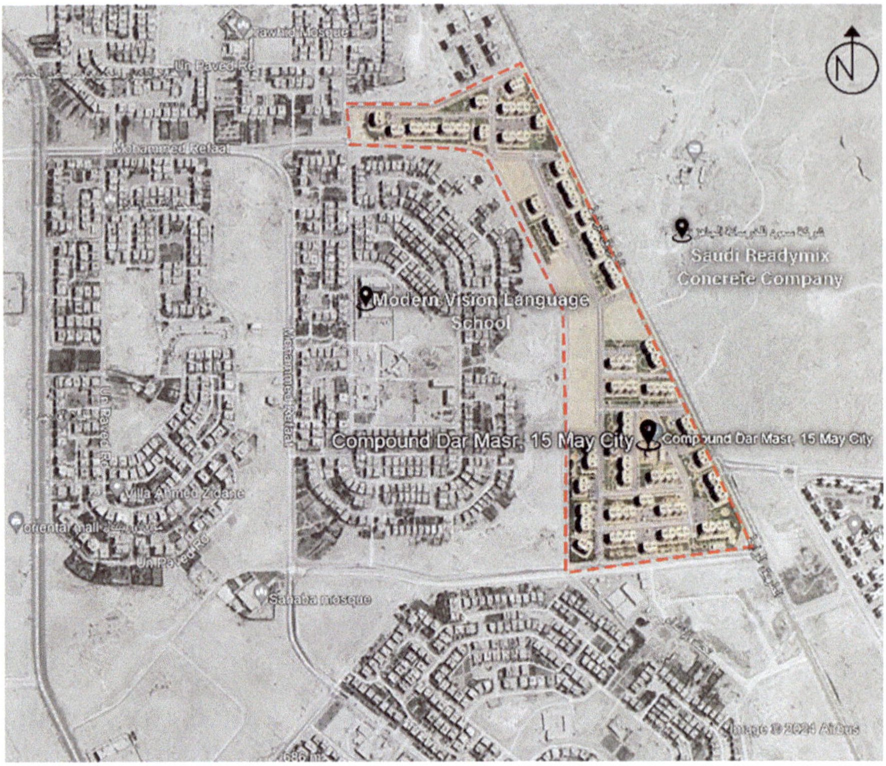

Fig. 15 Represents Dar Misr compound boundaries (*Source*: google earth pro)

Summers are extremely hot, with average high temperatures often exceeding 35 °C (95°F) in July and August. Winters are mild, with average temperatures rangingfrom 10 °C to 20 °C (50°F to 68°F) from December to February as shown in Fig. 16.

4.1.3 Current Water Management Systems

Currently, 15 May city is supplied with water through two main water stations with a design capacity of 65,000 m^3/day [44]:

- North Helwan Water Station.
- Tebeen Water Station as shown in Fig. 17.

The sewage treatment plant has been constructed with a capacity of 46,000 m^3/day, providing tertiary treatment for 30,000 m^3/day. The treated water is then directed to the Saf Canal via a discharge line of 13.4 km. The sewage networks extend 211.38 km throughout the city as shown in Fig. 18. Currently, sewage is being redirected to the excess discharge station, from where it is transferred to Helwan [45].

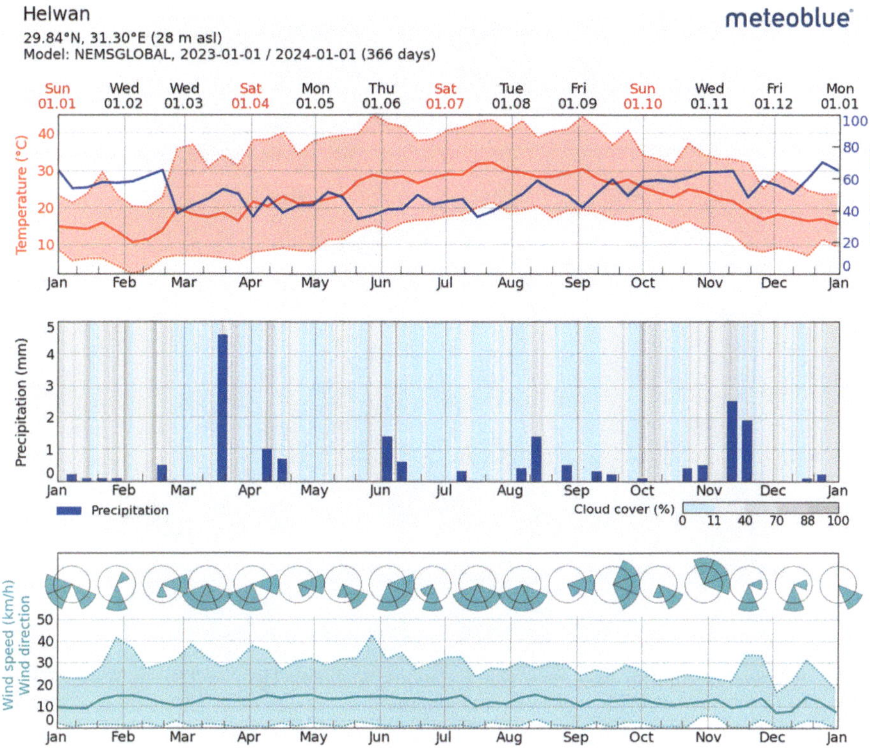

Fig. 16 Represents Helwan climate analysis [43]

The Tebeen Water Station's Primary Function: is to provide a reliable source of potable water to the residents of South Helwan and surrounding areas. It deals with the rising water demand brought on by population expansion and urbanisation.In order to guarantee that the water supplied satisfies health and safety regulations, the station is outfitted with cutting-edge water treatment technologies. This usually entails procedures like chemical treatment, filtration, and disinfection to get rid of pollutants and impurities [44].

At the northern entrance to the city a new water lifting station has been established capable of producing 78,000 m^3 per day. This station provides the city with roughly 70,000 m^3 per day which helps and encourages the city to expand the Tebeen Water Stations water network.

A main water supply line has been installed from the Tebeen Water Station expansions with a diameter of 1000 mm and a length of 13 km. The city's water network spans 250.3 km, and the irrigation networks cover 96.5 km. One of the projects in the 2015 National Action Plan of Egypt (NAP) is the proposed upgrade and extension of the Helwan WWTP, which is highly prioritised by the Egyptian government. It supports both Egypt's efforts to combat climate change and the country's effective and sustainable management of its water resources. The plan is

Fig. 17 Represents Helwan two main water stations: a) South Helwan Tebeen Water Station [44] and b) Helwan Northen Sewage treatment plant [45]

Fig. 18 Represents the location of each water treatment station in Helwan (*Source* Google earth pro)

to expand and upgrade Helwan (Arab Abo Sa'ed) WWTP, increasing its advanced treatment capacity from 550,000 m^3/day to 800,000 m^3/day.

The project will help clean up the Al-Saff Canal and the southern part of Greater Cairo. It will also add another water supply, which will boost agriculture's bottom line. The project will benefit the residents of the Cairo Governorate in terms of their health and the environment. The project will help with both energetically optimised and environmentally sound sludge treatment as well as environmentally sound effluent and sludge disposal and utilisation [46]. This proposed project was designed to redistribute the wastewater flows to decrease discharges from existing collectors and divert them to the WWTP in Arab Abo Sa'ed through the construction of a new collector, pump stations, and force-mains.

4.1.4 Water Use in Dar Misr

At Dar Misr Project, water usage is controlled to satisfy resident needs and maintain sustainability. Water use is broken down as follows: Domestic Water Supply: This category includes water used for drinking, cooking and bathing. Plumbing systems in buildings supply residents with clean water to use on a daily basis. Irrigation of the Dar Misr projects green areas and landscaping: Water-efficient irrigation techniques such as drip irrigation or rainwater harvesting can reduce water consumption while preserving the neighborhoods aesthetic appeal Cooling systems like evaporative coolers and air conditioners frequently use water because they require a steady flow of water to operate properly. Sprinkler systems in Dar Misr is one kind of fire safety measure, are installed in buildings to ensure safety in the case of an emergency. According to these percentages, domestic use uses up the majority of water in the Dar Misr Project, with smaller amounts going towards irrigation during construction and other uses [47].

Understanding grey water's potential for reuse, management, and environmental impact in the context of Dar Misr Compound in Helwan on May 15th is crucial. In Dar Misr, the main sources of grey water are as follows: Showers and Baths: Water used for personal hygiene, Laundry, Water utilised for cleaning clothing, Kitchen Sinks and Water used for food prep and dishwashing. Depending on the size of the home and how much water is used, the amount of grey water produced can vary, but it usually makes up a sizeable amount of all the wastewater produced in the home [48].

4.1.5 Location

Helwan's lowest point is 58 m, and its highest point is 167 m above sea level, east of the Nile River. Many soil types including clay and sandy soils are found in the city. The design of green areas and infrastructure projects can be influenced by the type of soil. For the purpose of planting or creating green projects soil may need to be amended. The proportion of cultivated land is 7% of both the total amount of

Fig. 19 Represents the green area between Dar Misr compound units [41]

green space and the urban mass as shown in Fig. 19, accounting for 1.3 km^2. Its irrigation system is mostly dependent on outside resources such as city water systems. In order to reduce water consumption green projects might need to use drip irrigation or other effective irrigation systems [49].

The compound contains public parks, undeveloped areas, roadside areas, and around residential buildings. It typically features a variety of residential unit types, including apartments and villas, catering to different family sizes and preferences. The designs often emphasize modern architecture and functionality. The housing sector in Dar Misr is supported by comprehensive infrastructure development, including roads, utilities, and public services, ensuring that residents have access to essential amenities (Fig. 20).

4.1.6 Residential Units

4525 plots are allocated with areas ranging from 209 to 180,000 m^2 including 2074 extension; 128 block in district number 24, 81 block in district number 33, 22 block for investment and 507 block for family housing. 3 compounds, 19 villas, 1038 social buildings: 2 beside the electronic bakery and 622 plots for social buildings as shown in Fig. 25 [50].

The number of service buildings implemented by New Urban Communities Authority is 309; 54 social buildings: (35 nurseries—1 club—2 youth centers—2 orphan institutions—1 Special needs—2 playing courts—11 open courts), 18 administrative buildings: (13 post office—1 Governmental Authorities1 Building—1 central—1 bus station—1 microbus station - Month mortgage -Department of Veterinary Medicine), 10 security buildings: (Police Department—4 Police Station—Public Security Bureau) as shown in Fig. 21.

49 Commercial market 44 educational buildings: (20 schools- 4 high schools—3 industrial high School- 5 Azhari—5 experimental schools—6 private schools—1 special high-Institute of Engineering), Health buildings: (2 hospitals—3 Ambulance buildings—health insurance—1 health centers), 36 religious buildings: (33

Fig. 20 Represents aerial view of Dar Misr compound [41]

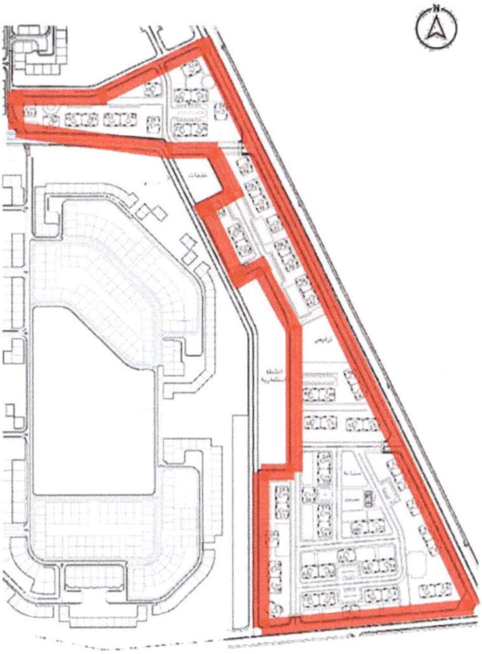

Fig. 21 Dar Misr compound master plan [50]

mosques, 3 churches) and cultural buildings: (1 a public library—1 museum—Art institution—Cinema Halls).

Dar Misr Compound on May 15 consists of residential units (Model B), which contains Residential units consisting of one residential building (4 apartments) and other units consisting of two attached residential buildings (8 apartments) [50] (Figs. 22, 23, 24).

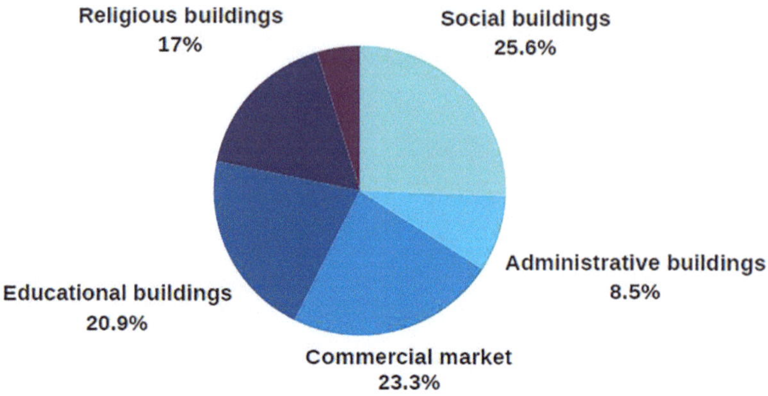

Fig. 22 Chart showing percentages of different types of buildings [50]

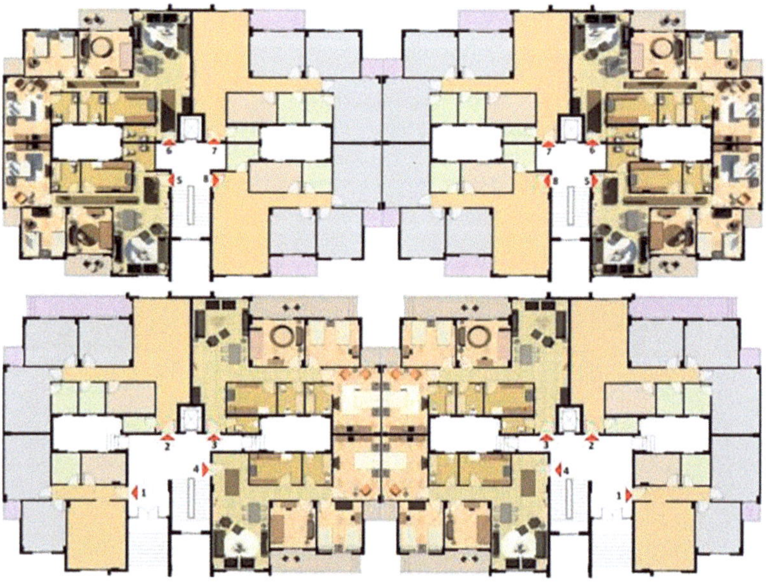

Fig. 23 Represents the architectural plan of Dar Misr Compound: **a** The ground floor of two attached residential buildings (100–150 m^2), **b** The last and repeated floor of two attached residential buildings (130–140 m^2) [41]

4.1.7 Water Challenges at Dar Misr

With several obstacles affecting both the residents and the water infrastructure of the compound, the management of water resources in Dar Masr is a difficult and urgent problem. These problems call for creative solutions that strike a balance between the needs of the users and the environment. Dar Misr can guarantee a more sustainable and resilient future for its citizens by comprehending

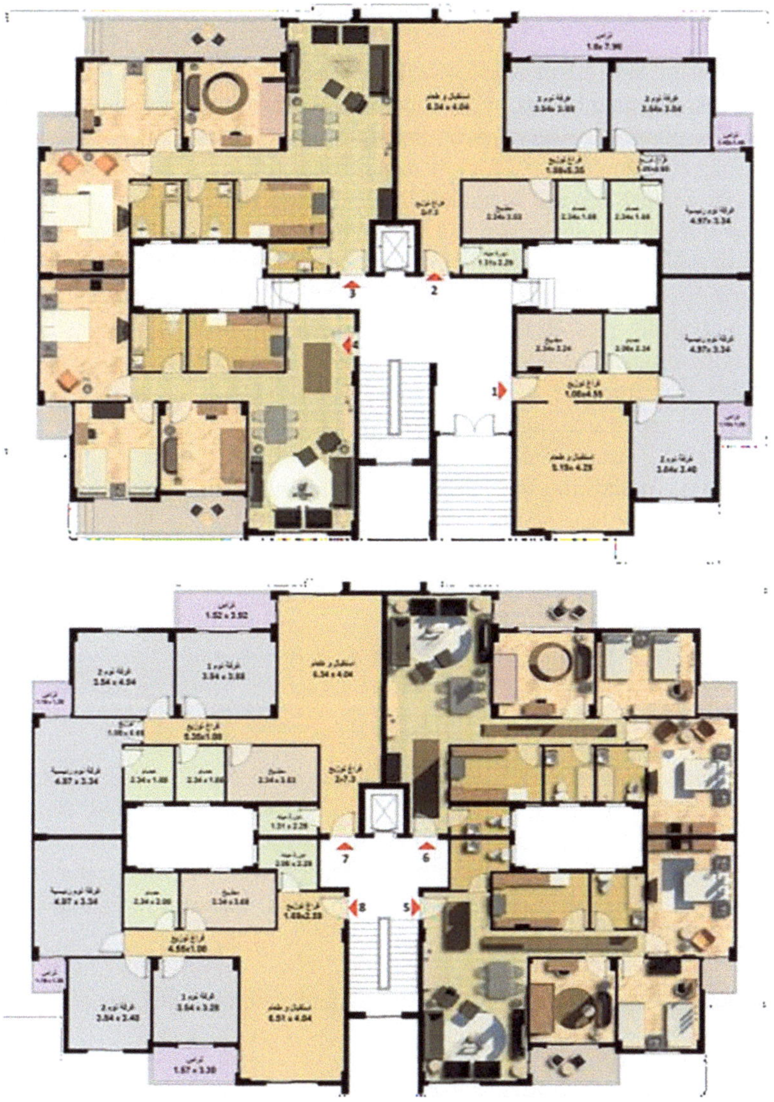

Fig. 24 Represents the architectural plan of Dar Misr Compound: **a** The ground floor of single residential building (100–150 m^2), **b** The last and repeated floor of single residential building (130–140 m^2) [50]

these issues and creating focused strategies to address them. Challenges related to water include: Limited capacity of existing wastewater treatment facilities to handle increasing urbanization and population growth; Poor upkeep of the current infrastructure causes frequent malfunctions; Insufficient public education regarding the significance of conserving water and utilizing water resources effectively. Insufficient involvement in water conservation efforts because of insufficient knowledge or rewards (Awad 2019).

4.1.8 Potentials of Adopting WSUD

The adoption of WSUD principles in Dar Misr Compound has the potential to improve the living conditions of inhabitants and tackle the problem of water scarcity. The compound can serve as an example of sustainable urban living in arid. These initiatives can support efficient water management while also boosting the urban environments general sustainability and resilience as shown in Table 1. For driveways and parking lot walkways permeable materials can be used which lowers surface runoff and encourages groundwater recharge. This makes it possible for any precipitation that does fall to be absorbed, which is particularly helpful in areas with little rainfall. Green roofs can increase insulation levels and lessen the effects of urban heat islands. In addition to adding more green space, vertical gardens can improve aesthetics and biodiversity in compact urban areas. Rainwater and greywater can be naturally treated by integrating bioretention areas and constructed wetlands into the landscape. These systems have the capacity to filter pollutants in addition to reducing runoff and offering habitat for wildlife. In addition to ensuring that plants receive adequate moisture and maximizing water use for gardens and landscaping intelligent irrigation systems can minimize waste (Awad 2019). These systems make use of sensors for soil moisture and meteorological data.

4.1.9 Design Calculations

The compound consists of 63 Residential Building: 11 Single Residential Building (4 appartements) each between (130–140 m^2), 5 floors and 52 Double attached residential buildings (4 appartements) each between (140–150 m^2) and 5 floors [50].

Average number of people per unit: 4 people, Average daily water usage per person: 150 L. Grey water sources: Sinks, showers, bathtubs, washing machines (typically 60–75% of total usage).

Total water per household = Number of residents* amount of Liter consumed per person = 4 people × 150 L/day = 600 L/day.

Table 1 Represents applicable WSUD technologies in Dar Misr (By Authors)

Available sources	Type of WSUD	Applicability
Several open spaces and parks	Wetlands, Permeable Pavements	Applicable
Existing of hardy plants like Sedums, succulents, and others	Green Roofs	Applicable
The city features various soil types, including sandy and clay soils	Green Open Spaces	Applicable
The city receives limited rainfall throughout the year, Climate Characterized as hot desert, with long, hot summers and mild winters	Storm Water Harvesting System, Rain Catchement System	Not Applicable

Total water for all households = 1260 households × 600 L/day = 756,000 L/day.

Total grey water produced = 756,000 L/day × 70% = 1,529,200L/day as shown in Fig. 27. Treatment requires 2 stages. Primary Treatment (Filtration): The collected grey water should be filtered to remove large particles, hair, and debris using mesh or screen filters. Secondary Treatment (Settling or Biological Treatment) uses settling tanks or biological treatment (like constructed wetlands or biofilters) to reduce organic matter and pathogens.

Total Area of Green Roofs: Appartement area * no. of appartements in the last floor * no. of residential buildings: [(140 * 4) * 11]+[(150*4) * 52] = 6160 + 31,200 = **37,360** m^2

Total Area of Constructed Wetlands: Each Double Attached Residential Buildings contain 3 circular greenery area with paths connected directly to the blocks [50] = (153m^2), 52 Building * 153 m^2 Greenery = 7956 m^2
Unconstructed land in front of Masjed Al Aziz of area = 2025.61 m^2, A constructed wetland of 800m2 can be constructed beside dense trees for increasing air quality and protecting the wetland, **Total area of constructed wetlands will be = 8756** m^2 as shown in Fig. 25 to be attached directly with the residential blocks through that waste water can be filtered and re used (Fig. 26).

Fig. 25 The proposed area for the constructed wetlands in blue (By Authors)

Fig. 26 Represents diagram of proposed implemented WSUD technologies in Dar Misr compound, **a** Represents the green roof area = 37,360 m^2, **b** Represents the permeable pavements area = 5000 m^2, **c** Represents the constructed wetland area = 8756m^2 and **d** Represents the Master Plan of Dar Misr Compounds (By Authors)

5 Dar Misr Recommended Proposal

Implementation of Green roofs on Residential Units (Single and double attached).

Exploitation of the open spaces and parks in constructing Wetlands to enhance water quality and increase water irrigation source.

Increasing green spaces between the units and Fertilizing the Soil to be applicable for various plants.

Constructing Permeable pavements as water passes through the permeable surface, it can be filtered naturally, reducing pollutants and sediments.

Awareness campaigns for residentials to not waste drinking/Potable water in other purposes than its intended purpose, as there aren't enough wastewater treatment plants in Helwan.

Separating Gray water from blue water to be re-used in irrigation and green roofs.

Install a separate plumbing system to direct grey water from these sources to a storage tank or treatment system, keeping it separated from black water (from toilets) then it will be connected to wetlands through filtration pipes and water can be re used in irrigation of green open spaces and green roofs as shown in the following diagram. Also Full depth permeable pavement (FDPP) consists of multiple permeable layers and the subsurface layers have the ability to capture and store stormwater until it infiltrates into the subgrade native soil and eventually into the groundwater.

Figure 27 illustrates the proposed water cycle for the Dar Misr Compound, which integrates all these different approaches toward better sustainability and efficiency in operations. This scheme shows the flow of water through the different components of the system: rainwater harvesting, treatment of greywater through constructed wetlands, and reuse of treated water in irrigation and landscape use. It shows the difference between gray water and black water, depicting a dual plumbing system that helps in increasing the recycling of water resources. The water cycle proposed gives importance to natural processes and innovative technologies in efficient management, waste reduction, and creation of a resilient urban ecosystem. This holistic approach meets the principles of WSUD, which tries to achieve sustainability and an eco-friendly community amidst challenges of water scarcity.

5.1 *Incorporating AI*

AI Prompt was used as follows *Implement water treatment systems within the wetlands to remove pollutants and contaminants from stormwater runoff. Urban wetlands and natural areas refer to restored or created ecosystems that provide habitat for wildlife, improve water quality, and enhance community livability*. Treatment lakes, Green Roofs, WSUD Development, Permeable Pavements,

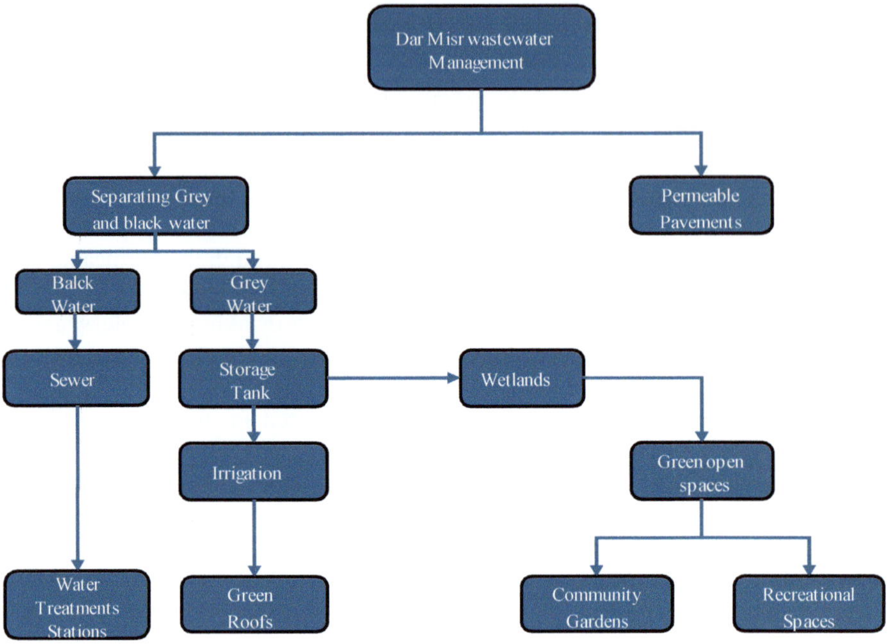

Fig. 27 Represents water cycle in Dar Misr as recommended proposal (By Authors)

Constructed Wetlands, Natural Urban Solutions, improving sustainability in Helwan, Install green roofs on buildings to reduce stormwater runoff and provide insulation. We used PromtAI for generating the photos. Grey water coming from sinks will be separated from the black water (flushing toilet water) through pipes then grey water will be collected to path through grey water filter, it will be then stored in underground storage tanks to be filtered again in the constructed wetlands this water can be re used in irrigation and the black water will go to the sewer where the waste water stations can filter it.

Figure 28 shows built wetlands and green roofs at the Dar Misr Compound, planned in the right place for environmental improvement of quality and to serve better in water management. While the constructed wetlands are sited to naturally clean stormwater and grey water, enhance biodiversity, and improve air quality, the design puts a heavy emphasis on the application of green roofs on residential buildings for insulation and reduction of stormwater runoff. The design shows commitment in using WSUD principles to realize a resilient and sustainable urban environment (Fig. 29).

Fig. 28 Represents Dar Misr wetlands and green roofs. (Prome AI)

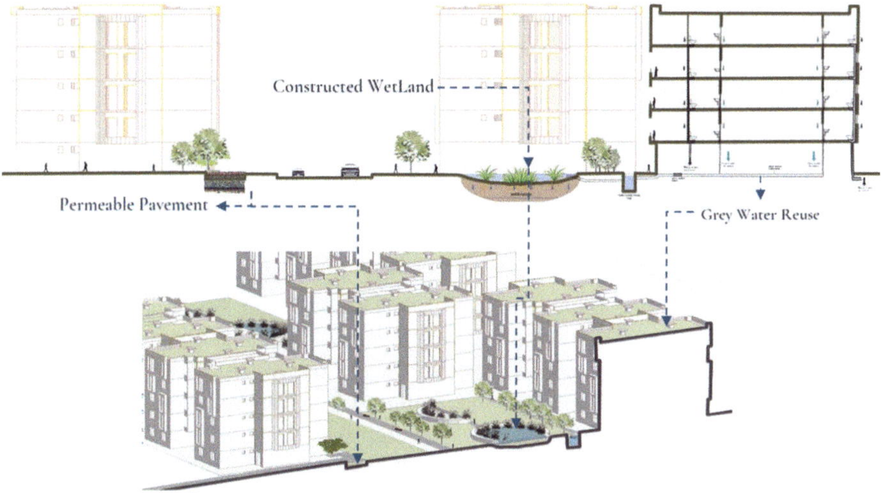

Fig. 29 Represents illustrating section for the water cycle in Dar Misr compound (By Authors)

6 Conclusion

In conclusion, Water scarcity is threatening Egypt due to the increasing population and an increase in demand which has resulted in a decrease in the availability of water per person. The country is currently using more water than that which is supplied by its internal renewable resources and is expected to use even more in the short term. In 2018 Egypt's annual per capita share of water fell to 570 m^3/year, below international standards. With a rapid urban growth rate of 2%, this means that cities have to absorb close to 1 million new entrants every year. Egypt's urbanization is taking place at a very fast rate of 2%. Although Egypt is creating new residential areas to satisfy the growing need for housing, it will face acute water problems by 2025 due to the unavailability of quality water management systems as well as droughts resulting from climate change.

Water scarcity is also worsened by the impacts of climate change, increased urban demand, land use change and environmental needs. The Ethiopian Renaissance Dam being the most prominent challenge, water quality challenges continue in urban centers and the Egyptian delta.

The implementation of Water Sensitive Urban Design strategies is an approach to eco-friendly management of the urban water cycle. It coordinates stormwater, drainage, sewage systems, and potable water supplies with urban design. WSUD aims to realize regenerative and equitable (as well as resilient) water management through propagating sustainable use and management of water within urban environments. Integrating WSUD into Dar Misr Compound showed many benefits to enhance environmental conditions and combat water scarcity. Using permeable surfaces, green infrastructure, green roofs on residential buildings, making use of open spaces and parks to create constructed wetlands, increasing the green space between units and soil fertilization for different kinds of plants are all applicable in the compound, preserving the scarce water resources and making the compound more sustainable and green.

7 Relevant SDGs

Water Scarcity and WSUD in Dar Misr Compound address a number of highly relevant SDGs, including but not limited to, SDG 6: Clean Water and Sanitation-assuring availability and sustainable management of water and sanitation for all. It would surely provide an evident upgrade in the improvement of water quality, promotion of water-use efficiency, and stimulation of approaches of sustainable water management within urban settings through the adoption of WSUD practices. It also has a relevance to SDG 11: Sustainable Cities and Communities, since it requires resilience in urban planning, underpinned by green infrastructure and sustainable practices to meet challenges arising from climate change and urbanization. The focus of this chapter is based on SDG 13: Climate Action at the level of community engagement toward environmental sustainability, where the emphasis is on adaptive strategies to address climate-related challenges.

8 Recommendations

Public Awareness Improvement: Educating the residents through campaigns on water conservation and benefits toward WSUD practices will improve the level of community involvement in the sustainable management of water.

Invest in Infrastructure: Invest in the creation and development of better water treatment facilities and infrastructure to be able to cater to the demand generated from the urban population.

Incentivize the development of green infrastructure, such as green roofs and permeable pavements, including artificial wetlands in new residential areas with better water management and environmental quality.

Grey Water Recycling Systems Installation: Develop grey-water separation and reuse systems to be used for irrigation and landscaping to generally lower water consumption for better sustainability.

Regular Monitoring: Develop a framework that allows for regular assessment and monitoring of urban water management practices to accommodate adaptive management strategies as environmental conditions and community needs change.

References

1. ElZein Z, Ahmed Milad N, Sherif Mohamed A, Kotb NM, Abdo N, Negm AM (2024a) Role of water, energy and food nexus in sustainable development of communities: case studies in the MENA Region. In: Negm AM, ElZein Z (eds) Integration of core sustainable development goals in rural areas, earth and environmental sciences Library. Springer Nature Switzerland, Cham, pp 15–66. https://doi.org/10.1007/978-3-031-60149-1_2
2. ElZein Z, Milad NA, Mohamed AS, Mahmoud N, Abdo N, Abdel-Ghafar HM (2024b) Sustainable development on the basis of the WEF nexus in arid coastal areas for climate change mitigation: case study of Rabia community in Matrouh, Egypt. Energy Nexus 14:100299. https://doi.org/10.1016/j.nexus.2024.100299
3. ElZein Z, Abdou A, Säumel I (2022) Lessons learned from water-scarce cities: proposed policies toward an integrated urban water management in Egypt. Front Water 4: 981261. https://doi.org/10.3389/frwa.2022.981261
4. Abdou A, Gawad I, ElZein Z (2017) Sustainable water treatment methods to be used in urban communities. In: Negm A (ed) Unconventional water resources and agriculture in Egypt. The handbook of environmental chemistry, vol 75. Springer, Cham
5. Kazemi F, Barton H (2021) Water sensitive urban design (WSUD) approach for mitigating groundwater depletion in urban geography. Water Supply 21(7):334–347. https://doi.org/10.2166/ws.2021.001
6. Elzein Z (2024) Sustainable built environments and the water, energy, and food nexus: toward an interdisciplinary approach to urban resilience. Chn J Urb Environ Stud 12:2450004. https://doi.org/10.1142/S2345748124500040
7. UNEP (2016) A snapshot of the world's water quality: towards a global assessment. United nations environment programme, Nairobi, Kenya. pp 162
8. Elshaarawy MK, Armanuos AM (2025) Predicting seawater intrusion wedge length in coastal aquifers using hybrid gradient boosting techniques. Earth Sci Inf 18(2):1755. https://doi.org/10.1007/s12145-025-01755-7

9. Climate Diplomacy (n.d) Security implications of growing water scarcity in Egypt. Available at: https://climate-diplomacy.org/case-studies/security-implications-growing-water-scarcity-egypt
10. Abd Ellah RG (2020) Water resources in Egypt and their challenges, Lake Nasser case study. Egypt J Aquat Res 46(1):1–12. https://doi.org/10.1016/j.ejar.2020.03.001
11. Gado TA, El-Agha DE (2021) Climate change impacts on water balance in Egypt and opportunities for adaptations. In Springer water, (pp 13–47). https://doi.org/10.1007/978-3-030-78574-1_2
12. Honorio V, Jara EP (2023) The dispute over the Nile River involving the Grand Ethiopian Renaissance Dam. Jejaring Administrasi Publik 15(2):79–92. https://doi.org/10.20473/jap.v15i2.52096
13. Wood J (2024) Desalination: How can it help tackle water scarcity? World Economic Forum. https://www.weforum.org/agenda/2024/04/desalination-drinking-water-water-scarcity/
14. Ministry of Water Resources and Irrigation, Arab Republic of Egypt (2019, October 21) مصر والمياه: حقائق وأرقام [Egypt and water: Facts and figures]. Ministry of Water Resources and Irrigation.
15. WWF (2021) Waterways to resilience: nature-based solutions for adaptation in Africa. World Wide Fund for Nature. https://wwfint.awsassets.panda.org/downloads/wwf_report_waterways_to_resilience_f_web.pdf
16. EPA (2016) Green infrastructure strategic agenda 2016. US environmental protection agency. https://19january2021snapshot.epa.gov/sites/static/files/2015-10/documents/2013_gi_final_agenda_101713_0.pdf
17. Masi F, Bresciani R, Rizzo A, Conte G (2017) Constructed wetlands for combined sewer overflow treatment: ecosystem services at Gorla Maggiore, Italy. Ecol Eng 98:427–438. https://doi.org/10.1016/j.ecoleng.2016.03.043
18. IEA (2018) Renewables 2018. IEA Publications. ISSN: 2520-2774. ISBN: 978-92-64-30684-4. https://iea.blob.core.windows.net/assets/79e1943b-9401-478f-9f60-5e8c2bff9342/Market_Report_Series_Renewables_2018.pdf
19. URBANET (2019) Urbanisation and urban development in Egypt: infographic. URBANET. Available at: https://www.urbanet.info/urbanisation-and-urban-development-in-egypt/
20. Sallam OM, El Shewy MA, Dawoud MA (2014, October). New reclamation mega projects and increasing the pressure on water system in the Nile valley and Delta in Egypt. Conference paper presented at the WSTA 11th Gulf Water Conference. Water in the GCC … Towards Efficient Management. Muscat, Sultanate of Oman. 1. https://doi.org/10.13140/RG.2.1.1845.4564
21. Sugita M, Matsuno A, El-Kilani RMM, Abdel-Fattah A, Mahmoud MA (2017) Crop evapotranspiration in the Nile Delta under different irrigation methods. Hydrol Sci J 62(10):1618–1635. https://doi.org/10.1080/02626667.2017.1341631
22. Sutcliffe J, Hurst S, Awadallah AG, Brown E, Hamed K (2015) Harold Edwin Hurst: the Nile and Egypt, past and future. Hydrol Sci J 61(9):1557–1570. https://doi.org/10.1080/02626667.2015.1019508
23. Abdelhaleem FS, Helal EY (2015) Impacts of grand Ethiopian Renaissance dam on different water usages in upper Egypt [Original Research Article]. Br J Appl Sci Technol 8(5):461–483. https://doi.org/10.9734/BJAST/2015/17252
24. National Academies of Sciences, Engineering, and Medicine (2021) Quality water from every tap: proceedings of a workshop—in brief. The National Academies Press. https://doi.org/10.17226/26069
25. Fisher-Jeffes L, Carden K, Armitage N (2017) A water sensitive urban design framework for South Africa. Town Reg Plann 71(1). https://doi.org/10.18820/2415-0495/trp71i1.1
26. Wu W, Jamali B, Zhang K, Marshall L, Deletic A (2023) Water Sensitive Urban Design (WSUD) spatial prioritisation through global sensitivity analysis for effective urban pluvial flood mitigation. Water Res 235:119888. https://doi.org/10.1016/j.watres.2023.119888
27. Kumareswaran K, Jayasinghe GY (2023) Urbanization and sustainable urban planning. In Springer eBooks, pp 99–144. https://doi.org/10.1007/978-3-031-37081-6_3

28. Taylor A, Lang A (2023) Urban green infrastructure: an introduction—WeADAPT. Available at: https://weadapt.org/knowledge-base/cities-and-climate-change/urban-green-infrastructure-an-introduction/. Accessed August 28, 2023)
29. Bioretention (2020) Bioretention systems: the different types & benefits. River Sands. Available at: https://www.riversands.com.au/understanding-bioretention-systems
30. Herrera Environmental Consultants Inc (2020) Bioretention media blends to improve stormwater treatment: final phase of study to develop new specifications. Final Report. Available at: https://www.ezview.wa.gov/Portals/_1962/Documents/SAM/Bioretention%20Study%20Final%20Report.pdf
31. The Constructor (2019) Bioswale: design, applications and advantages. The Constructor. Available at: https://theconstructor.org/water-resources/bioswale-design-applications-advantages/34380/
32. Constructing constructed wetlands (2024). ArcGIS StoryMaps. Available at: https://storymaps.arcgis.com/stories/59fa4a7d31d64a2d9dba62f8ac662f96
33. ElZein Z, Abdou A, ElGawad IA (2016) Constructed wetlands as a sustainable wastewater treatment method in communities. Procedia Environ Sci 34:605–617. https://doi.org/10.1016/j.proenv.2016.04.053
34. Martien (n.d.) Types of constructed wetlands—Kilian water. Killan water Available at: https://kilianwater.nl/en/types-of-constructed-wetlands/
35. Ilyas H, van Hullebusch ED (2020) Performance comparison of different types of constructed wetlands for the removal of pharmaceuticals and their transformation products: a review. Environ Sci Pollut Res 27(13):14342–14364. https://doi.org/10.1007/s11356-020-08165-w
36. Shukla V, Hiwrale I, Dhodapkar RS, Pal S (2024) Constructed wetlands: insights and future directions in sustainable approach for wastewater treatment. In: Earth and environmental sciences library, pp 303–343. https://doi.org/10.1007/978-3-031-63046-0_13
37. Minnesota Stormwater Manual (2023) Overview for stormwater and rainwater harvest and use/reuse. Minn Pollut Control Agency. Available at: https://stormwater.pca.state.mn.us/index.php/Overview_for_stormwater_and_rainwater_harvest_and_use/reuse
38. Minnesota Pollution Control Agency (2016) Overview for stormwater and rainwater harvest and use/reuse. Minn Stormwater Manual. Available at: https://stormwater.pca.state.mn.us/index.php/Overview_for_stormwater_and_rainwater_harvest_and_use/reuse
39. AdriAdapt (2022) Water sensitive urban and building design. AdriAdapt. Available at: https://adriadapt.eu/adaptation-options/water-sensitive-urban-and-building-design/
40. Sydney Water (2022) Water use and conservation. Sydney Water. Available at: https://www.sydneywater.com.au/education/drinking-water/water-use-conservation.html
41. Cairo Governorate (2019) Districts details. Cairo.gov.eg. https://www.cairo.gov.eg/en/Culture/Pages/Districts_detials.aspx?ID=2
42. Power Technology (2021) South Helwan supercritical power plant. Power Technol. Available at: https://www.power-technology.com/projects/south-helwan-supercritical-power-plant/
43. Meteoblue (n.d) Weather archive helwan. Available at: https://www.meteoblue.com/en/weather/historyclimate/weatherarchive/helwan_egypt_355795?fcstlength=1y&year=2023&month=9. (Accessed September 5, 2025)
44. ICAT (2019) Helwan wastewater treatment plant and irrigation pumping station. ICAT. Available at: https://icat.com.eg/portfolio/helwan-wastewater-treatment-plant-and-irrigation-pumping-station/
45. The Arab Contractors (2013) Helwan wastewater treatment plant (Contract 5). The Arab Contractors. Available at: https://www.arabcont.com/english/project-518
46. Qarani A, Sabah M (2022) Wastewater sludge characteristics, treatment techniques and energy production. Reciklaža i Održivi Razvoj 15(1):9–26. https://doi.org/10.5937/ror2201009a
47. Abulkasem MT, Nagy S (2021) Development of drainage and water supply systems to reduce water consumption in residential buildings in Egypt. Eng Res J 170:47–62

48. Ghaitidak DM, Yadav KD (2013) Characteristics and treatment of greywater: a review. Environ Sci Pollut Res 20(5):2795–2809. https://doi.org/10.1007/s11356-013-1533-0
49. Topographic Maps (2024) Helwan topographic map. Topographic map. Available at: https://en-gb.topographic-map.com/map-6s314/Helwan/?center=29.91007%2C31.26343&zoom=10
50. New Urban Communities Authority (2024) Booklet of conditions for reserving residential units in Dar Misr project—Second Offer. In Dar Misr—Second Offer. New Urban Communities Authority. https://reserve.newcities.gov.eg:8084/Files/booklet-dar-masr.pdf
51. Nassar UA, El-Samaty H, Soussa H (2024) Water sensitive urban design: a sustainable design approach for reforming open spaces in low-income residential rehabilitation projects in Egypt. In: Russo F, Poponi S (eds) Smart and sustainable cities: policies, programs, and strategies for urban development. Springer, pp 235–252. https://doi.org/10.1007/978-3-031-50132-6_12
52. Egypt's 2021 Voluntary National Review (2021) United Nations Sustainable Development. https://sustainabledevelopment.un.org/content/documents/279512021_VNR_Report_Egypt.pdf
53. Climate Change Impacts on Water Balance in Egypt and Opportunities for Adaptations—Scientific Figure (n.d.) *ResearchGate*. https://www.researchgate.net/figure/Water-resources-and-water-consumption-in-Egypt-Data-source-50_fig4_355186497
54. Aziz, M. (2020). Egypt's water challenges: Beyond the dam saga. *Ahram Online*. https://english.ahram.org.eg/NewsContent/1/64/359272/Egypt/Politics-/Egypts-water-challenges-Beyond-the-dam-saga-.aspx
55. Gad W (2017) Water scarcity in Egypt: Causes and consequences. IIOAB J 8:40–47. https://www.mewf.de/uploads/1549104207849.pdf
56. Elsaie Y, Soussa H, Gado M, Balah A (2022) Water desalination in Egypt: literature review and assessment. Ain Shams Eng J 14(7):101998. https://doi.org/10.1016/j.asej.2022.101998
57. Cohen Y (2021) Egypt has a water problem—and no, it's not only the GERD. Atlantic Council. https://www.atlanticcouncil.org/blogs/menasource/egypt-has-a-water-problem-and-no-its-not-only-the-gerd/
58. Kim S, Song SK (2019) Sustainable stormwater management using bioretention systems: a review. Environ Technol Innov 14:100325. https://doi.org/10.1016/j.eti.2019.100325
59. Razzaghmanesh M, Beecham S, Brien CJ (2014) Developing resilient green roofs in dry climates. Sci Total Environ 490:579–589. https://doi.org/10.1016/j.scitotenv.2014.05.073
60. Chui TFM, Liu X, Zhan W (2016) Assessing cost-effectiveness of specific WSUD techniques in enhancing infiltration and reducing runoff. Sci Total Environ 548–549:253–263. https://doi.org/10.1016/j.scitotenv.2015.12.140
61. Furlong C, Phelan K, Dodson J, Sun L (2019) Water sensitive urban design: Rethinking general approaches to urban water management. Cities 95:102397. https://doi.org/10.1016/j.cities.2019.102397
62. Lucke T, Nichols PWB (2015) The pollution removal and stormwater reduction performance of street-side bioretention systems. Water Sci Technol 71(1):80–86. https://doi.org/10.2166/wst.2014.488
63. Hamdy A, Aly A (2014) Water crisis and food security in Egypt: current situation and future perspective. Int J Water Resour Arid Environ 3(1):12–22. https://www.psipw.org/attachments/article/83/IJWRAE_3(1)12-22.pdf
64. Shalash S (2022) Water crisis in Egypt: current situation and future challenges. Afr J Environ Sci Technol 16(3):85–94. https://doi.org/10.5897/AJEST2021.3041
65. Ministry of Water Resources and Irrigation (MWRI) - Egypt (2017) National Water Resources Plan for Egypt 2017–2037. http://extwprlegs1.fao.org/docs/pdf/egy190563.pdf
66. El-Fadel M, El-Sayegh Y, Bou-Zeid E (2003) The Nile River Basin: a case study in surface water conflict resolution. J Nat Resour Life Sci Educ 32(1):107–117. https://doi.org/10.2134/jnrlse.2003.0107
67. Nour El-Din M (2013) Egypt's water crisis and degeneration of water quality: an overview and recommendations for future policies. Sustain Water Resour Manag 2(1):1–12. https://doi.org/10.1007/s40899-015-0004-7

68. World Bank (2020) Egypt: Enhancing Water Efficiency. https://www.worldbank.org/en/news/feature/2020/03/22/egypt-enhancing-water-efficiency
69. United Nations Development Programme (UNDP) (2014) Water Governance in Egypt: Challenges and Innovative Approaches. https://www.undp.org/arab-states/publications/water-governance-egypt
70. El-Gafy I (2017) Water–food–energy nexus index for sustainable development in Egypt. Water Resour Manage 31(10):3121–3135. https://doi.org/10.1007/s11269-017-1660-0
71. Abd-Elaty I, Mohamed MF, Badr M (2020) Urban water management and climate change adaptation: strategies for Egypt. J Water Clim Change 11(S1):166–177. https://doi.org/10.2166/wcc.2019.275
72. Yehia M, Mahmoud SH (2020) Assessment of water security and crop production under climate change impacts in Egypt. Sci Total Environ 735:139378. https://doi.org/10.1016/j.scitotenv.2020.139378
73. Mahmoud SH, Gan TY (2018) Impact of climate change and human activities on water resources in the arid Nile River Basin. Sci Total Environ 639:217–227. https://doi.org/10.1016/j.scitotenv.2018.05.155
74. FAO (2021) Egypt: Water Scarcity Overview. https://www.fao.org/egypt/programmes-and-projects/water-scarcity/en/

Urban Greening in Cities

Green Space Expansion Using Haloculture Technology to Combat Climate Change and Global Warming in MENA Region: Cases, Strategies, and Pedagogical Content Knowledge (PCK) Approaches

Amir Parnian, Mina Mahbod, Akbar Soliemanzadeh, Ali Momenpour, Maryam Golbaz, Zeina ElZein, Hossein Beyrami, and Hamid Reza Boostani

Abstract Climate change and global warming pose critical challenges to the Middle East and North Africa (MENA) region, adversely affecting natural resources, food security, health, and urban stability. This region, already grappling with the severe impacts of rising temperatures, faces acute drought and salinity issues that undermine the stability and expansion of urban green spaces. Haloculture, a pioneering technology that utilizes saline soil and water, offers a promising solution for managing and

A. Parnian (✉) · A. Momenpour · H. Beyrami
National Salinity Research Center (NSRC), Agricultural Research Education and Extension Organization (AREEO), Yazd, Iran
e-mail: a.parnian@areeo.ac.ir

A. Momenpour
e-mail: a.momenpour@areeo.ac.ir

M. Mahbod
Department of Educational Sciences, Farhangian University, Tehran, Iran
e-mail: mina.mahbod@cfu.ac.ir

A. Soliemanzadeh
Soil and Water Research Department, Hormozgan Agricultural and Natural Resources Research and Education Center, Agricultural Research Education and Extension Organization (AREEO), Bandar Abbas, Iran

M. Golbaz
Imam Khomeini Higher Education Center, Agricultural Research, Education and Extension Organization (AREEO), Karaj, Iran

Z. ElZein
Architecture Department, Faculty of Fine Arts, Helwan University, Cario, Egypt
e-mail: Zeina.ElZein@f-arts.helwan.edu.eg

H. R. Boostani
College of Agriculture and Natural Resources of Darab, Shiraz University, Darab, Iran
e-mail: Hr.boostani@shirazu.ac.ir

Z. ElZein and A. Negm (eds.), *Recent Approaches of Sustainable Architecture in Arid and Semi-arid Cities*, Urban Sustainability,
https://doi.org/10.1007/978-981-95-1645-2_12

expanding green spaces under these challenging conditions. This chapter employs the PRISMA approach to explore the effects of climate change on MENA cities, examining the risks to communities and the exacerbation of urban heat islands. It discusses the potential of haloculture as a sustainable technology, highlights suitable plants, and presents case studies for green space expansion. Furthermore, it proposes haloculture strategies and pedagogical content knowledge (PCK) approaches to enhance green space management, adaptation, and expansion in response to climate change in the MENA region.

Keywords Climate change · Haloculture · Green space · Adaptation · MENA · Pedagogical content knowledge

1 Introduction

Human-caused climate change and global warming, which is initiated by the continuous increase of carbon emissions from industries, transportation, livestock, and agriculture activities, are modifying Earth's temperature and precipitation patterns. Climatic emissions have been decreasing since 2000 according to the [1]. However, BP [2] has noted that energy consumption emissions went back to pre-pandemic levels in 2021, showing the global economic recovery from the COVID-19 crisis. These events have severe and mass implications, which especially put the polar regions and coastal areas at risk. Ice caps that are melting and rising sea levels are detrimental to the living conditions and food security of coastal residents, who depend on fertile and productive lands for their social activities, the economy, and their well-being. For that reason, it is really important to take immediate action by both mitigating and adapting to climate change and global warming to safeguard the coastal human populations.

Climate change affects inland areas differently in comparison to coastal areas, especially in arid and semi-arid regions. It causes water scarcity, which harms agricultural production and enhance environmental salinization (soil and water resources). It also leads to desertification, land degradation, and dust storms, which disrupt food and goods supply chains, increase involuntary migration, disruption in sustainable power supply, increase industrial sustainability and investment risks, increase health problems, and weaken political and economic stability. These risks and disadvantages endangered the sustainability of local communities [3, 4].

2 Methodology

The Preferred Reporting Items for Systematic Reviews and Meta-Analyses (PRISMA) approach guided a bibliographic research study on "green space expansion using haloculture technology to combat climate change and global warming in the MENA region" from 1993 to 2024, utilizing databases like Scopus, Web of Science, and Google Scholar. Search terms included "climate change," "haloculture," and "green space." The review was limited to peer-reviewed journals and books, with the author reviewing the titles and abstracts to ensure relevance to the study's scope. The Context-Intervention-Mechanism-Outcome (CIMO) framework helped in selecting the most pertinent articles, and snowball sampling was employed to incorporate additional relevant cited articles.

3 Climate Change Effects on MENA Cities and Residential Locations

Climate change significantly influences MENA cities, by altering weather patterns, increasing temperatures, exacerbating water scarcity, and put them in risk of soil/ water resources salinization. These changes threaten urban infrastructure, health, and livelihoods, necessitating adaptive strategies for residential areas [3, 4]. Understanding these effects is crucial for developing resilient cities capable of withstanding and mitigating the adverse impacts of a changing climate [5].

3.1 Climate Change Poses a Severe Challenge to Communities

Climate change is a critical problem that makes food safety and job market face serious difficulties, especially to the communities in coastal and arid regions such as the Middle East and North Africa (MENA) and southern Europe, which are already dealing with its harmful effects (see Table 1).

3.2 Global Climate Change Affects Communities

Climate change endangers human populations and their surrounding ecosystems in various aspects, including food, water, energy, and safety, though not all communities and environments are equally affected (see Table 2). Some, like indigenous communities, are more vulnerable due to their strong reliance on local natural resources for livelihoods and culture, experiencing significant impacts on their food systems and lands. Similarly, low-income groups, minorities, women, children, and the elderly face heightened risks from extreme weather events such as floods, droughts, heat waves, and storms, compounded by their limited access to resources and services to manage the effects of climate change [4].

Table 1 Climate change negative impacts reports

Number	Location	Climate change possible/present effects	Reference
1	Nile Delta, Egypt	Increasing agriculture and tourism risk	[6]
2	Jordan River Basin	Water scarcity	[7]
3	Maghreb region	Droughts, heat waves, floods, storms, and desertification problems	[8]
4	Lake Urmia, Iran	Urmia lake diminishing	[9]
5	Red Sea coral reefs, Saudi Arabia	Local ecosystem services are threatened by climate change	[10]
6	Mesopotamian marshes, Iraq	Reduce precipitation, increased evaporation, and altered river flows	[11]
7	Indus River Delta, Pakistan	Intensification of water and soil salinization	[12]
8	Doha Bay, Qatar	Sea level rise, storm, coastal flooding, heat stress, and air pollution	[13]

Table 2 Some climate change negative effects and challenges

Number	Effects and challenges	References
1	Forest fires	[14]
2	Hurricanes	[15]
3	Droughts	[16]
4	Heat waves	[17]
5	Air pollution	[18]
6	Food insecurity	[19]
7	Soil salinization	[20]

3.3 Climate Change Silent Movement is Behind Social and Psychological Problems

Climate change has severe effects on human health, the physical and mental, and leads to a number of common health problems in various parts of the world [21]. It influences the mental health through various natural disasters such as floods, droughts, heat waves, and storms, which can undermine mental health and social relationships. The emerging health risks including climate anxiety and the accompanying emotional reactions are to be addressed with clinical attention, despite the fact that they can be interpreted as either adaptive or maladaptive. It is of the utmost importance that individual mental health should not become the only focus while we as a society address global warming [22]. The effects of climate change on human mental health are of different types and durations, from direct and immediate to indirect and

Table 3 Some climate change social and psychological problems

Number	Effects	References
1	Climate anxiety in children and young	[23]
2	Anger, depression and anxiety	[24]
3	Post-traumatic stress disorder, depression, anxiety, psychotic symptoms, suicidal ideation and suicide completion	[25]
4	Ecological anxiety ecological grief	[26]

long-lasting. The most dangerous events can cause people to suffer from traumatic stress disorders, which can also become the matter for the common mental disorders, and the manifestation of chronic events can produce effects that are delayed and/or are intergenerationally carried on like, for example, posttraumatic stress disorders [21] (Table 3).

Climate change has serious environmental (air, water and soil) effects [20], which lead to health impacts and violate human rights and social justice, especially for low-income societies and poor people in high-income ones. Climate change causes temperature rise, precipitation change, harsh weather, and sea level rise, which could damage agriculture productivity, water quality and quantity, job market, and land degradation. Regarding the matter of the facts, it forces many individuals to migrate their homelands. Climate change also prompts heat stress, infectious diseases, allergies, malnutrition, violence rises, and mental disorders. These impacts affect the rights to life, food, water, health, security, shelter, and culture of vulnerable groups, such as the poor, minorities, women, children, elderly, disabled, sick, and workers. Moreover, low-income countries emit less greenhouse gases but suffer more from climate change than high-income countries, and have less ability to adapt. Therefore, climate action should protect human rights, promote social justice, and support vulnerable populations [27, 28]. Short-term changes, especially crucial changes in climate conditions, trigger a significant response in and by human society. The anthropogenic climate changes have long-term effects on societies. Regarding the deep socio-economic impacts of these changes, policy makers are advised to develop optimal adaptation strategies [29].

4 Heat Island Effect, Energy Sustainability, and Prevention Methods

The heat island effect is a phenomenon found in urban areas, where temperatures are higher than those in nearby suburban regions. This temperature rise increases energy consumption for air conditioning and negatively impacts human health, activities, and work performance [30]. Driven by rapid urban population growth, socio-economic development, and changes in land use, this effect significantly alters land surface temperatures. Addressing the urban heat island effect is essential for sustainable urban development [31].

Several strategies have been suggested to mitigate the heat island effect and lower energy consumption,, including altering the thermal properties of roofs and paved surfaces, using more reflective materials, and increasing green spaces and plant cultivation [30]. Yet in the dry and semi-dry regions, the growth of greenery and the application of old-school farming can bedifficult because of scarce water endowments and the issues of land salinization and water salinity [32]. Hence, haloculture technology could be a promising solution for green space development in these climate conditions.

4.1 Heat Island Effects

The increase in land surface temperature is driving significant changes in urban ecological systems, disrupting material and energy flows, and altering their structures and functions. This affects urban climate, hydrology, soil characteristics, atmospheric conditions, biological patterns, and public health. Alterations in wind patterns, precipitation, air quality, evaporation, runoff, and water infiltration may lead to heightened greenhouse gas emissions, ozone formation, and increased particulate matter. Additionally, the heat island effect raises resource consumption, waste production, and heat-related health issues, necessitating urgent action from urban planners, policymakers, and residents [33].

The heat island effect, as mentioned earlier on, is a term used to describe the phenomenon whereby the urban areas boast of higher temperatures than the neighboring suburbs. This temperature increase is mostly ascribed to the heat absorption and retention by buildings, roads, and other infrastructure [34]. The heat island area should be considered in terms of removing the natural cover of the earth (plants, soil, and water) and replacing it with anthropogenic surface items (concrete, asphalt, etc.), reducing evaporation and transpiration, creating and importing heat sources, and reducing wind flow in urban areas. It is important to note that heat island effects can significantly affect energy consumption, human health, and overall urban livability. This heat island effect increases energy consumption for communities and cities [35].

High temperatures significantly impact daily life and activities, especially in large cities where heat-absorbing surfaces from buildings and roads can double the temperature increase. Urban activities like transportation contribute further to this rise. As heat absorption grows, so does the surrounding temperature. Planning, design, and construction methods can hinder air and wind flow, reducing wind speed and causing air stagnation, which in turn exacerbates pollution and raises air temperatures. In warmer regions, these high temperatures lead to economic issues, such as increased maintenance costs and damage to urban infrastructure, as well as social problems, including a rise in certain physical and mental health conditions [36].

4.2 Heat Island and Energy Consumption in Municipal Areas

Heat islands directly affect the energy consumption in urban areas and the main way is through the increased use of air conditioning of buildings in these regions. This rise in cooling demands, not only is the increased energy consumption but also the increase in carbon emissions [34] that accompany it. Besides the rising energy needs, the heat island areas also put great pressure on the electrical grid, which can cause power cut-offs and a subsequent increase in the city's energy demand.

Urban heat islands significantly influence building energy consumption by raising cooling demand and lowering heating demand. However, the effects of these heat islands on energy use remain understudied due to the challenges in quantifying temperature changes. On average, cooling energy consumption increases by 19.0%, while heating energy consumption decreases by 18.7%. Notably, inter-city variations show that cooling energy consumption can rise between 10 and 120%, while heating consumption can drop from 3 to 45%. Intra-city variations are particularly pronounced in urban centers [37]. Both urban heat islands and global warming contribute to rising ambient temperatures, significantly affecting electricity consumption in the building sector. For every degree increase in temperature, peak electricity demand can rise by 0.45% to 4.6%, leading to an additional electricity penalty of 21 watts per degree per person. Moreover, total electricity demand can increase by 0.5% to 8.5% for each degree of temperature rise [38]. These findings underscore the critical need to address urban heat islands and global warming to mitigate their impact on electricity consumption in buildings. The increases in peak and total electricity demand per degree of temperature highlight the urgency for sustainable energy solutions and efficient cooling systems in urban areas.

Cities, occupying only 2% of the Earth's surface, consume 75% of the world's energy resources. Strategies in urban design can be modified to reduce energy consumption and CO2 emissions, like green spaces, trees, albedo, pavement surfaces, vegetation, and building design and types, which can directly affect city temperatures on a local scale, assisting professionals in urban planning and design [39]. Implementing sustainable energy solutions and efficient cooling systems in urban areas can help alleviate the strain on electricity demand during peak times and reduce overall energy consumption. Additionally, incorporating strategies such as green spaces, trees, and building design can not only mitigate the heat island effect but also contribute to a more environmentally friendly and livable city.

4.3 Heat Island Prevention Methods

The urban heat island has an impact on global health and well-being. It happens because man-made materials and heat production are increasing, which make cities hotter than other areas. This effect also means we need more energy, which heats up cities even more and leads to problems for the environment and people's health [40]. Mitigating the urban heat island effect involves implementing strategies that can lower land surface temperatures and enhance the urban environment. Improvements in energy efficiency, urban landscape optimization, green roof construction, high-reflectivity materials, and green land cultivation can significantly alleviate this effect. These approaches help decrease heat absorption and generation, promote heat dissipation and reflection, and enhance cooling and shading in urban areas. Ultimately, they offer multiple benefits, including energy savings, reduced emissions, improved air quality, increased biodiversity, and enhanced human health and well-being [41].

In this regard, asphalt concrete, a common pavement material, contributes significantly to urban heat islands due to its low albedo and high volumetric heat capacity. To mitigate this, cool pavements, either reflective or evaporative, have been suggested. However, these strategies should be based on urban geometry and local atmospheric conditions, as they can increase the urban heat island effect. Furthermore, urban green space elements like street trees, parks, and roof gardens reduce urban heat island effects and enhance cooling from water and wind sources. Combining these techniques is optimal for effective urban heat island mitigation. Likewise, there are many possible sources, and alternative measures for mitigation have been described [40]. To mitigate the heat island effect and reduce energy consumption in arid regions, haloculture can be implemented as a solution. Haloculture refers to cultivating plants using saline water and soil, which can help overcome the limitations of freshwater availability in arid and semi-arid areas. Using haloculture techniques, green spaces can be developed and maintained without relying on freshwater resources.

5 Green Space, Climate Change, and Community Risks

Green spaces are essential for mitigating climate change impacts and reducing community risks in the MENA region. They help regulate temperatures, improve air quality, and enhance flood resilience, offering significant environmental and social benefits. Expanding urban greenery is vital for creating resilient and sustainable cities in this climate-vulnerable area (Table 4).

6 Saline Water and Soil Resources in MENA Region

A saline environment refers to areas where land is impacted by high levels of salt, such as coastal regions or places with naturally salty soil and water. The plants in these environments have evolved special adaptations, like adjusting their internal water balance and isolating excess salt in specific cells, allowing them to survive in harsh conditions. Despite these adaptations, extreme salinity can still stunt plant growth, reduce productivity, and ultimately disrupt biodiversity, causing an ecological imbalance. To maintain the health of these unique ecosystems—and the human

Table 4 Green space expansion/development can protect cities climate change consequences

Number	Impact	Description	References
1	Temperature regulation/heat island mitigation	Green spaces help mitigate the urban heat island effect by providing shade and cooling through evapotranspiration, reducing the need for air conditioning	[42]
2	Flood mitigation	Vegetated areas absorb rainwater, reducing runoff and lowering the risk of urban flooding during heavy rainfall events	[43]
3	Air quality improvement	Plants in green spaces filter pollutants from the air, improving overall air quality and reducing health issues related to air pollution	[44]
4	Dust control	Green space and green belts can control dust storms and wind erosion	[45]
5	Carbon sequestration	Trees and plants act as carbon sinks, absorbing CO2 from the atmosphere and helping to mitigate greenhouse gas emission	[46]
6	Mental health healing	Green space development and expansion can mitigate the adverse impact of climate change on residents mental health	[47]
7	Biodiversity support	Green spaces provide habitats for various species, promoting biodiversity and enhancing ecosystem resilience against climate change	[48]

communities that rely on them—it's essential to manage and protect saline soils and water resources effectively [49, 50].

The cooling effects of green spaces were affected significantly by the saline surrounding environment. Saline water bodies, like lakes or ponds, enhance cooling through evaporative processes and help moderate lower desert night temperatures. Additionally, the soil composition and moisture content in green spaces affect their heat dissipation capacity; soils with higher moisture can absorb and release heat more effectively, contributing to overall temperature balance [51]. Therefore, incorporating these factors in the design and management of green spaces is crucial for maximizing their potential to mitigate the urban heat island effect.

The saline environment, characterized by high soil and water salinity, presents various economic opportunities through haloculture development. This sustainable approach to producing both biological and non-biological products leverages engineering and scientific methods to enhance food security, generate water and energy, conserve biodiversity, and support rural development. However, the saline environment faces challenges such as climate change, land degradation, water scarcity, pollution, and socio-economic issues. Therefore, haloculture must be supported by integrated and adaptive management strategies that address environmental, social, and economic factors, helping communities mitigate environmental disorders, enhance resilience, and promote sustainability in salt-affected regions [52].

Table 5 Some of the potential applications of haloculture

Number	Haloculture potential application	Reference
1	Mitigate the negative impacts of pandemics on the environment, society, and economy	[52]
2	Incorporate saline soil and water resources to improve local society's sustainability	[53, 54]
3	A proper strategy for sustainable management of saline resources, improvement of environmental quality, and improving bio-economy in salt-affected areas	[55]
4	Promoting food security while also supplying water and energy to improve living conditions in salt-affected regions	[53, 54]
5	Medicinal plants production	[49]

7 Haloculture

Haloculture is a salt-based method that utilizes saline soils and water for economic activities in salt-affected regions. This includes aquaculture, agriculture, horticulture, poultry, and livestock using unconventional resources like sea water, saline lakes and surface waters, saline groundwater, salty wastewater, and drainage. Haloculture provides communities with food, vegetable oil, biofuels, biomass, wood, and new jobs, while also helping cities overcome environmental problems by maintaining green space and expanding the tourism industry [53, 54]. It has great potential in environmental protection and food production (Table 5).

8 Haloculture Use in Green Space Development

Many halophytes have the potential to be cultivated in green spaces, parks, and green belts (Table 6). Haloculture techniques can help develop green space in saline soils and also use saline waters. Haloculture is a technique that involves cultivating salt-tolerant plants in saline soils and utilizing saline waters for irrigation. This approach involves growing salt-tolerant plants, which can increase green space in salt-affected areas and contribute to the production of medicinal plants and vegetables. Such cultivation promotes a bio-economy based on urban agriculture, providing local communities with valuable resources [52]. Furthermore, sustainable management of saline resources can improve environmental quality, positively impacting both society and the economy.

Many halophytes can be cultivated in green spaces, parks, and green belts, making haloculture a valuable technique for developing green areas in saline soils while utilizing saline water for irrigation. This method involves growing salt-tolerant plants, which can enhance green space in salt-affected regions and contribute to the production of medicinal plants and vegetables. Such cultivation promotes a

Table 6 Potential halophytes for green space, parks, and green belts

Species	Life Form	Maximum tolerance
Acacia stenophylla	Tree	61 dS m^{-1}
Alhagi camelorum	herbaceous perennial	30 dS m^{-1}
Atriplex nummularia	Shrub	60 dS m^{-1}
Atriplex paludosa	Shrub	62 dS m^{-1}
Calotropis procera	Tree, Shrub	54 dS m^{-1}
Limonium iranicum	Shrub	31 dS m^{-1}
Lotus pedunculatus	Herbaceous perennial	34 dS m^{-1}
Lycium chinense	Chaemaephyte	51 dS m^{-1}
Prosopis juliflora	Tree	57 dS m^{-1}
Schoenoplectus californicus	Herbaceous	27 dS m^{-1}
Sonneratia apetala	Tree	40 dS m^{-1}
Suaeda nigra	Shrub	54 dS m^{-1}
Suaeda vera	Nano-chamaephyte	54 dS m^{-1}
Tamarix chinensis	Tree, Shrub	54 dS m^{-1}
Tamarix indica	Tree	112 dS m^{-1}
Trifolium fragiferum	Herbaceous perennial	42 dS m^{-1}
Typha domingensis	Herbaceous perennial	30 dS m^{-1}
Typha latifolia	Herbaceous perennial	29 dS m^{-1}
Vigna marina	Herbaceous	40 dS m^{-1}

bio-economy based on urban agriculture, providing local communities with valuable resources [52]. Furthermore, sustainable management of saline resources can improve environmental quality, positively impacting both society and the economy.

Haloculture enhances biodiversity and ecological balance by cultivating halophytic plants and creating salt marshes and wetlands, serving as modern green spaces and green belts for cities [50]. Additionally, haloculture has significant potential for wastewater treatment and pollutant removal from saline effluents [53, 54], addressing saline water issues while generating green spaces and parks near urban areas. Halophytes can also be used for biomass and bioenergy production, contributing to a sustainable and renewable energy sector, which helps reduce carbon footprints and greenhouse gas emissions, thereby mitigating the heat island effect. Moreover, haloculture plays a crucial role in restoring degraded salt-affected areas, revitalizing ecosystems, and improving overall environmental health [55]. This approach has great potential for reducing heat island effects and energy consumption in urban environments.

8.1 Haloculture Effects on Energy Consumption in Arid Areas

Landscape structure—encompassing size, shape, and configuration—affects surface heat flux and temperature mitigation, with spatial complexity being vital for maintaining landscape stability across seasons. The relationship between landscape structure types may differ, as in some cities, the spatial arrangement of green spaces influences landscape variation more than their total area. Urban greening seeks to counteract the negative effects of impervious surfaces on urban temperatures, although the cooling mechanisms of green spaces can vary based on factors like vegetation activity and albedo [42]. Haloculture techniques, which involve cultivating salt-tolerant plants in saline soils and using saline water for irrigation [52], can significantly enhance green space development, especially in coastal cities or arid regions where freshwater is scarce. By employing haloculture, urban greening projects can expand into areas previously considered unsuitable for traditional vegetation. Additionally, cultivating halophytes can increase the diversity and resilience of urban green spaces, enriching their ecological and aesthetic values.

Arid regions experience high-temperature changes, which leads to high energy consumption for cooling and warming. As is obvious, green space development could help reduce energy consumption in arid regions, regarding the role of green space in mitigation and balancing the temperature in residential areas. Implementing haloculture techniques in urban greening can enhance ecological value and lower energy consumption. By cultivating salt-tolerant plants and halophytes, cities can create natural shade, minimizing reliance on artificial cooling during sweltering summers. Additionally, these plants help stabilize soil and combat desertification, increasing the resilience of urban green spaces in challenging climates. Ultimately, haloculture presents a viable strategy for establishing sustainable green areas that promote energy efficiency in arid regions with saline water and soil resources.

8.2 Haloculture Biomass Production in Cities

Salinity is a major issue affecting agricultural lands, leading to lean agricultural yields and barren lands. Halophytes, which can withstand environmental stress and biosynthesize nutrient-rich substances, are gaining attention for food security and saline-enriched landscapes. They can be commercially cultivated for biomass production and oil-rich seed production. Promoting halophyte propagation is an eco-friendly approach to reclaiming salt-enriched soils, and boosting agricultural production and green space development in salt-affected areas (Bhat and [56]).

Biofuel and biodiesel are terms used by politicians, economists, scientists, and the industrial community to describe the production of high-energy organic liquids from plants that can partially replace fossil fuels like gasoline or diesel. These "green" fuel substitutes recycle CO2 in the atmosphere, unlike fossil fuels that add CO2 by burning organic matter. To be economically, environmentally, and socially acceptable, biofuel production should co-exist with food and fiber production, rather than replace it [57]. Haloculture is expanding the capacity of bioenergy production in arid areas.

Halophytes, such as Halopyrum mucronatum, Desmostachya bipinnata, Phragmites karka, Typha domingensis, and Panicum turgidum, produce abundant biomass using saline resources like water and soil [58]. Furthermore, salt cedar trees (Tamarix spp.), known for their fast growth rates, can be used for syngas and biofuel production by converting cellulosic biomass into simple sugars and fermentation [57]. Morover, forage plant species, e.g., sesbania (Sesbania aculeata), kallar grass (Leptochloa fusca), millet rice (Echinochloa colona), and finger millet (Eleusine coracana), could be planted in saline fields to produce biomass [59]. These halophytes and many other similars have the potential to be utilized in the green belt and green space of cities and towns simultaneously, which, alongside biomass production, could expand the green space and mitigate heat island effects.

Biofuel and biodiesel are terms used by politicians, economists, scientists, and the industrial community to describe the production of high-energy organic liquids from plants that can partially replace fossil fuels like gasoline or diesel. These "green" fuel substitutes recycle CO2 in the atmosphere, unlike fossil fuels that add CO2 by burning organic matter. To be economically, environmentally, and socially acceptable, biofuel production should co-exist with food and fiber production, rather than replace it [57]. Haloculture is expanding the capacity of bioenergy production in arid areas.

Halophytes, such as Halopyrum mucronatum, Desmostachya bipinnata, Phragmites karka, Typha domingensis, and Panicum turgidum, produce abundant biomass using saline resources like water and soil [58]. Furthermore, salt cedar trees (Tamarix spp.), known for their fast growth rates, can be used for syngas and biofuel production by converting cellulosic biomass into simple sugars and fermentation [57]. Morover, forage plant species, e.g., sesbania (Sesbania aculeata), kallar grass (Leptochloa fusca), millet rice (Echinochloa colona), and finger millet (Eleusine coracana), could be planted in saline fields to produce biomass [59]. These halophytes and many other similars have the potential to be utilized in the green belt and green space of cities and towns simultaneously, which, alongside biomass production, could expand the green space and mitigate heat island effects.

The development of saline agriculture for bioenergy crops, based on haloculture technology in cities, requires identifying region-specific halophytes, improving seed accessibility, developing nurseries, optimizing cultural practices, establishing processing plants, and developing new landscape management systems. Plant breeders and molecular biologists must collaborate to incorporate desirable changes in plant makeup for fuel production, high-value products, and improved yield potential. Large-scale planting of halophytes in cities has not yet been attempted, so future research will focus on practical challenges and future research directions [60].

9 Some Potential Plants and Cases for Green Space, Management, Adaptation, and Expansion Using Haloculture Technology

Halophytes and other salt-tolerant plants offer promising alternatives for many developing countries. These plants thrive in saline or extremely salty environments due to their unique adaptations that help them cope with high salinity through various eco-physiological mechanisms. They naturally grow or can be cultivated in areas affected by salt, such as saline semi-deserts, mangrove swamps, marshes, degraded soils, and seashores. Their ability to flourish in these challenging conditions makes them an excellent potential source of livestock fodder [52].

Several species of halophytes, along with salt-tolerant shrubs, grasses, and legumes, could be effectively grown in saline lands to feed livestock. Examples include species like Kochia, Juncus, Leptochloa fusca, Acacia, Suaeda fruticosa, Nitraria retusa, Salsola, Atriplex, Paspalum distichum, and Scirpus litoralis. The quality of fodder from these plants depends on a mix of climate, soil, and plant factors. One of the key advantages of halophytes is their ability to tolerate high salt levels, which gives them economic potential in arid and semi-arid regions [53].

Recognizing their value, salt-tolerant forage crops have already been integrated into rangeland improvement programs in several salt-affected regions across the globe. Though the economic benefits of halophytes are still being explored, these plants are receiving more attention, especially in arid regions where soil salinity is a critical issue. On grazing lands, halophytes like Atriplex spp. serve as a supplementary source of nutrients to traditional feed, and species like Atriplex canescens have been widely used in ecological conservation and livestock production [3].

Halophytes also offer potential as nutraceutical sources. For instance, *Salicornia perennis* (also known as *Sarcocornia perennis subsp. perennis*) is used in human diets for its health-promoting components. Its leaves are rich in proteins (6.9 mg per 100 g dry weight), polyunsaturated fatty acids (with linolenic acid making up about 21% of PUFA), and essential minerals like calcium, magnesium, iron, and potassium. Another species, *Salicornia ramosissima*, is valued as gourmet food because it contains antioxidants such as phenolic compounds and vitamins like α- and γ-tocopherol [52, 53, 55].

Additionally, *Suaeda maritima* (also known as *Chenopodium maritimum*) is rich in calcium, magnesium, iron, zinc, and copper—providing up to 10% of the recommended daily intake—and is also a good source of vitamins like beta-carotene, lutein, and vitamins A and C. Many halophytic species, such as *Suaeda* and *Sarcocornia*, contain phenolic compounds like gallic acid, flavonoids like quercetin, and other beneficial antioxidants. *Suaeda vera*, in particular, has been noted for its antioxidant, anti-inflammatory, and anticancer properties. Its leaves also accumulate sugars like galactoarabinans, which have antioxidant effects and contribute to cellular stability [3].

In summary, halophytes offer not only agricultural and ecological benefits but also potential health benefits, making them valuable resources in regions with saline soils and harsh conditions.

10 Haloculture Strategies, and Pedagogical Content Knowledge (PCK) Approaches to Combat Climate Change Defect in MENA Region

Most of MENA cities face soil and water resources salinity problems. Moreover, haloculture is promising potential technology to establish/develop/expand/adapt green spaces in these cities inner/surrounding environment. Using haloculture to expand green spaces in terrestrial cities and coastal cities differs primarily due to the environmental conditions. Coastal cities naturally have saline conditions, making it easier to integrate halophytes in green space development. Additionally, coastal cities can leverage natural saline water sources, whereas terrestrial cities often need to manage saline groundwater or wastewater. On the other hand, terrestrial cities with salt-affected soils mostly have dry and warm environments. In addition, in terrestrial areas soil conditions and natural vegetation are different. Therefore, according to the nature of these environments, different potential haloculture strategies are possible to apply (Table 7).

Promoting haloculture for the sustainable use of saline soils in the MENA region necessitates the development of effective pedagogical content knowledge

Table 7 Some potential haloculture strategies for green space development

Zone	Potential Haloculture Strategies
Coastal cities	Wetland restoration; Restoring and creating wetlands using haloculture principles can serve as natural buffers against coastal pollution and degradation
	Green space development with seawater using halophytes, e.g., *Polygonum* spp, *Aster Tripolium*, etc
	Coastal park development in shallow waters using halophytes e.g., *Avicennia* spp. trees
Terrestrial cities with salt-affected soil and saline water resources	Establishment of green belt with halophytes, e.g., genus of *Tamarix* and genus of *Haloxylon* trees
	Green space development using halophytes, e.g., *Amaranthus* spp., *Panicum* spp., etc
	Phytoremediation/land restoration with halophytes; haloculture can be used a base tech to restore salt-affected and polluted soil and effluents

(PCK) pathways. These pathways can connect scientific principles with practical applications, enhancing understanding among both students and practitioners [5].

One promising PCK method involves integrating haloculture into environmental science and agricultural curricula tailored for local communities. This can be achieved through a blend of theoretical lessons and hands-on activities focused on haloculture. For example, students can study the biology, ecology, and cultivation of halophytes, as well as the challenges posed by saline soils and waters, while also learning the principles of developing green spaces based on halophyte cultivation. This theoretical framework can be enhanced through practical experiences, such as field trips to haloculture sites and laboratory experiments that investigate plant growth in saline conditions. Additionally, students can observe traditional cultivation practices and the utilization of halophytes, providing real-world context that reinforces their learning. Such an integrated approach not only deepens understanding but also equips communities with the skills needed to implement sustainable practices effectively.

Additionally, project-based learning (PBL) serves as an effective PCK pathway for integrating haloculture into education. Students can engage in projects that involve designing and implementing small-scale haloculture systems within their schools using native halophytes. These initiatives can encompass tasks such as selecting suitable halophyte species, developing irrigation plans utilizing saline water, creating simple designs, and exploring the environmental benefits of haloculture. By addressing real-world challenges, students can enhance their critical thinking and problem-solving skills while acquiring practical knowledge about sustainable haloculture practices, climate change, and green space development. Overall, incorporating haloculture into educational programs through PCK pathways can significantly increase awareness and adoption of sustainable practices. This approach not only contributes to the development of green spaces but also serves as a viable strategy for mitigating the impacts of climate change in the MENA region.

11 Conclusion

Climate change and global warming are among the most pressing issues of our time, affecting the environment, human well-being, and sustainable development in various ways. They pose serious threats and challenges to Cities of the MENA regions, which are already experiencing the negative impacts of climate change on their natural resources, food security, health, and energy. This chapter introduced and discussed "climate change effects on MENA cities, green space, climate change, community risks, and heat island, saline water and soil resources, haloculture is a technology, and also some potential plants and cases for green space expansion using haloculture technology". At the bottom line, in this chapter authors tried to propose haloculture for green space establishment, management, and expansion to combat climate change defects in the MENA region. Moreover, this chapter introduces some pedagogical content knowledge (PCK) approaches to promote haloculture technology for the use of haloculture in green space.

12 Recommendations

In light of the deliberations in this chapter concerning haloculture and its capacity to confront climate change threats in MENA cities, the following constructive and scientifically grounded suggestions can be proposed:

Develop Comprehensive Educational Programs: Identify and establish educational programs that include haloculture in existing courses such as environmental science and agriculture. These programs would include PBL exercises to help students construct plans and develop small saltwater agriculture systems. During such activities, learners will be getting practical knowledge and experience on the importance of halophytes and their role in developing sustainable agricultural systems and greening.

Conduct Field Research and Case Studies: Commence field studies for finding the scope and limits of the halophyte species performance within various saline-salt affected areas coverage in the cities of the MENA region. This research can provide several useful insights related to how fast halophytes grow, how much biomass they produce, and their ecologic service potentials, which would recommend the right approaches to be used in the incorporation of halophytes within the modern cities.

Implement Community-Based Haloculture Projects: Engage local communities in haloculture initiatives to bolster food security and promote sustainable land use. Establishing community gardens with native halophytes can provide educational opportunities while fostering community involvement and stewardship.

Advocate for Policy Support for Haloculture: Position policies that enable the practise of haloculture in the course of planning and constructing cities. This may entail providing a reward for the use of halophytes in beautifying the environment, green projects and combating heat in cities.

Enhance Public Awareness Campaigns: Mobilize the citizenry to understand that haloculture can solve food problems and climate change threats. These actions may showcase some of the existing projects and those in plan to make people sympathetic to the cause.

Collaborate with Stakeholders: Encourage collaboration among governments, educational institutions, non-governmental organizations and the private sector to make r a multidisciplinary approach to haloculture. Such collaborations might help in sharing knowledge, pooling resources and making coordinated attempts to practice sustainable methods.

Integrate Technological Innovations: Use advanced technologies like irrigation and plant-health remote sensing to assess the performance and growth of the halophytes systems. These measures can improve the management practices and enhance the efficiency of resources use in salinizing conditions.

Establish Monitoring and Evaluation Frameworks: Create a strong set of mechanisms for monitoring and evaluating the success of various haloculture initiatives. This would identify positive practices and those that require adjustment to assure long-term and widespread implementation of haloculture.

Thus, by implementing these recommendations, MENA cities can use haloculture to improve climate change urban resilience, provide additional green spaces, and efficiently address climate change-related challenges.

References

1. FAO (2020) Emissions due to agriculture. Global, regional and country trends 2000–2018. FAOSTAT Analytical Brief Series No. 18
2. British Petroleum (BP) (2022) Statistical Review of World Energy 2022 (71st ed.)
3. Parnian A, Parvizi H, Selmy S, Mushtaq Z (2024) Haloculture: a pathway to reduce climate change consequences for societies. In: Negm AM, ElZein Z (eds) Integration of core sustainable development goals in rural areas. Springer, Berlin, pp 1–15. https://doi.org/10.1007/978-3-031-60149-1_14
4. Mahbod M, Parnian A (2024) Sustainable development in rural areas: climate change impacts on water, agriculture, energy, and psychological health consequences. In: Negm AM, ElZein Z (eds) Integration of core sustainable development goals in rural areas. Springer, Berlin, pp 1–20. https://doi.org/10.1007/978-3-031-60149-1_3
5. Mahbod M (2025) Climate change and global warming effects on agricultural communities: Mitigation through pedagogical content knowledge (PCK) approach. Int J Appl Experim Biol 4(1):149–154. https://doi.org/10.56612/ijaaeb.v1i1.130
6. El-Nahry AH, Doluschitz R (2010) Climate change and its impacts on the coastal zone of the Nile Delta, Egypt. Environ Earth Sci 59:1497–1506
7. Al Qatarneh GN, Al Smadi B, Al-Zboon K, Shatanawi KM (2018) Impact of climate change on water resources in Jordan: a case study of Azraq basin. Appl Water Sci 8:1–14
8. Schilling J, Hertig E, Tramblay Y, Scheffran J (2020) Climate change vulnerability, water resources and social implications in North Africa. Reg Environ Change 20:1–12
9. Schulz S, Darehshouri S, Hassanzadeh E, Tajrishy M, Schüth C (2020) Climate change or irrigated agriculture–what drives the water level decline of Lake Urmia. Sci Rep 10(1):236
10. Hereher ME (2016) Vulnerability assessment of the Saudi Arabian Red Sea coast to climate change. Environ Earth Sci 75(1):30
11. Albarakat R, Lakshmi V, Tucker CJ (2018) Using satellite remote sensing to study the impact of climate and anthropogenic changes in the Mesopotamian marshlands. Iraq. Remote Sensing 10(10):1524
12. Jamali H, Jamali M, Hayat S, Iqbal J (2023) Indus river delta: contested narratives in the climate of change. Asian J Soc Sci
13. Ajjur SB, Al-Ghamdi SG (2022) Exploring urban growth–climate change–flood risk nexus in fast growing cities. Sci Rep 12(1):12265. https://doi.org/10.1038/s41598-022-16410-0
14. Aponte C, de Groot WJ, Wotton BM (2016) Forest fires and climate change: causes, consequences and management options. Int J Wildland Fire 25(8):i–ii

15. Appendini CM, Meza-Padilla R, Abud-Russell S, Proust S, Barrios RE, Secaira-Fajardo F (2019) Effect of climate change over landfalling hurricanes at the Yucatan Peninsula. Clim Change 157:469–482
16. Stewart IT, Rogers J, Graham A (2020) Water security under severe drought and climate change: disparate impacts of the recent severe drought on environmental flows and water supplies in Central California. J Hydrol X 7:100054
17. Hopke JE (2020) Connecting extreme heat events to climate change: Media coverage of heat waves and wildfires. Environ Commun 14(4):492–508
18. Weiand L, Schmitz S, Becker S, Niehoff N, Schwartzbach F, von Schneidemesser E (2019) Climate change and air pollution: The connection between traffic intervention policies and public acceptance in a local context. Environ Res Lett 14(8):085008
19. Dasgupta S, Robinson EJ (2022) Attributing changes in food insecurity to a changing climate. Sci Rep 12(1):4709
20. Okur B, Örçen N (2020) Soil salinization and climate change. In: Climate change and soil interactions. Elsevier, Amsterdam, pp 331–350
21. Cianconi P, Betrò S, Janiri L (2020) The impact of climate change on mental health: a systematic descriptive review. Front Psych 11:74
22. Clayton S (2020) Climate anxiety: psychological responses to climate change. J Anxiety Disord 74:102263
23. Hickman C, Marks E, Pihkala P, Clayton S, Lewandowski RE, Mayall EE, van Susteren L (2021) Climate anxiety in children and young people and their beliefs about government responses to climate change: a global survey. Lancet Planet Health 5(12):e863–e873
24. Stanley SK, Hogg TL, Leviston Z, Walker I (2021) From anger to action: Differential impacts of eco-anxiety, eco-depression, and eco-anger on climate action and wellbeing. J Clim Change Health 1:100003
25. Burke M, González F, Baylis P, Heft-Neal S, Baysan C, Basu S, Hsiang S (2018) Higher temperatures increase suicide rates in the United States and Mexico. Nat Clim Chang 8(8):723–729
26. Cunsolo A, Harper SL, Minor K, Hayes K, Williams KG, Howard C (2020) Ecological grief and anxiety: the start of a healthy response to climate change? The Lancet Planet Health 4(7):e261–e263
27. Grundmann R (2016) Climate change as a wicked social problem. Nat Geosci 9(8):562–563
28. Levy BS, Patz JA (2015) Climate change, human rights, and social justice. Ann Glob Health 81(3):310–322
29. Stehr N, Von Storch H (1995) The social construct of climate and climate change. Clim Res 5(2):99–105
30. Zou Z, Yan C, Yu L, Jiang X, Ding J, Qin L, Qiu G (2021) Impacts of land use/land cover types on interactions between urban heat island effects and heat waves. Build Environ 204:108138. https://doi.org/10.1016/j.buildenv.2021.108138
31. Qiu J, Li X, Qian W (2023) Optimizing the spatial pattern of the cold island to mitigate the urban heat island effect. Ecol Ind 154:110550. https://doi.org/10.1016/j.ecolind.2023.110550
32. Abulibdeh A (2021) Analysis of urban heat island characteristics and mitigation strategies for eight arid and semi-arid gulf region cities. Environ Earth Sci 80(7):259
33. Karimi A, Mohammad P, García-Martínez A, Moreno-Rangel D, Gachkar D, Gachkar S (2022) New developments and future challenges in reducing and controlling heat island effect in urban areas. Environ Dev Sustain 25(10):10485–10531. https://doi.org/10.1007/s10668-022-02530-0
34. Hirano Y, Fujita T (2012) Evaluation of the impact of the urban heat island on residential and commercial energy consumption in Tokyo. Energy 37(1):371–383. https://doi.org/10.1016/j.energy.2011.11.018
35. Liu Y, Li Q, Yang L, Mu K, Zhang M, Liu J (2020) Urban heat island effects of various urban morphologies under regional climate conditions. Sci Total Environ 743:140589. https://doi.org/10.1016/j.scitotenv.2020.140589
36. Guo A, Yang J, Xiao X, Xia J, Jin C, Li X (2020) Influences of urban spatial form on urban heat island effects at the community level in China. Sustain Cities Soc 53:1072. https://doi.org/10.1016/j.scs.2019.101972

37. Li X, Zhou Y, Yu S, Jia G, Li H, Li W (2019) Urban heat island impacts on building energy consumption: A review of approaches and findings. Energy 174:407–419. https://doi.org/10.1016/j.energy.2019.02.183
38. Santamouris M, Cartalis C, Synnefa A, Kolokotsa D (2015) On the impact of urban heat island and global warming on the power demand and electricity consumption of buildings—a review. Energy and Buildings 98:119–124. https://doi.org/10.1016/j.enbuild.2014.09.052
39. Gago E, Roldan J, Pacheco-Torres R, Ordóñez J (2013) The city and urban heat islands: A review of strategies to mitigate adverse effects. Renew Sustain Energy Rev 25:749–758. https://doi.org/10.1016/j.rser.2013.05.057
40. Mohajerani A, Bakaric J, Jeffrey-Bailey T (2017) The urban heat island effect, its causes, and mitigation, with reference to the thermal properties of asphalt concrete. J Environ Manage 197:522–538. https://doi.org/10.1016/j.jenvman.2017.03.095
41. Yang L, Qian F, Song DX, Zheng KJ (2016) Research on urban heat-island effect. Procedia Eng 169:11–18. https://doi.org/10.1016/j.proeng.2016.10.002
42. Sun R, Chen L (2017) Effects of green space dynamics on urban heat islands: Mitigation and diversification. Ecosyst Serv 23:38–46. https://doi.org/10.1016/j.ecoser.2016.11.011
43. Kim HY (2021) Analyzing green space as a flooding mitigation–storm Chaba case in South Korea. Geomat Nat Haz Risk 12(1):1181–1194
44. Venter ZS, Hassani A, Stange E, Schneider P, Castell N (2024) Reassessing the role of urban green space in air pollution control. Proc Natl Acad Sci 121(6):e2306200121
45. Sheng Q, Guo Y, Lu J, Song S, Li W, Yang R, Zhu Z (2024) A study on the dust retention effect of the vegetation community in typical urban road green spaces—in the case of Ying Tian Street in Nanjing City. Sustainability 16(7):2656
46. Fan L, Wang J, Han D, Gao J, Yao Y (2022) Research on promoting carbon sequestration of urban green space distribution characteristics and planting design models in Xi'an. Sustainability 15(1):572
47. Astell-Burt T, Feng X (2019) Association of urban green space with mental health and general health among adults in Australia. JAMA Netw Open 2(7):e198209
48. Lai H, Flies EJ, Weinstein P, Woodward A (2019) The impact of green space and biodiversity on health. Front Ecol Environ 17(7):383–390
49. Yazdani Biouki R, Khajahhosseini S, Rad M (2020) The potential of using medicinal herbs in haloculture, a case study of the caper plant (Capparis spinosa L.). Acta Hort 1287:9–14
50. Gheyi HR, Sandhu D, de Lacerda CF (2023) Fields of the future: pivotal role of biosaline agriculture in farming. Agriculture 13(9):1774. https://doi.org/10.3390/agriculture13091774
51. Lim J, Skidmore M (2020) Heat vulnerability and heat island mitigation in the United States. Atmosphere 11(6):558. https://doi.org/10.3390/atmos11060558
52. Pirasteh-Anosheh H, Parnian A, Spasiano D, Race M, Ashraf M (2021) Haloculture: A system to mitigate the negative impacts of pandemics on the environment, society and economy, emphasizing COVID-19. Environ Res 198:111228. https://doi.org/10.1016/j.envres.2021.111228
53. Khorsandi F (2022) Haloculture: strategy for sustainable utilization of saline land and water resources. Iran J Earth Sci 8(2):164–172
54. Khorsandi F (2022) Haloculture and Haloengineering in the Context of Water-Energy-Food Nexus. Environ Sci Proc 16(1):63
55. Khorsandi F, Siadati SH, Rastegary J (2020) Haloengineering as a vital component of sustainable development in salt-affected ecosystems. Environ Dev 35:100545. https://doi.org/10.1016/j.envdev.2020.100545
56. Bhat RA, Hakeem KR (2020) Biomass production of various halophytes. In: Handbook of halophytes. Springer, Berlin, pp 1–13. https://doi.org/10.1007/978-3-030-17854-3_90-1
57. Santi G, D'Annibale A, Eshel A, Zilberstein A, Crognale S, Ruzzi M, Petruccioli M (2014) Ethanol production from xerophilic and salt-resistant Tamarix jordanis biomass. Biomass Bioenerg 61:73–81. https://doi.org/10.1016/j.biombioe.2013.11.024
58. Abideen Z, Ansari R, Khan MA (2011) Halophytes: Potential source of ligno-cellulosic biomass for ethanol production. Biomass Bioenerg 35(5):1818–1822. https://doi.org/10.1016/j.biombioe.2011.01.023

59. Qadir M, Qureshi RH, Ahmad N, Ilyas M (1996) Salt-tolerant forage cultivation on a saline-sodic field for biomass production and soil reclamation. Land Degrad Dev 7(1):11–18
60. Sharma R, Wungrampha S, Singh V, Pareek A, Sharma MK (2016) Halophytes as bioenergy crops. Front Plant Sci 7:1372. https://doi.org/10.3389/fpls.2016.01372

The Impact of Green Roofs on Buildings' Thermal Performance in the Hot-Arid Climate

Nesma A. Zakaria Elian

Abstract Cities nowadays are facing several complex challenges due to the rapid increase in urbanization and shortage of green spaces proportion, resulting in negative impacts on all aspects (environmentally, socially, and economically). Green roofs are one of the sustainable tools for healthy cities. They represent an innovative approach to rebalancing the Urban Heat Island phenomena, which is considered a major problem causing high temperatures in cities. Green roofs play a significant role to reduce energy consumption, improve air quality, and decrease stormwater runoff. Despite green roofs' benefits, they still have limited usage in Egypt. In a densely populated city like Cairo, the ratio of green areas per inhabitant is considered one of the lowest proportions in the world. Cairo has a huge number of unutilized rooftops, these spaces could be regarded as a solution to the city's lack of green areas, Additionally, In the face of climate change and human urbanization, the need for more eco-friendly buildings is very essential nowadays. A broad overview of green roofs' benefits and typologies is represented in the paper, analyzing various green roof applications. The main objective of the study is to investigate the impact of green roofs on buildings' thermal performance in hot-arid climates, Moreover, discusses the challenges facing green roof usage in Egypt. Finally, the research proposes an action plan to increase the green areas in Cairo.

Keywords Green roofs · Thermal performance · High temperature · Hot-arid climate · Green areas

---

N. A. Z. Elian (✉)
Architecture Department, Faculty of Fine Arts, Helwan University, Cairo, Egypt
e-mail: nesma-ahmed@f-arts.helwan.edu.eg

Z. ElZein and A. Negm (eds.), *Recent Approaches of Sustainable Architecture in Arid and Semi-arid Cities*, Urban Sustainability,
https://doi.org/10.1007/978-981-95-1645-2_13

1 Introduction

Egypt has been suffering recently from rapid urbanization, high rates of pollution, and a serious lack of green and open areas in both distribution and quantity which has led to several environmental, economic, and social problems, such as Urban Heat Island (UHI) phenomena which causes the high temperature of cities compared to the surrounding spaces and increase in energy consumption [1].

Globally the building sector consumes approximately 40% of total energy consumption, while in Egypt the building sector is responsible for 62% of electricity consumption, and 70% of CO2 emissions [2]. Recently, Energy and water consumption issues have gained great attention due to their direct effect on the budget of the government and individuals [3].

Egypt is currently suffering from losing agricultural lands every day, due to many reasons like desertification and people's ignorance, this has resulted in a lack of greenery areas, a shortage of food insecurity, and struggling to manage water storms. In a densely populated city like Cairo, the ratio of green areas per inhabitant is considered one of the lowest proportions in the world, and it's even lesser as compared to other cities worldwide, which is one of the major problems [4].

In the face of climate change and human urbanization, the need for more eco-friendly and sustainable buildings is becoming more widely recognized worldwide. Adopting green roofing in building sectors is one of the green strategies for mitigating the effects of rapid urbanization and reducing UHI impact, they are considered an effective way to contribute to environmental sustainability and provide several benefits to cities, communities, and citizens. Planting buildings' roofs is a civilized manifestation and a new form of agriculture inside cities to reclaim green areas as a natural breath for residents [1].

Green roofs (GRs) have environmental, economic, and social benefits, these systems can provide ecosystem services in urban areas; improve air quality, mitigate building temperature, increase biodiversity, absorb stormwater, and decrease noise levels as shown in Fig. 1 [5]. Moreover, these systems are common in different climates, mostly for rainwater harvesting and to improve stormwater management, by reducing the peak of urban runoff. Green roofs are not only an architectural element but also an element of sustainable urban ecology and agriculture.

Green roofs are known as (living, vegetative, or eco) roofs that have vegetation planted on the top of the growth medium, adding plants and soil to underutilized rooftop surfaces is frequently regarded as a valuable strategy for converting buildings to sustainable buildings.

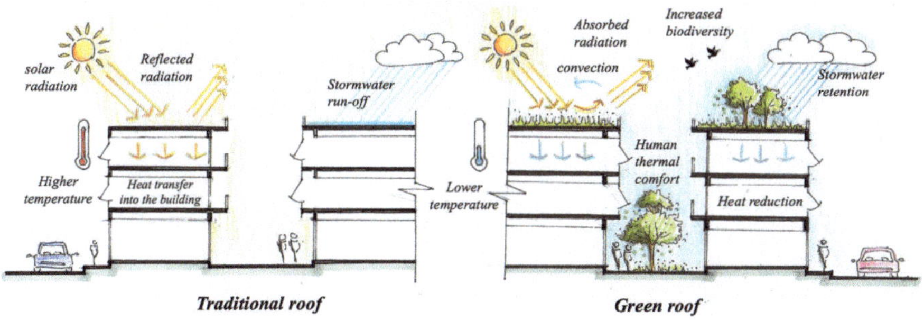

Fig. 1 Schematic showing the environmental impact of green roofs (*Source* by author)

Green roofs are divided into three divisions; (Extensive, Intensive, and Semi-intensive), the main differences between each of these divisions are the plant's height, layer thickness, weight, and maintenance requirements [6]. The green roof system is a very complicated process that depends on various aspects including policies, climatic, environmental, cultural, economic, social, and technical aspects affecting the widespread use of green roofs currently and need to be considered in future studies [7].

A green roof typically consists of various components from the top to the bottom including vegetation, growing media, filtration layer, drainage layer, protection layer, root barrier layer, waterproofing membrane, and structural layer. Green roofs are distinguished from traditional roofs due to their additional components. These components include layers that protect the roofs from leakages, allow excess water to drain, retain sufficient water to support plants, and prevent roots from penetrating the roof membrane [8]. Green roofs' components can improve building performance in their local ecosystems and reduce building energy consumption [9]. There is a significant need to transform greenery areas into livable and attractive places to contribute to the development of sustainable cities [10].

Cairo has a population of more than 22 million people and is among the most densely populated and overcrowded cities in the Middle East, and the decrease in greenery shares occurred because of rapid growth which may have outpaced the need for greenery. A study indicates that the current proportion of the world's population living in urban regions was only 30% in 1950 but reached 55% in 2018 and is anticipated to rise to 68% by 2050. Roughly 90% of this increase is happening in Africa and Asia [11]. At the local, regional, and international levels, these profound changes have a significant influence on people, the environment, and development. Changes include air and water pollution, land use, loss of biodiversity, and ecosystem degradation [12], while carbon emissions from cities and accompanying land use changes are becoming a major contributor to climate change [13].

Cairo has a nearly low share of green areas which is less than the international green spaces standards as shown in (Fig. 2), Between 2017 and 2020, Cairo city lost 910,894 m^2 of its limited green areas. Table 1 illustrates how, as the population grew, the individual proportion of green space decreased from 0.87 m^2 to 0.74 m^2 per capita

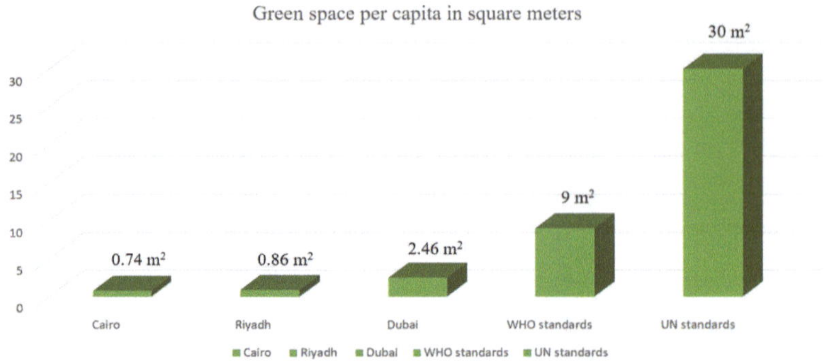

Fig. 2 Schematic representing space per capita in different cities compared to the international standards. (*Source* by author, based on data from [15] and [70])

[4]. Other studies mention that open space has reached 1.5 m^2 per capita but, when Cairo is compared to even more hot-arid cities such as Dubai and Riyadh, it's still lesser [10]. Furthermore, this proportion per capita isn't evenly distributed across Cairo's population, as more than half of the city's inhabitants only have 0.5 m^2 per capita [14]. According to the World Health Organization (WHO), it recommended that 9 m^2 per capita is the minimum ratio, and the ideal is 50 m^2 per capita [15].

The rooftop can be used in a variety of ways including installing solar panels, cultivating plants, and creating reactional spaces. In Egypt, most rooftops were unutilized and regarded as a waste of opportunity [16], roofs were sometimes used to store old building materials, garbage, and building services such as water tanks and satellite dishes as shown in Fig. 3. These spaces could be regarded as a solution to the city's lack of green areas. Green roofs aren't widely used in Egypt, but researchers have recommended that green roofs will solve the shortage of green areas and increase the individual share of greenery in the future.

The goal of this study is to investigate the impact of green roofs on buildings' thermal performance in hot-arid climates, discuss the challenges and limitations facing green roof usage in Egypt, and finally, propose an action plan to increase the green areas in Cairo city by applying green roof systems.

Table 1 Summary of green areas per capita in Egypt between 2017 and 2020 (*Source*: modified by author from [4])

Year	Green spaces area m^2	Population number	Individual share of green spaces m^2/ individual
2017	7,834,388	9,030,128	0.87
2020	6,923,494	9,414,766	0.74

Fig. 3 Rooftops of buildings in Cairo City showing the use of rooftops (*Source* Photo by author)

2 Methodology

The research methodology is based on two approaches; Firstly, the theoretical approach highlights green roofs' definitions, components, typologies, and benefits. The study discusses the ratio of green spaces in Cairo city. Additionally, the paper investigates the impact of green roofs on a building's thermal performance in hot-arid climates. Secondly, the descriptive analytical approach analyzes local and regional case studies of green roof applications, using different research tools. The current research discusses the Challenges facing GRs' application in Egypt, and finally, the research proposes an action plan to increase the green areas in Cairo city see Fig. 4.

3 Literature Review

3.1 Green Roof (GR) Concept

A "Green roof" (GR) is defined as a living vegetation system planted on the building's rooftops [9]. It's known as "Living, Vegetative, or Eco-roofs" because of their ecological benefits [17]. Another definition of GR is a building's roof that is partially or completely covered by vegetation. It can be a sloped roof surface, or a flat roof designed to sustain vegetation cover besides working as a fully functioning roof [18]. Adding a green roof system has been recognized as a sustainable and ecological approach to enhancing the urban and built-up environment [19].

GRs include different layers that preserve the support and enhance the performance of the system. They typically include a vegetation layer, growing media, filter layer, and drainage layer [17]. The idea of GR was designed to promote vegetation

Fig. 4 The research methodology starting by literature review and ends of by results and recommendations. (*Source* Author)

growth over a building's rooftop and that way provides environmental, economic, social, and aesthetical benefits [5].

A "Green roof" is a horizontal living system, that assists in reducing various ecological problems [18]. "Living roof systems" (LR) have been shown as an efficient nature-based solution (NBS) that uses empty urban rooftops to help reduce various environmental, economic, and social problems [20].

Nowadays, cities are using green roofs because of their sustainable ability to resolve many environmental problems, improve building energy efficiency, and limit thermal stress. The most significant goal of planting green roofs in Egypt is to compensate for a lack of green areas, mitigate heat gain, reduce energy consumption, improve the air-purifying of pollutants, enhance the building's thermal performance, and increase the production of the essential needs of fruits, vegetables, and plants, and finally, creating peaceful retreats for people and animals [16].

3.2 *Green Roof Benefits*

Green roofs have several potential benefits; environmental, economic, and social benefits, also GRs can contribute positively towards issues such as mitigation of climate change, biodiversity, flood prevention, and increasing green spaces ratio in urban.

3.2.1 Environmental Benefits

- Boosting thermal performance: green roofs can reduce the ambient temperature, different studies found that the surface of an experimental GR was 2–5 °C cooler during daytime (depending on GR depth and typologies) and was commonly cooler than the ambient temperature [21].
- Mitigation of Urban heat island impact, through shading and cooling buildings.
- Reducing (CO_2) emissions: Green roofs can significantly decrease greenhouse gas emissions, particularly CO2 emissions into the atmosphere [22].
- Insulating properties: A green roof acts as a sound barrier to buildings. It absorbs sound resulting in a quieter environment. Green roofs with deeper substrates can reduce noise by up to 10 dB [21].
- Pollution reduction: the vegetation and growing medium in GR can act as a natural cleaning filter, remove certain airborne particles and pollutants from the atmosphere.
- Increasing biodiversity and preservation of habitat, A well-designed green roof can enhance existing habitats or even provide new habitats for both fauna and flora, resulting in increased biodiversity [23].
- Stormwater management: different studies indicated that green roofs could intercept 50% of the annual rainfall volume from a roof through retention and evapotranspiration [21]. Therefore, it can reduce stormwater runoff, stormwater can be filtered especially by the intensive green roof type, which can help to reduce overload to city sewage treatment systems and decrease the risk of flooding.
- Increasing rooftop agriculture [24], see Fig. 5.

3.2.2 Economic Benefits

- Improve Energy Efficiency: green roofs can act as a heat-transfer-reduction method, resulting in lower energy requirements in buildings in any climate, in either summer or winter [25]. According to [26], applying green roof systems in residential buildings offers significant economic benefits (energy savings ranging from 13 to 39% when compared to un-isolated and traditional roofs).
- Increase the roof's lifespan and protect the building's membrane: When GR is compared to conventional roofs, it is expected that the GR's life span may increase up to 40 years, due to the protective impact of the vegetation from external factors, for example, the sun, wind, rain, and temperature fluctuations [27].

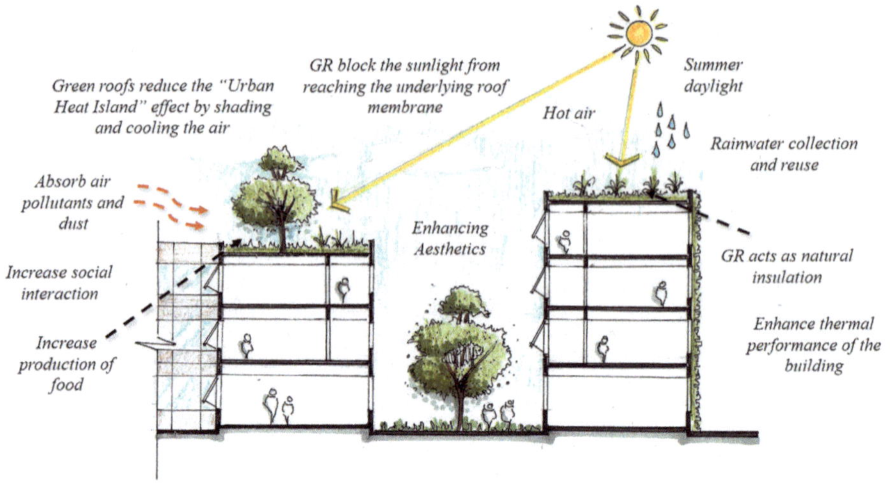

Fig. 5 Schematic showing green roof benefits (*Source*: by author)

- Growing food in cities has a positive impact on local economies by bringing production closer to consumption which saves money and energy.
- Rooftop agriculture also supports local economies by enhancing buildings' economic value [24], The initial cost of green roofs is more expensive to build than traditional roofs, but they are more economical in the long run by increasing food production than urban agriculture [28].

3.2.3 Social Benefits

- Increase social interaction: living or working in a greenery environment brings people together, which improves social cohesion. also, aggression, vandalism, and violence are less of a hindrance in areas with greener areas [29].
- Green spaces play an important role in improving public open space.
- Enhancing Aesthetics: GRs have a natural appearance and make buildings more attractive [17].
- Create job opportunities, in green roof projects, more people are required at different stages of the project including installation workers, maintenance staff, and drivers [30].
- Improve human health: lack of greenery impacts mental and psychological health because people living in cities have fewer opportunities to be physically active. Also, greenery reduces stress and offers relaxation [31].
- Provides a healing environment: green spaces encourage patients to recover faster, which results in a shorter stay in the hospital, a person's pain tolerance increases in a green environment, it is referred to as a "healing environment".

3.3 Green Roof Typologies

Green roofs are categorized into three different types; Extensive, Intensive, and Semi-intensive as shown in (Fig. 6) [17]. The main difference between each of these types is the depth of the substrate, the vegetation used (the thicker layer of substrate allowing for a greater plant variation), maintenance requirements, cost, and construction factors.

3.3.1 Extensive Green Roof

Extensive GR is known as "eco-roofing".It is one of the cheapest roof types to install. It's thin, lightweight, usually panted into a layer of shallow soils, has a growing media with a depth between (60–200 mm), uses small and low- growing plant species, and insignificant needs of maintenance and irrigation [32]. Extensive GR weighs from 60 to 150 kg/m^2, and finally, it's suitable for large flat roof buildings and retrofit projects [33].

3.3.2 Intensive Green Roof

Intensive GR is known as a "roof garden", it's generally heavier mainly because it contains a deeper layer of growing media with a depth between (150 to 400 mm), and it offers the widest variety of vegetation that can include trees, shrubs, and flowers [5], but it's more expensive to install and maintain than the extensive GR, it needs an irrigation system, and it weighs from 180 to 500 kg/m^2 [33]. The construction

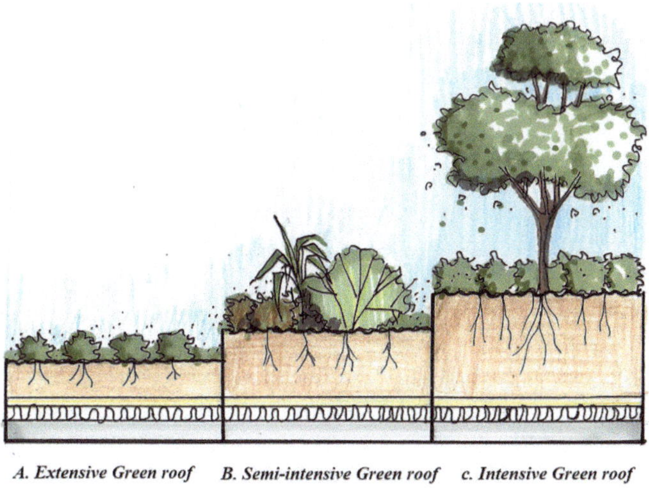

Fig. 6 Typologies of green roofs, (*Source* by author based on [33])

process is more complex compared to the previous type [28]. Intensive GR is a full landscape garden built directly on structures with large weight load capacity. It's easily accessed by people using the roof for different activities. Intensive GR must take into consideration live and dead loads before building construction and it can be a problem to apply it to existing buildings due to these loads.

3.3.3 Semi-Intensive Green Roof

Semi-intensive GR known as "simple intensive" has intermediate characteristics of extensive and intensive GRs [34]. It has a thicker growing media than the extensive roof with a depth between (120–250 mm) and is thinner than the intensive roof, it requires periodic irrigation when needed and a little maintenance. Semi-intensive weighs from 120 to 200 kg/m [33]. Semi-intensive GRs generally accommodate a wide variety of vegetation types due to their substrate depth which includes shrubs, grass, and herbs. Figure 7 shows the comparison between green roof typologies.

3.4 Green Roof Components

Green roof's structure components have eight layers including the roof structure support which is generally a deck. Other layers as shown in (Fig. 8) are a high-quality waterproof membrane, root barrier system, protection fabric, roof drainage layer, filter fabric, a lightweight growing media, and vegetation [16].

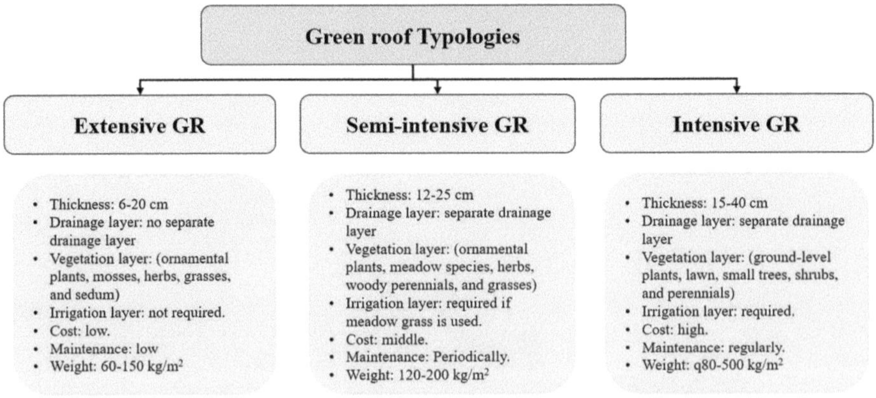

Fig. 7 Comparison between green roof typologies. (*Source* by author from [33])

Fig. 8 Green roof's structural components. (*Source* by author based on [16])

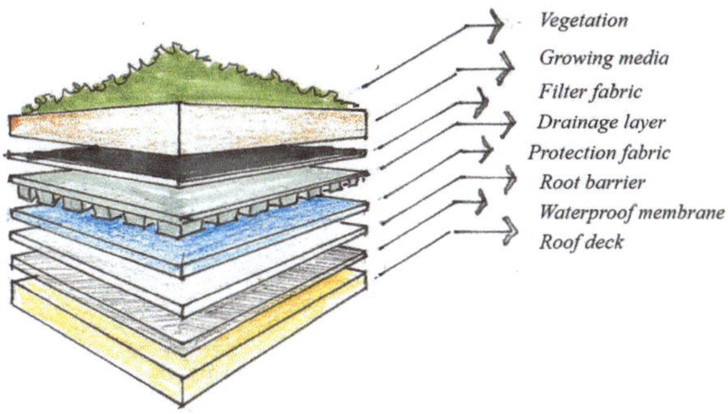

3.4.1 Vegetation Layer

Plants constitute the highest layer that adds life to the rooftop. The success of GRs depends on the plant's health, and the choice of vegetation generally depends on GR typologies (intensive or extensive) and the structural loads. Also other factors must be taken into consideration such as geographical location, growing media, cost, humidity, sun and wind exposure, rainfall intensity and biodiversity benefits. Water is considered a limiting factor for the roof environment. Also, building regulations restrict soil's depth and the soil needs nutrients to maintain plant life performance, so vegetation must have some characteristics such as; a) it does not need frequent irrigation, b) it must have short roots, c) it easily available, d) it must capable to withstand drought and suitable to climate conditions [35].

Different types of plants are suitable for green roofs such as sedums, shrubs, turfs, and herbaceous perennials. It is possible to cultivate various types of plants in Egypt including (a) leafy greens such as; cabbage, spinach, and lettuce, (b) fruiting vegetables such as; eggplant, zucchini, pepper, and cantaloupe, (c) ornamental plants, (d) medicinal and aromatic plants like; basil, rosemary, marjoram, and mint [16].

3.4.2 Growing Media

It is considered one of the main components of the GR system. It's known as the crucial layer that has a direct impact on plant growth and the performance of green roofs, it has various benefits such as a high proportion of natural minerals for increasing vegetation, lightweight, water quality enhancement, runoff reduction, and thermal advantages. The growing media should be designed alongside the design process of the GR project. The load-bearing strength of the roof will influence the design weight and the growing media's depth, and as a result, the selection of planting as well as the green roof system, whether extensive or intensive [35].

3.4.3 Filter Layer

It is used to prevent tiny particles of the vegetation layer from entering the drainage layer, this maintains the function of the drainage layer, filter fabrics have high tensile strength and high water permeability to flow water into the drainage layer [36].

3.4.4 Drainage Layer

It gathers the excess water in the holes therein and drains the rest out of the roof, protects the water-resistant membrane and the hydro insulation from the roots, and improves the thermal properties of the green roof. The selection of a proper drainage layer depends significantly on cost, vegetation species, construction requirements, and GR scale [35].

3.4.5 Protection Layer

The protection layer is located between the drainage and waterproofing layer. It is generally a nonwoven geotextile that protects the waterproofing layer from mechanical damage and root growth.

3.4.6 Root Barrier

The root barrier is an important component of the GR system because it prevents vigorous plant species' roots from penetrating the waterproof membrane. The root barrier type used can be determined by the GR system and plant species.

3.4.7 Waterproofing Membrane

It is considered the most critical layer of the green roof system, due to its ability to prevent water from penetrating the building while also allowing for run-off. Because the moist soil raises the chances of leakages of green roofs, it must be installed by a professional who has experience in GR applications as this layer is critical to the system's success, this layer is made of a material that can withstand hydrostatic pressure (ponding water) for long periods. In hot arid climates, the membrane's durability must be considered during storm events where a large amount of precipitation may fall in a short period [36].

Fig. 9 Illustrative cross-section for soil-based green roof system and an example photo (*Source* photos by author)

3.5 *Green Roof Installation System*

There are different installation systems for green roofs.

3.5.1 Soil-Based Green Roof System

The Soil-based system is the typical type of GR systems that use soil as a growth medium for vegetation. This system consists of three layers; soil, drainage material, and plants. The soil layer is normally about 15 cm deep and it is comprised of a mixture of soil, compost, and other materials suitable for plant growth. The plants used in this type are hardy species that grow in a limited space and are exposed to the conditions of the rooftop environment. Figure 9 shows the details of the Soil-based system and an example photo [1].

3.5.2 Modular Green Roof System

The modular GR system is a modern energy-efficient construction technology, it is comprised of individually pre-planted trays that are arranged on the rooftop, most of these trays include engineered soil blends and plans depending on the regional climate of the building's area, modular systems are characterized by quick installation, easily interchangeable, flexible design, and giving an instant greening effect. Most modular systems are extensive GR their growing media depths are less than 6 inches and plant palettes consist of low-lying ground covers. Tray systems are the most encountered in practice, plastic interlocking containers are frequently filled with growing media, drainage layer, and vegetation, and tray systems are easily replaced or removed without causing any damage to other plants or the original structure. Figure 10 shows the details of the Modular green roof system [37].

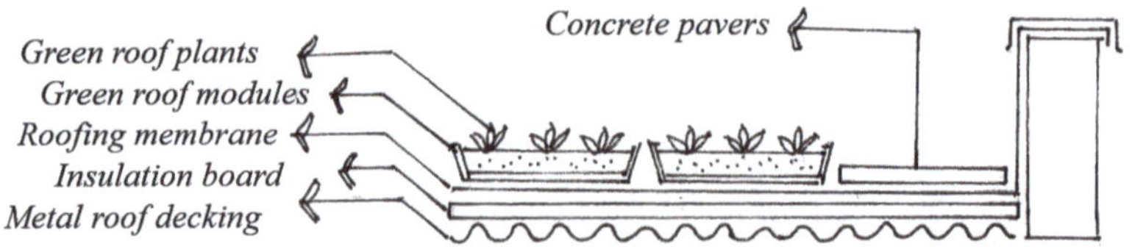

Fig. 10 Illustrative cross-section for modular green roof system (*Source* by author)

3.5.3 Nutrient Film Technique Green Roof System

The hydroponic farming system is a type of agriculture that includes growing plants using only water nutrients and growing media. This method of gardening doesn't use soil. This system takes up less space and water than the traditional agriculture systems, uses recycled water, does not require pesticides, and can be used in arid regions.

The Nutrient Film Technique (NFT) is a type of hydroponic growing system that uses a thin film of nutrient-rich water constantly over the roots of the plants, it is commonly used to grow small fast-growing plants including leafy greens, herbs, and lettuce. The NFT system comprises a narrow channel or trough filled with nutrient solution, as well as a sloping surface that allows the solution to flow over plant's roots. The main benefits of applying an NFT system for a GR are that it enables plants to grow in a small space with little water and nutrient inputs and increases the growth rate of most plants by at least 25% more than their counterparts within the soil. Figure 11 shows the details of the NFT system and an example photo [1, 38].

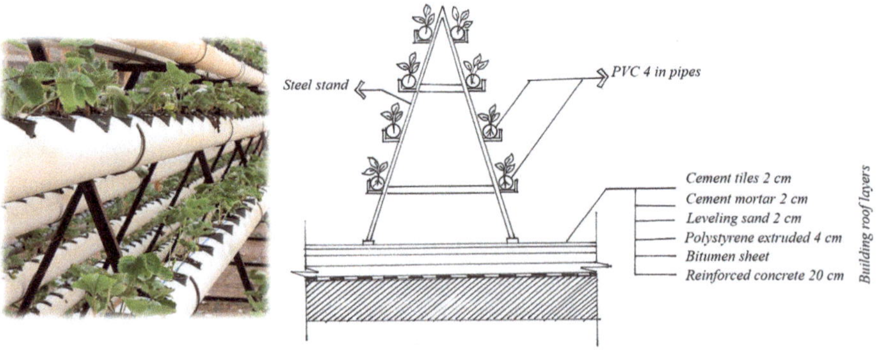

Fig. 11 Illustrative cross-section for NFT green roof system and an example photo (*Source* photos by author)

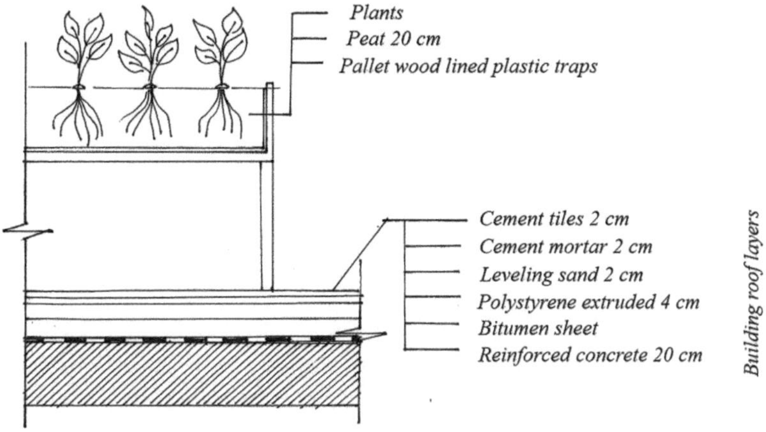

Fig. 12 Illustrative cross-section for DWC green roof system, (*Source* by author)

3.5.4 Deep Water Culture Green Roof System

The Deep Water Culture (DWC) is a hydroponic growing system that cultivates plants in a submerged environment using a nutrient-rich water solution, in this system plants are suspended in water-filled, nutrient-filled airtight containers, with their roots allowed to grow downward into the water. The plants absorb the oxygen they require to survive from an air stone or other oxygenation device. Growing media is used to sustain the plants and give them the nutrients they require. The main benefits of applying a DWC system for a GR are that it makes it possible to grow the plants in a very small area with minimal nutrients and water, requires little maintenance, low cost, and is easy to set up. It is an appropriate solution for growing medium-sized fruits and vegetables. Its disadvantage is that the vast amount of water can serve as a breeding ground for mosquitoes. Figure 12 shows the details of the DWG system [1].

3.6 Thermal Performance of Green Roofs

Different studies conducted all over the world have investigated the thermal performance of green roof systems. In general, it has been observed that the GR thermal performance differs among the climatical conditions and depends on the materials used to implement the GR system and their associated thermal properties [39, 40]. Another study observed that the green roof's thermal performance is influenced by many factors, including the form, kind, and diversity of vegetation, the coverage ratio, plant height, and plant biological processes such as photosynthesis, respiration, and transpiration. Furthermore, the temperature and heat flux of green roofs can be influenced by the physical qualities of the growing media, such as thickness, water content, and mass given by plants. Numerous studies have investigated how green roof characteristics (green canopy, soil, and roof support) affect the temperature of the building's surface, the number of heat fluxes in a building, and the total amount of energy required when applying a green roof system [41].

According to [41] the experimental result confirmed that the vegetative canopy on buildings reflected almost 13% of incident solar radiation and absorbed 56% so the solar radiation entered the system can be approximately 31%. Another study showed that the amount of greenery coverage has a big effect on the temperature. The ambient temperature decreased by increasing the greenery coverage and plant height, [42].

There are various variables also that affect the thermal performance of GRs; soil depth, water balance, and organic matter, first, the soil acts as an inertial mass with a high heat thermal capacity, it can absorb and retain solar heat, several studies proved that increasing the soil thickness improve the thermal comfort in summer in hot-arid and humid climates. [43, 44]. In the winter, 0.15 m of soil layer or more can improve thermal comfort because the green roof is warmer than the bare roof [45]. Some studies found that using natural ventilation could improve the green roof's impact on indoor conditions in different climates [46, 47]. In terms of thermal insulation, non-insulated GRs reduce the cooling energy demand in hot climates [48]. However, insulated GRs are more effective at decreasing winter heating energy consumption [49]. Regular irrigation improved energy efficiency and enhanced indoor thermal comfort in hot climes during the summer [50]. Several studies use numerical simulation methods to analyze saving energy and thermal comfort levels. A research study conducted in KSA used self-design methodologies and a simulation program called DesignBuilder to prove that an extensive GR with a soil depth of less than 15 cm can reduce energy consumption by approximately 8% [51]. According to (Schweitzer and [39]) the water requirements of plant species range from 2.6 to 9.0 L/m2 per day, depending on plant species and the climatic conditions. Moreover, a study by [52] found that grey water and roof runoff can be recycled for usage at home, when grey water is used for irrigation and toilet flushing, a residential home can save 31% of its average potable water consumption.

3.7 *The Climatic Characteristics of Cairo City*

Cairo city is described by a hot arid climate, high sunlight, and low rainfall. Daytime temperature in summer ranges between 37 °C to 40 °C, the temperature reaches its maximum in August at 42 °C, and humidity ranges from 38% in May to 65% in January. The average low temperatures vary from 9.5 °C in wintertime to 23 °C. The city occasionally experiences dust storms known as "khamsin" which carry dust and sand across the northern coast of Africa, these khamsin storms come from the Southwest of Egypt (between the end of March and the middle of May) and can raise the temperature by 20 °C in two hours and stay for many days [53]. Different studies have investigated how temperatures would vary in the MENA region over the twenty-first century. They concluded that even if the average global increase is within 2 °C, the temperature in these regions will more than double during the summer. They predict that by mid-century, throughout the hottest periods, temperatures at night will not fall below 30° and will rise to 46 °C during the daylight, with an average temperature increase of approximately 5 °C throughout the summer [54].

4 Case Studies

Different case studies of green roof applications are studied in the last part of the research study, the case studies were selected according to different criteria:

The geographical location: the case studies of GRs were chosen in the MENA region. Examples include three different projects in Cairo, Riyadh, and Kuwait cities, which are situated in major urban centers characterized by rapid urban growth and population increase.

The climatic conditions: the case studies fall under the same climate region, and green roofs' performance is influenced by the surrounding climatic conditions.

Project construction age: The chosen case studies belong to the category of contemporary green roof projects, which are still under construction.

Application of different typologies of GR systems: Careful consideration was given to choosing examples that include different types of green roofs.

4.1 Olaya Metro Station, Riyadh City, KSA

The design of "Olaya Metro Station" is a part of the KSA plan to develop the Riyadh metro station. The project was designed by "Gerber Architekten" who won the first prize in the Metro-Station competition in Riyadh city in 2012. The project is located on one of the most essential traffic junctions between Olaya Street and is still under construction. The following case study was chosen to study the application of the green roof system and investigate its thermal performance. The roof's undulating shape is inspired by the beauty of Saudi Arabia's dunes, which create a dialogue between the future and tradition. The external form of the project creates a unique rooftop park that rises in gentle waves and is planted with palm trees in a grid configuration over the entire area of the station, this park includes green areas, water elements, and gathering spaces with an innovative design as shown in (Fig. 13 and 14). The natural interplay of landscape and architecture provides a communication oasis in the public area for the residents of the city. The underground floors include multilevel car parking, a shopping Centre, and two subway lines. The project offers Riyadh City the amenities of the main transportation hub altogether with public gardens [55]. The main objectives of this project are to provide attractive urban

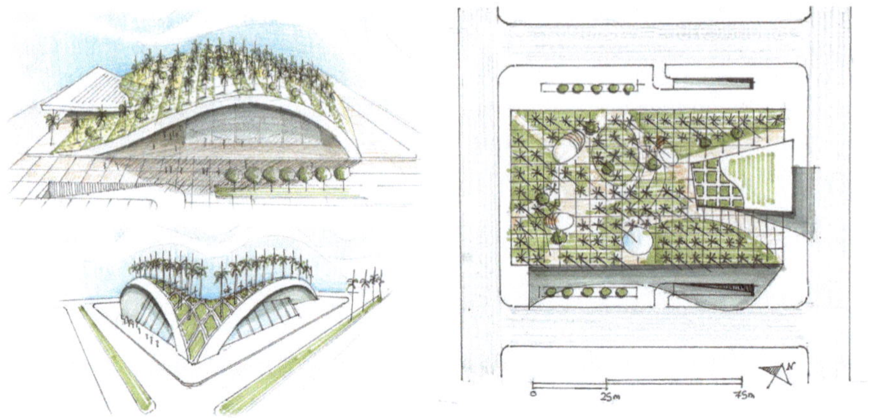

External views of Olaya station showing the rooftop gardens over the building structure *Top view of the Olaya metro station, Riyadh, KSA*

Fig. 13 Olaya Metro Station project, Riyadh, KSA (*Source* by author based on Gerber [55])

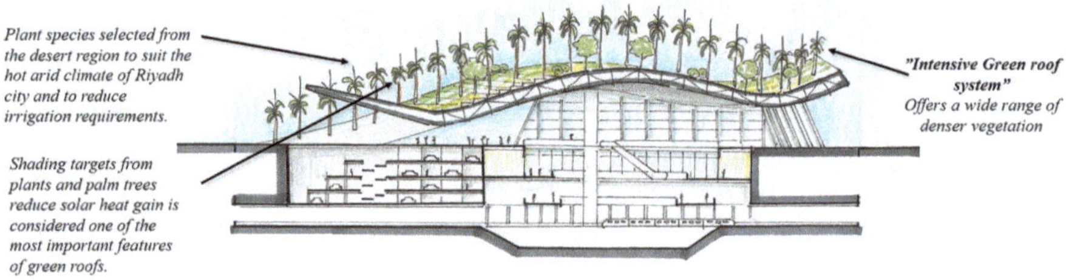

Fig. 14 Cross-section of Olaya Metro Station (*Source* Analysis by author based on Gerber [55])

public space for the residents of Riyadh to enhance their quality of living and to create a highly attractive and visible Station building for the residents [56].

Riyadh city is described by its high temperature and arid climate. Daytime summer temperatures vary from 34.7 to 44 °C, and the main objective of the project was to transform the metro station from a conventional building to a sustainable green building by applying a green roof system which contributes to reducing heat gain, thus minimizing energy demand also green roofs add esthetic value to the metro station.

This project applied the "Intensive" type of green roof which offers a full landscape garden built directly on the structure and provides the widest range of vegetation opportunities. The surface planting on the rooftop was chosen to endure the harsh climate conditions of the KSA and to simulate the oasis such as; Palm trees, Aloe vera, shrubs, and Perennials, resulting in a major reduction in the internal temperature of the building, also, absorbing dust in the building's immediate surroundings and produces habitats for wildlife [55].

Therefore, the most important challenges of the project were enhancing the thermal performance of the building using sustainable development, showing the integration between landscape and project structure, providing attractive urban public space for the residents of Riyadh to improve their quality of life, creating a unique landmark to make the Olaya Metro Station easily recognizable and highly attractive.

4.2 Al-Shaheed Park, Kuwait City, Kuwait

Kuwait City is described by a harsh arid climate that can reach temperatures between 45 °C to 53.2 °C in the summer period making it among the hottest places on earth. Dusty days in Kuwait are the highest in the region compared to other Gulf countries. Climate change is one of the most challenging issues that Kuwait faces due to the country's climate characteristics, high CO_2 emission rates, and heavy dependence on unsustainable energy sources. Al-Shaheed park is mainly designed to reduce air pollution, purify the weather, protect the city from sandstorms, improve the microclimate, and increase the rates of greenery areas.

Al-Shaheed Park is a historical and cultural park located on the periphery of the capital "Kuwait City". The park was officially opened by the Kuwait Amiri Diwan in the early 1960s, to celebrate the city's modernization. It was named "Al-Shaheed" after Saddam's Invasion to commemorate the victims of the first Gulf 1990 war. Nowadays, the park acts as an ecological haven for Kuwait's inhabitants. The current case study was selected to identify the influence of green roof applications on the building's thermal efficiency and the fact that project faces climatic circumstances similar to Egypt. The park is considered the most essential green infrastructure project in Kuwait with an area of 20 hectares as a part of the city's green belt as shown in Fig. 15, The park has the largest overstructured green roofs in the MENA region, The project was redesigned by the architect "Ricardo Camacho" who has carried out the reconstruction of the old park, The work was carried out in two phases to date. The park is situated in the middle of a heavily built-up area with a network of busy roadways crisscrossing it. On the other hand, the green belt is built in such a way that all the individuals' areas in the park are connected. The park was designed as a reflection of paradise according to the Islamic tradition. It is an organized landscape with plenty of vegetation and water bodies. The park consists of 200,000 m^2 of greenery, featuring beautiful botanical gardens, two museums, a congress center, exhibition areas, restaurants, a visitor center, outdoor theatres, a large underground garage with 800 parking lots, walkways, and jogging tracks [57]. As a part of the redesign of the park, an artificial lake (9000 m^3) was designed in the heart of the park, not only as a landscape feature but also as a reservoir ensuring the park has a water supply used for irrigation. Almost all buildings, including the ceiling of the underground parking, were covered with green roofs with a total area of 35,000m^2 during the first phase of the construction (2013–2015). In the second phase of the construction (2017), the park's Green Roof area was doubled [58].

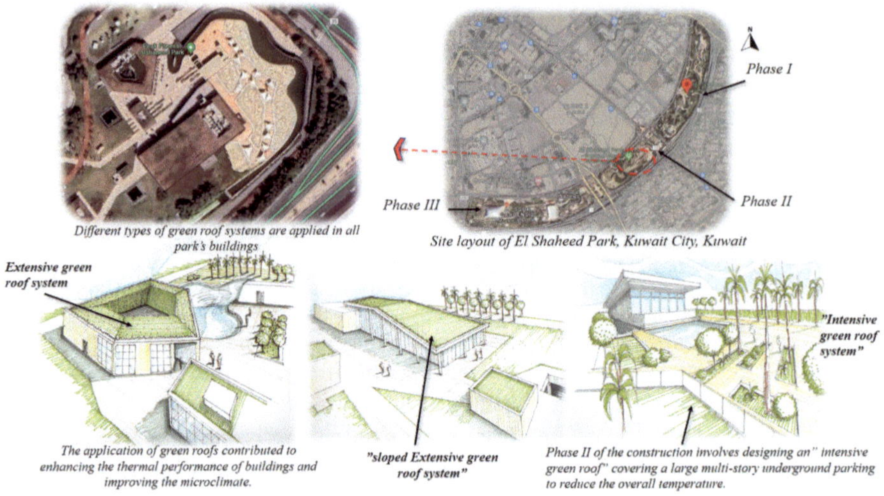

Fig. 15 Al-Shaheed Park, Kuwait City, Kuwait [71] (*Source* Analysis by author from Google Maps "Al-Shaheed Park, Kuwait city")

The project applied the "Extensive and Intensive" type of green roof on most of the park's buildings and over the underground garage. The vegetation layer of the green roofs consists of native plants that are suitable for the harsh climate of Kuwait to minimize irrigation requirements like shrubs such as *Oleander, Agave, Desert rose*, and the *Euphorbia Tirucalli.* Also, 25 various plant species were used and are irrigated by an automated irrigation system like Herbs such as *Carissa, Rosemary, and Portolacaria.* Moreover, 1500 trees were planted such as *Date trees, Lemon trees, Acacia Arabia, Agave, Euphorbia Tirucalli, Desert Rose, and Prosopis Chilensis* as shown in (Fig. 16 and 17) [58].

In summary, the current study shows that the sustainable strategies used in Al Shaheed Park; Green roofs, living walls, vegetation, and building configurations have all been combined to improve the microclimate, purify the air, enhance the thermal performance of buildings, and mitigate energy demand [59].

4.3 The Gate Residence, Cairo, Egypt

The Gate Residence is a private compound situated in a densely populated area in the Heliopolis district in Cairo. The project was designed by Vincent Callebaut Architectures with a total area of about 450,000 m^2, the construction started in 2015 till date. The project is considered the biggest green residential and mixed-use compound in Egypt and it was designed according to the highest standards of green building technology. The project consists of 1000 residential units spread across nine levels with commercial and retail spaces. These units are arranged in rectangular buildings

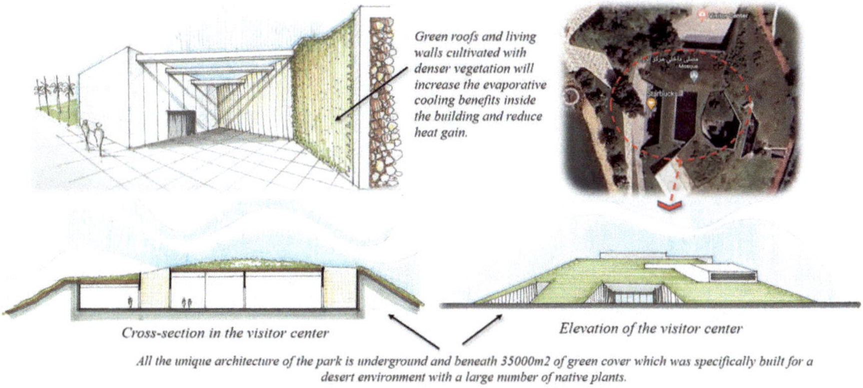

Fig. 16 Green roofs and living walls application in Al-Shaheed Park, [71] (*Source* Analysis by the author from Google Maps "Al-Shaheed Park, Kuwait city")

Fig. 17 Plant species and irrigation resources in Al-Shaheed Park (*Source* photos by author)

(U-shaped) and are connected to the central street. The main idea of the project is to establish a sustainable landmark in Cairo city by changing this efficient building mass into a huge urban oasis and raising awareness of sustainable architecture. The project uses Passivhaus principles and renewable energy technologies to achieve a 50% reduction in energy usage. The building has been designed with bioclimatic rules in mind, like; prevailing wind direction, the solar cycle, and plant species that are suited to the hot-arid region. There are different green architectural features integrated into the project, as shown in (Fig. 18 and 19). (a) The rooftop community garden covers the whole building to enhance the roof's thermal inertia. Moreover, increasing food production and provides a recreational area for the inhabitants, (b) the windcatchers transferred into mega trees in the heart of the courtyard to ventilate naturally all basement spaces, (c) the green walls are situated along the mega trees and building facades to reduce the internal temperature, (d) solar water heating tubes (SWH) are designed on the roof to deliver hot water for all residences throughout

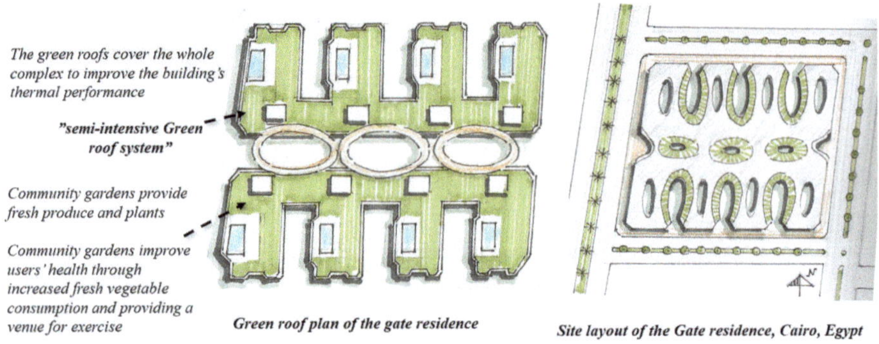

Fig. 18 The gate residence in Cairo City, Egypt (*Source* Analysis by author based on [60])

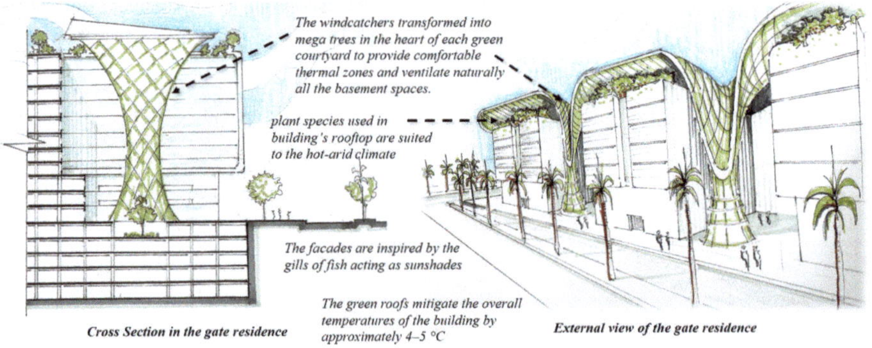

Fig. 19 The green architectural features integrated into the project (*Source* Analysis by author based on [60])

the year, (e) solar photovoltaic cells are covering the roof, west and east facades to generate electricity [60].

In conclusion, the main idea of the gate residence is to transform the city into an ecosystem, the project objective is to mitigate the demand for artificial lighting and mechanical ventilation by enhancing natural lighting systems, natural ventilation of buildings, and reducing heat gain. Green roofs are considered the most essential system applied in this project for achieving better thermal performance of the buildings, they can reduce the internal air temperature of buildings from 4 to 5 °C during daytime and at night 2 °C lower than the bare roof.

5 Contribution of Green Roofs in the Achievement of Sustainable Development Goals (SDGs)

The sustainable development goals (SDGs) of the United Nations were initially put forth at the 1992 Rio de Janeiro Earth Summit. After more than 12 years of effort by different countries. "The 2030 Agenda for Sustainable Development Goals" was adopted by all UN members in 2015. At its heart are the 17 SDGs, which recognize that ending poverty and other deprivations must go together with policies addressing climate change, protecting the ocean and forests, lowering inequality, enhancing health and education, and promoting economic growth [61].

The City of Sydney assessed the alignment of green roofs with the SDGs and found that they contribute to 11 of these goals. After further investigation, we agreed with this conclusion and identified an additional benefit: SDG 14 (Life Below Water). Green roofs play a crucial role in managing stormwater runoff, helping to prevent pollution from reaching water bodies, which in turn protects aquatic ecosystems [62].

The widespread implementation of green roofs could substantially increase green spaces in urban areas, help mitigate the urban heat island effect and improve overall livability by making cities more sustainable and resilient to climate change, green roofs support the achievement of SDG 11 (Sustainable Cities and Communities) [63]. Green roofs reduce climate change's effects by increasing building energy efficiency (via improved insulation) and absorbing CO2, thus contributing to lower carbon footprints (SDG 13: Climate Action) [64].

Water stored in large green roofs can be harnessed to generate power using turbines. Furthermore, green roofs help lower the energy demand for cooling. Also, Local food production reduces the need for long-distance transportation, thus limiting CO2 emissions. This also helps lower energy consumption for heating and cooling, contributing to SDG 7 (Affordable and Clean Energy). Harvested rainwater can be reused for non-potable domestic uses, such as toilet flushing and washing. This helps reduce pressure on clean water supplies and decrease the load on sanitation systems (SDG 6: Clean Water and Sanitation). Additionally, managing water quantity and quality from green roofs helps limit eutrophication, preserves the quality of receiving water bodies, and supports aquatic life, including phytoplankton and fish. As a result, green roofs also contribute to achieving SDG 14 (Life Below Water). The plants on green roofs offer mental health benefits (SDG 3: Good Health and Well-being). They also create habitats for a variety of plants, insects, and birds, enhancing biodiversity (SDG 15: Life on Land) [63].

The growing green roof industry provides opportunities for green jobs, from implementation and maintenance to design, construction, and research. This benefits local economies, increases investment, and helps to transition to a green economy (SDG 8: Decent Work and Economic Growth). Local food production in cities can contribute to lowering the pressure on the food supply system, and multi-layered green roofs can support food production further advancing SDG 2 (Zero Hunger) and SDG 12 (Responsible Consumption and Production) [65]. Green roofs promote innovation and contribute to climate-resilient infrastructure (SDG 9: Industry, Innovation, and Infrastructure) [64]. Collaboration and partnerships, both nationally and internationally, are essential for sharing knowledge and improving green roof technology (SDG 17: Partnership to achieve goal) [63].

6 Challenges Facing Green Roof Application in Egypt

Green roofs are an innovative approach for solving environmental, social, and economic problems in hot arid climates, green roofs have various advantages such as mitigating the overall temperature, reducing emission levels, and improving air quality. Although green roofs offer an opportunity to reduce pollution and try to restore natural hydrology in urban areas, there are still challenging issues preventing their widespread application in Egypt, as follows lack of guidelines standards, climatic suitability, Construction barriers, Cost, Maintenance, etc.

6.1 Lack of Governmental Policies and Regulations

Egypt does not have a scientific guideline or national design standards related to the proper application of green roof systems. Clear design guidelines from governmental bodies can encourage the integration of GRs into building codes and national planning policies [66].

6.2 High Initial Cost

The initial cost of implementing green roofs can be relatively high, different aspects affect the cost including materials, labor, maintenance, quality of roof membrane underneath, irrigation system, and the type of green roof system. The intensive green roof type is the most expensive one due to the high maintenance required. Moreover, there is a Lack of financial support from governmental or nongovernmental sources. Although green roofs have long-term advantages including stormwater management and energy savings, it can be difficult to persuade property owners of their economic viability [28].

6.3 Lack of Awareness

Despite the benefits of green roof systems, there is still a lack of awareness among architects, citizens, and property owners. Most of them are unaware of their implementation techniques and the environmental issues involved. Citizens and property owners are still worried that the added green rooftop will lead to overload problems in their buildings and water leakage. Building expertise through training and education programs is important to promote the widespread adoption of green roofs in Egypt [66].

6.4 Climatic Suitability

The successful installation of green roofs in Egypt is affected by the climatical conditions. Egypt is characterized by a hot-arid climate with limited rainfall which requires careful selection of plant species and irrigation systems to ensure the survival and functionality of green roofs. [67].

6.5 Water Scarcity

The most critical challenge facing the future of green roofs in Egypt is the shortage of water resources, leading to difficulties in maintaining GRs that require regular irrigation. Efficient water management strategies, such as the Greywater recycling system, which recognized recently as a substantial alternative source of water [68].

6.6 Construction Barriers

Several construction and design barriers to building rooftops can delay the green roof application in Egypt, such as construction weight, low quality of rooftop membrane, and insulation system. It is very essential to study the relationship between vegetated areas and other roof-mounted structures or purposes before GR applications (mechanical ventilation equipment, satellite dishes, RC columns meant for future building vertical expansion, etc.). Existing buildings may not be structurally designed to sustain the extra weight of green roofs. Retrofitting buildings with green roofs can be costly, and structural examinations are required to assure safety and stability.

6.7 *Maintenance*

Proper maintenance is vital to the long-term success of green roof systems. However, a lack of qualified maintenance team and resources may make it difficult to ensure regular maintenance, such as pruning, fertilization, pest control, and cleaning the irrigation system. The maintenance mostly depends on green roof typologies; Intensive GR requires more intense and frequent maintenance than Semi-intensive and Extensive ones. The extensive type needs maintenance once or twice a year [69].

Addressing these challenges needs coordinating efforts from governmental bodies, policymakers, architects, stakeholders, and research institutions that can help promote the sustainable application of green roofs in Egypt. A suitable ratio of green spaces can be attained by overcoming these challenges and seizing opportunities. Once the appropriate green roofs have been constructed and utilized a ratio of 4 to 5 m^2 per person can be provided.

7 Conclusion

Increasing the number of green spaces all over the world will stop the "Global warming" phenomenon, which is nowadays a serious threat facing the whole world. Most cities throughout the world are overpopulated, which causes resource depletion and desertification, particularly in developing nations like Egypt, which suffers from an increase in the population every year and lack of green spaces and urban agriculture, especially in Cairo city which is considered one of the most polluted and densely cities in the whole world. Therefore, green roofs could be a vital solution to rebuild the lost agriculture again after the great demolishing occurred by humans. Green roofs have several benefits such as improving the quality of life, enhancing the building's thermal performance in a hot climate, creating a balanced ecosystem, and providing buildings with remarkable spaces where people can work, relax, and grow food.

Despite the benefits of green roof systems, their application is still limited in Egypt due to; the lack of environmental awareness among citizens, limited leadership, shortage of practical plans, and effective programs for increasing the green areas in Cairo city whether in the long or short term.

Green roof applications in Egypt are facing some critical challenges such as the lack of Egyptian authoritative standards, legal aspects, high cost, construction barriers, Maintenance, Water Scarcity, and lack of information. No green roof implementation can be designed or constructed without understanding the various aspects that influence it.

Green roofs protect the building's surface and its users from the surrounding harsh climate, especially in the summer period, whereas they can minimize the internal air temperature of the building by approximately 4 °C to 5 °C during the daytime and at night 2 °C lower than the bare roof, thus, improve building thermal performance, reducing energy consumption and saving money.

There is an intermediate relationship between the thickness of the vegetation layer of the green roof and the reduction in the building's surface temperature. Hence the thicker the vegetation layer the higher the reduction.

Green roofs' performance in a hot-arid environment is strongly affected by the availability and shortage of water as well as the limited varieties of plants that can tolerate these harsh weather conditions.

On the urban scale, connecting green roofs could provide the city with a network of accessible open spaces and be considered an extension of spaces with different uses.

8 Recommendations

In response to Cairo's ecological challenges and growing urban density, an innovative action plan has been proposed to increase the city's green areas through the application of GR systems. This approach aims to transform underutilized rooftop spaces into vibrant green areas, offering various benefits such as improved air quality, enhanced buildings' thermal performance, reduced UHI effects, and increased aesthetic appeal. By incorporating GRs into both new and existing buildings, this initiative seeks to address the pressing need for sustainable urban solutions while fostering a more resilient and eco-friendly urban environment. Through collaboration between government agencies, the private sector, architects, and developers, this plan aspires to create a network of green roofs that contribute to a healthier, more sustainable Cairo.

The action plan encompasses multiple stages, beginning with the assessment and planning phase, followed by policy support, public awareness efforts, stakeholder engagement, technical and design considerations, and culminating in the implementation and monitoring phase (Fig. 20).

By systematically implementing these stages, Egypt can gradually increase green areas through green roofs, contributing to environmental sustainability and enhancing urban resilience to climate change.

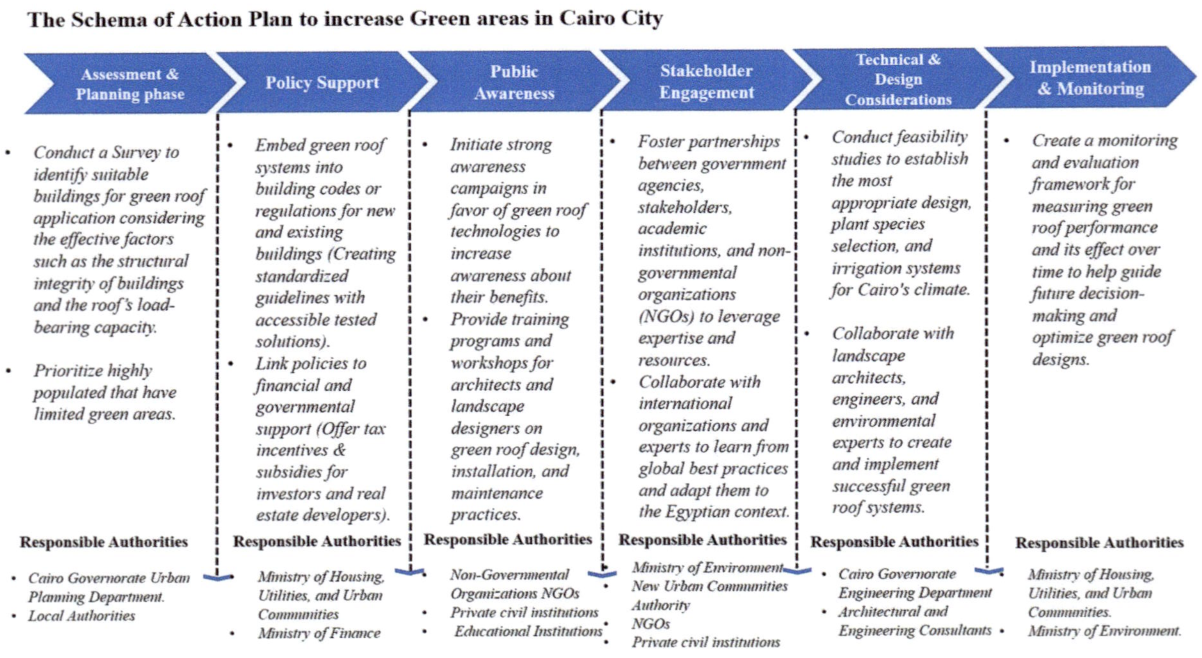

Fig. 20 An action plan to increase the green areas in Cairo city by applying green roof systems (*Source* Author)

References

1. Desouki M, Madkour M, Abdeen A, Elboshy B (2024) Multicriteria decision-making tool for investigating the feasibility of green roof systems in Egypt. Sustain Environ Res 34:2. https://doi.org/10.1186/s42834-024-00207-z
2. Abdelsalam GM, El-Razzaz ZM, Elnekhaily F (2020) Effectiveness of high energy efficiency to minimize energy consumption for residential buildings in Egypt. Eng Res J. https://doi.org/10.21608/erj.2020.131818
3. United Nations, Economic and Social Commission for West Asia (ESCWA) (2011) Green economy in the context of sustainable development and poverty eradication principles, opportunities, and challenges in the Arab region. In: Paper presented at the Arab Regional Preparatory Meeting for the United Nations Conference on Sustainable Development (Rio+20) Cairo, 16–17 October 2011, League of the Arab States Egypt
4. Aly D, Dimitrijevic B (2022) Public green space quantity and distribution in Cairo, Egypt. J Eng Appl Sci 69:15. https://doi.org/10.1186/s44147-021-00067-z
5. Manso M, Teotonio I, Silva MC, Cruz CO (2021) Green roof and green wall benefits and costs: A review of the quantitative evidence. Renew Sustain Rev. https://doi.org/10.1016/j.rser.2020.110111
6. Kazemi M, Courard L (2021) Modeling hygrothermal conditions of unsaturated substrate and drainage layers for the thermal resistance assessment of green roof: effect of coarse recycled materials. Energy Build 250:111315. https://doi.org/10.1016/j.enbuild.2021.111315
7. Ziaee S, Gholampour Z, Soleymani M, Doraj P, Eskandani OH, Kadaei S (2022) Optimization of energy in sustainable architecture and green roofs in construction: A review of challenges and advantages. Complexity 2022:1–15. https://doi.org/10.1155/2022/8534810
8. Kamarulzaman N, Hashim S, Hashim H, Saleh A (2014) Green roof concepts as a passive cooling approach in tropical climate—an overview. E3S Web Conf 3:1. https://doi.org/10.1051/e3sconf/20140301028
9. Joshi MY, Teller J (2021) Urban integration of green roofs: current challenges and perspectives. Sustainability 13(22):12378. https://doi.org/10.3390/su132212378
10. Kafafy N, Al-Betawi Y (2011) Urban green space benefits and the pivotal role of conservation, Cairo's case – Egypt. In: The 2nd conference on conservation of architecture, urban areas, nature, and landscape: towards a sustainable survival of cultural landscape, Zarqa
11. United Nations, Department of Economic and Social Affairs, Population Division (2018) World Urbanization Prospects: The 2018 Revision (ST/ESA/SER.A/420). United Nations, New York
12. Li X, Stringer LC, Dallimer M (2022) The impacts of urbanization and climate change on the urban thermal environment in Africa. Climate 10(11):164. https://doi.org/10.3390/cli10110164
13. Wigginton NS, Fahrenkamp-Uppenbrink J, Wible B, Malakoff D (2016) Cities are the future. American Association for the Advancement of Science, Washington, DC, USA. https://doi.org/10.1126/science.352.6288.904
14. Hamdy O (2022) Using remote sensing techniques to assess the changes in the rate of urban green spaces in Egypt: a case study of greater Cairo. Int Des J 12(3):53–64. https://doi.org/10.21608/IDJ.2022.128280.1039
15. World Health Organization (WHO) (2012) Health indicators of sustainable cities in the context of the Rio+20 UN Conference on Sustainable Development. WHO, Geneva, Switzerland
16. Abd El-Hafeez M, Elmokadem A, Megahed N, EL-Gheznawy D (2016) Green roof analysis and design software calculator in Egypt. Port Said Eng Res J 20(1):12–20. https://doi.org/10.21608/pserj.2016.33626
17. Shafique M, Kim R, Rafique M (2018) Green roof benefits, opportunities, and challenges—a review. Renew Sustain Rev 90:757–773. https://doi.org/10.1016/j.rser.2018.04.006
18. Abass F, Ismail LH, Wahab A, Elgadi AA (2020) A review of green roof: definition, history, evolution and functions. IOP Conf Ser: Mater Sci Eng 713:012048. https://doi.org/10.1088/1757-899X/713/1/012048

19. Shafique M, Luo X, Zue J (2020) Photovoltaic-green roofs: a review of benefits, limitations, and trends. Sol Energy 202:485–497. https://doi.org/10.1016/j.solener.2020.02.101
20. Stefanakis A, Calheiros C, Nikolaou L (2021) Nature-based solutions as a tool in the new circular economic model for climate change adaptation. Circ Econ Sustain 1:303–318. https://doi.org/10.1007/s43615-021-00022-3
21. Grant G, Gedge D (2019) A decade of green roofs and green walls in London. FCIEEM. Infrastructure. P8. https://cieem.net/wp-content/uploads/2021/12/InPractice105sept2019.pdf#page=8
22. Seyedabadi MR, Eicker U, Karimi S (2021) Plant selection for green roofs and their impact on carbon sequestration and the building carbon footprint. Environ Challenges 4:100119. https://doi.org/10.1016/j.envc.2021.100119
23. Nadeeshani M, Ramacgandra T, Gunatilake S, Zainudeen N (2021) Carbon footprint of green roofing: a case study from Sri Lankan construction industry. Sustainability 13(12):6745. https://doi.org/10.3390/su13126745
24. Appolloni E, Orsini F, Specht K, Thomaier S, Sanyé-Mengual E, Pennisi G, Gianquinto G (2021) The global rise of urban rooftop agriculture: a review of worldwide cases. J Clean Prod 296:126556. https://doi.org/10.1016/j.jclepro.2021.126556
25. United States General Services Administration (2011) The benefits and challenges of green roofs on public and commercial buildings; GSA: Colorado. CO, USA, p 2011
26. Abdel Salam A (2012) The future of green Roofs in Egypt: the economic and environmental benefits when installing green roofs on a residential building in Cairo. CU Theses 29 July 2016, from http://www.cu.edu.eg/data_journals/2/articles/3337/submission/original/3337-6553-1-SM.pdf
27. Calheiros CS, Pereira SI (2023) Resilience of green roofs to climate change. Adapt Built Environ Clim Change. https://doi.org/10.1016/B978-0-323-95336-8.00008-1
28. Faisal Z, Elsaadany AS (2022) Green roof awareness, opportunities, and challenges in Egypt. IOP Conf Ser: Earth Environ Sci. 1113:012006. https://doi.org/10.1088/1755-1315/1113/1/012006
29. Nguyen Dang HA, Legg R, Khan A, Wilkinson S, Ibbett N, Doan AT (2022) Social impact of green roofs. Front Built Environ 8:1047335. https://doi.org/10.3389/fbuil.2022.1047335
30. Mazzucco G, Canepa M, Perini K (2023) Application of social-life cycle assessment in urban settings: social impact assessment of green roofs. Buildings 2023(13):1659. https://doi.org/10.3390/buildings13071659
31. Rezaei M, Emmanuel N, Kim J, Kim DH (2021) Analyzing the impact of green roof functions on the citizens' mental health in metropolitan cities. Iran J Publ Health 50(5):900. https://doi.org/10.18502/ijph.v50i5.6107
32. Perivoliotis D, Arvanitis I, Tzavali A, Papakostas V, Kappou S, Andreakos G, Fotiadi A, Paravantis JA, Souliotis M, Mihalakakou G (2023) Sustainable urban environment through green roofs: a literature review with case studies. Sustainability 15(22):15976. https://doi.org/10.3390/su152215976
33. Fernandez-Cañero R, Emilsson T, Fernandez-Barba C, Machuca MÁH (2013) Green roof systems: a study of public attitudes and preferences in southern Spain. J Environ Manag 128:106–115. https://doi.org/10.1016/j.jenvman.2013.04.052
34. Shahmohammad M, Hosseinzadeh M, Dvorak B, Bordbar F, Shahmohammadmirab H, Aghamohammadi N (2022) Sustainable green roofs: A comprehensive review of influential factors. Environ Sci Pollut Res 29(52):78228–78254. https://doi.org/10.1007/s11356-022-23405-x
35. Vijayaraghavan K (2016) Green roofs: A critical review on the role of components, benefits, limitations and trends. Renew Sustain Energy Rev 57:740–752. https://doi.org/10.1016/j.rser.2015.12.119
36. Dauda I, Alibaba HZ (2020) Green roof benefits, opportunities, and challenges. Int J Civil Struct Eng Res 7:106–112
37. Korol E, Shushunova N (2016) Benefits of a modular green roof technology. World multidisciplinary civil engineering-architecture-urban planning symposium, pp 1820–1826. https://doi.org/10.1016/j.proeng.2016.08.673

38. Goddek S, Delaide B, Mankasingh U, Ragnarsdottir KV, Jijakli H, Thorarinsdottir R (2015) Challenges of sustainable and commercial aquaponics. Sustainability 7(4):4199–4224. https://doi.org/10.3390/su7044199
39. Schweitzer O, Erell E (2014) Evaluation of the energy performance and irrigation requirements of extensive green roofs in a water-scarce Mediterranean climate. Energy Build 68:25–32. https://doi.org/10.1016/j.enbuild.2013.09.012
40. Zhao M, Tabares-Velasco PC, Srebric J, Komarneni S, Berghage R (2014) Effects of plant and substrate selection on the thermal performance of Green roofs during the summer. Build Environ 78:199–211. https://doi.org/10.1016/j.buildenv.2014.02.011
41. D'Orazio M, Di Perna C, Di Giuseppe E (2012) Green roof yearly performance: a case study in a highly insulated building under temperate climate. Energy Buildings 55:439–451. https://doi.org/10.1016/j.enbuild.2012.09.009
42. Liu TC, Shyu GS, Fang WT, Liu SY, Cheng BY (2012) Drought tolerance and thermal effect measurements for plants suitable for extensive green roof planting in humid subtropical climates. Energy Build 47:180–188. https://doi.org/10.1016/j.enbuild.2011.11.043
43. He Y, Yu H, Ozaki A, Dong N (2020) Thermal and energy performance of green roof and cool roof: a comparison study in Shanghai area. J Clean Prod. https://doi.org/10.1016/j.jclepro.2020.122205
44. Jamei E, Chau HW, Seyedmahmoudian M, Stojcevski A (2021) Review on the cooling potential of green roofs in different climates. Sci Total Environ. https://doi.org/10.1016/j.scitotenv.2021.148407
45. Wei T, Jim CY, Chen A, Li X (2021) A random effects model to optimize soil thickness for green-roof thermal benefits in winter. Energy Build. https://doi.org/10.1016/j.enbuild.2021.110827
46. Moghtader Gilvaei Z, Haghighi Poshtiri A, Mirzazade Akbarpoor A (2022) A novel passive system for providing natural ventilation and passive cooling: Evaluating thermal comfort and building energy. Renew Energy 198:463–483. https://doi.org/10.1016/j.renene.2022.07.151
47. Tuck NW, Zaki SA, Hagishima A, Rijal HB, Yakub F (2020) Affordable retrofitting methods to achieve thermal comfort for a terrace house in Malaysia with a hot–humid climate. Energy Build. https://doi.org/10.1016/j.enbuild.2020.110072
48. Yazdani H, Baneshi M (2021) Building energy comparison for dynamic cool roofs and green roofs under various climates. Sol Energy 230:764–778. https://doi.org/10.1016/j.solener.2021.10.076
49. Bevilacqua P, Bruno R, Arcuri N (2020) Green roofs in a Mediterranean climate: energy performances based on in-situ experimental data. Renewable Energy 152:1414–1430. https://doi.org/10.1016/j.renene.2020.01.085
50. Abdalazeem ME, Hassan H, Asawa T, Mahmoud H (2022) Review on integrated photovoltaic-green roof solutions on urban and energy-efficient buildings in hot climates. Sustain Cities Soc 82:103919. https://doi.org/10.1016/j.scs.2022.103919
51. Al-Tamimi N (2022) Passive design strategies for energy-efficient buildings in the Arabian Desert. Frontiers in Built Environment 7:805603. https://doi.org/10.3389/fbuil.2021.805603
52. Chang NB, Rivera BJ, Wanielista MP (2011) Optimal design for water conservation and energy savings using green roofs in a green building under mixed uncertainties. J Clean Prod 19(11):1180–1188. https://doi.org/10.1016/j.jclepro.2011.02.008
53. USAID (2018) Climate risk profile—Egypt. Fact Sheet. https://www.climatelinks.org/sites/default/files/asset/document/2018_USAID-ATLAS-Project_Climate-Risk-Profile-Egypt.pdf
54. Khalil HAEE, Ibrahim A, Elgendy N, Makhlouf N (2018) Could/should improving the urban climate in informal areas of fast-growing cities be an integral part of upgrading processes? Cairo case. Urban Climate 24:63–79. https://doi.org/10.1016/j.uclim.2018.01.007
55. Gerber Architekten (2015) Metro Station in Riyadh. https://www.gerberarchitekten.de/en/project/olaya-metro-station/. Accessed 13 July 2024
56. German architects (2017) Olaya Metro Station, Gerber Architekten. https://www.german-architects.com/en/gerber-architekten-dortmund/project/olaya-metro-station. Accessed 13 July 2024

57. Al-Nakib F (2022) Kuwait's Changing Landscape. Struggles for political change in the Arab World, p 70. https://doi.org/10.1353/book.103145
58. ArchDaily (2015) Al Shaheed Park/Ricardo Camacho. https://www.archdaily.com/608253/al-shaheed-park-ricardo-camacho. ISSN 0719–8884. Accessed 13 July 2024
59. Green Roofs (2022) Al Shaheed Park. https://www.greenroofs.com/projects/al-shaheed-park/. Accessed 13 July 2024
60. Quddus S (2014) ArchDaily, "Vincent Callebaut designs sustainable mixed-use complex for Cairo". https://www.archdaily.com/565550/vincent-callebaut-designs-sustainable-mixed-use-complex-for-cairo. ISSN 0719–8884. Accessed 13 July 2024
61. United Nations, Department of Economic and Social Affairs. The 17 Goals (2024) https://sdgs.un.org/goals. Accessed on 26 December 2024
62. City of Sydney (2024) Green Roofs and Walls Policy. https://www.cityofsydney.nsw.gov.au/policies/green-roofsand-walls-policy. Accessed on 27 December 2024
63. Rahman MA, Alim MA, Jahan S, Rahman A (2022) Vegetated roofs as a means of sustainable urban development: a scoping review. Water 14(19):3188. https://doi.org/10.3390/w14193188
64. Vourdoubas J (2024) The contribution of green roofs in the achievement of sustainable development goals. Eng Technol J 9(10):1. https://doi.org/10.47191/etj/v9i10.03
65. Cristiano E, Deidda R, Viola F (2021) The role of green roofs in urban water-energy-food-ecosystem nexus: a review. Sci Total Environ 756:143876. https://doi.org/10.1016/j.scitotenv.2020.143876
66. Gawad IO (2014). The rise of rooftop gardens in informally developed areas in Egypt: exploring the abilities and boundaries. ARCHCAIRO: Responsive Urbanism in Informal Areas, pp 208–223
67. Ragab A, Abdelrady A (2020) Impact of green roofs on energy demand for cooling in Egyptian buildings. Sustainability 12(14):5759. https://doi.org/10.3390/su12145729
68. Morsi AAG, Elian NAZ (2021) The vertical greenery systems significant role in achieving better architectural and urban thermal performance. MEJ-Mansoura Eng J 46(2):102–115. https://doi.org/10.21608/bfemu.2021.188836
69. Abada H (2023) Evaluation of green roof technology in Egypt (Master's thesis, The American University in Cairo (Egypt)). https://fount.aucegypt.edu/cgi/viewcontent.cgi?article=3165&context=etds
70. Almayouf A (2013) Preserving the green in hot-arid desert environments: the case of Riyadh, Saudi Arabia. King Saud University. Archit Plan 25:39–49. https://cap.ksu.edu.sa/sites/cap.ksu.edu.sa/files/imce_images/jap_ksu_jan2013_e4_0.pdf
71. Google Maps (n.d.) [Google Maps directions for Al Shaheed Park, Kuwait City]. Retrieved July 24, 2024, from https://www.google.com/maps/place/Al+Shaheed+Park/@29.367041,47.970404,2998m/data=!3m1!1e3!4m10!1m2!2m1!1sal+shaheed+park!3m6!1s0x3fcf84a59ee1ca37:0xdd5589fb20e82577!8m2!3d29.3693379!4d47.9937309!15sCg9hbCBzaGFoZWVkIHBhcmtaESIPYWwgc2hhaGVlZCBwYXJrkgEEcGFya-ABAA!16s%2Fg%2F11fy94q05n?entry=ttu
72. Martens R, Bass B, Alcazar SS (2008) Roof-envelope ratio impact on green roof energy performance. Urban Ecosyst 11(4):399–408. https://doi.org/10.1007/s11252-008-0053-z

Energy Transition Toward Carbon Neutral and Green and Smart Cities

Application of Renewable Energy in the Built Environment: Towards Sustainable Development Through Community Centers

Merna Abdelhamid, Donia Mostafa-Kamel, Sama Abdelghany, Mariam ElSayed, Mariam Hassan, Momen Ghonim, Sara Mohamed, Donia Muhammad, and Zeina ElZein

Abstract The transition to sustainable energy sources is crucial for addressing the pressing challenges of climate change and energy security. Renewable energy (RE) plays a vital role in this transition, offering environmentally friendly alternatives that reduce greenhouse gas emissions and reliance on finite resources. The chapter offers an in-depth examination of the renewable energy (RS) scenario, identifying the major hurdles faced by Egypt and the demographic issues affecting its growth. It starts by discussing the energy demand landscape that is likely to grow with trend toward urbanization and subsequent population increase straining India's

M. Abdelhamid · D. Mostafa-Kamel · S. Abdelghany · M. ElSayed · M. Hassan · M. Ghonim · S. Mohamed · D. Muhammad · Z. ElZein (✉)
Architecture Department, Faculty of Fine Arts, Helwan University, Cairo, Egypt
e-mail: Zeina.ElZein@f-arts.helwan.edu.eg; Zeinabaher@gmail.com

M. Abdelhamid
e-mail: Merna.Moataz.2021-03142@f-arts.helwan.edu.eg; mernaa.moatazz@gmail.com

D. Mostafa-Kamel
e-mail: Donia.hatem.2021-05322@f-arts.helwan.edu.eg; doniahatem2903@gmail.com

S. Abdelghany
e-mail: Sama.Mohamed.2021-02982@f-arts.helwan.edu.eg; sama.abdelghany04@gmail.com

M. ElSayed
e-mail: Mariam.Ahmed.2021-00725@f-arts.helwan.edu.eg; eng.mariamahmedezzat@gmail.com

M. Hassan
e-mail: Mariam.Hussein.2021-00448@f-arts.helwan.edu.eg; mariamhussein2407@gmail.com

M. Ghonim
e-mail: MOMEN.ADEL.2022-00834@f-arts.helwan.edu.eg; ghonimmomen@gmail.com

S. Mohamed
e-mail: sara.salah.2021-04861@f-arts.helwan.edu.eg; SaraAldieb26@gmail.com

D. Muhammad
e-mail: Donia.Ashraf.2021-00744@f-arts.helwan.edu.eg; donia.alhassan.ahmed@gmail.com

Z. ElZein and A. Negm (eds.), *Recent Approaches of Sustainable Architecture in Arid and Semi-arid Cities*, Urban Sustainability,
https://doi.org/10.1007/978-981-95-1645-2_14

existing power infrastructure. A detailed analysis of energy consumption in buildings follows, covering residential, retail, mixed-use developments, office buildings, industrial facilities, public institutional structures and hospitality buildings. Finally, a case study is presented to illustrate the practical implementation of RE solutions in Egypt. This section analyzes a specific project by examining its successes, challenges and lessons learned. The case study serves as an example of how RE can be effectively integrated into Egypt's energy landscape while also highlighting the obstacles that must be overcome for adoption. By comprehensively addressing these interconnected topics, this report aims to provide a view of Egypt's energy challenges and opportunities with a specific emphasis on the potential for RE to transform the country's built-environment and energy sector.

1 Introduction

As Egypt faces an increase in energy demands driven by rapid urbanization and population growth, it is important to analyze energy usage across different building types and investigate RE options to ensure the countries sustainable development [1]. The massive increase in energy demand is due to Egypt's expanding population and urbanization. Roughly 75% of the population as shown in Figs. 1 and 2 is expected to live in cities by 2050. Urbanization boosts energy use by 5.6% per year till at least that time [2].

Big cities like Cairo will get much bigger. The government is even building new cities to fit everyone. As more people move to cities, fewer will live in the countryside. This big change will affect many things. Cities will need more houses, roads and services. They might offer more jobs, but they will also use more energy. Meanwhile, rural areas will change too.

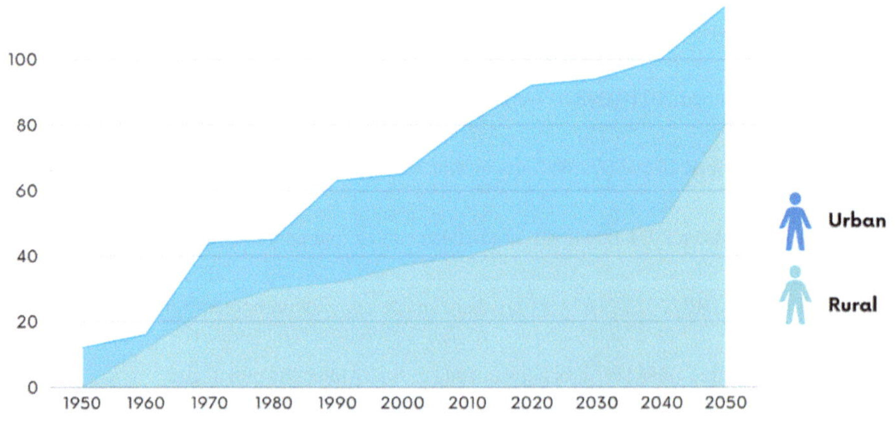

Fig. 1 Urbanization in Egypt (modified by authors, based on [3])

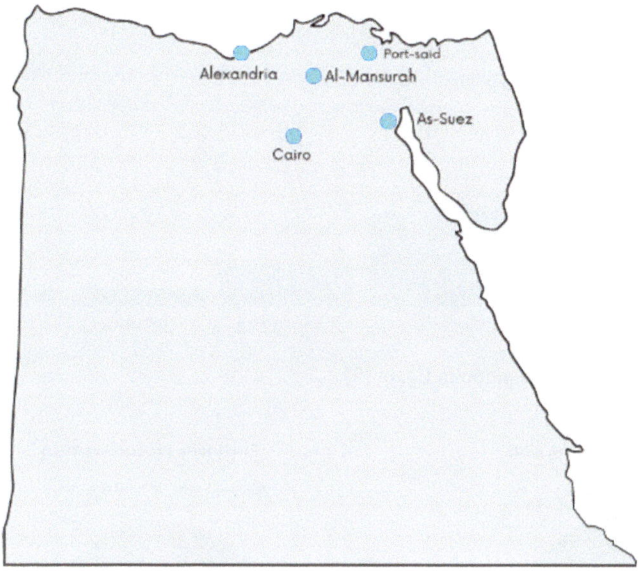

Fig. 2 Urban development in Egypt (by author(s), based on [3])

They will have fewer people but will still need to grow a lot of food. This shift from countryside to city life will shape how Egypt grows and develops over the next few decades.

Despite these impressive improvements, Egypt's energy infrastructure faces substantial obstacles. This includes very high transmission losses in the country's national grid of up to 15%, for instance. Access to electricity is available for just 40% of rural areas in comparison with cities, indicating a chasmic contrast between urban and rural when it comes access energy [4]. These infrastructural difficulties present obstacle to the further growth of RE sources in Egypt. The rural–urban divide in accessibility and efficiency cannot be bridged without big investments, better technology and policy focus on the issues as shown in Fig. 3 The total consumption of electricity is consumed by the built environment and residential areas.

Egypt has a very complex regulatory system for its RE sector. While [5] has the share that government aspires to until 2035 is around 42% and IRECT, where it would increase to 61% by 2050 [5].

Financing is still one of the principal hurdles impeding a transition to an energy regime reliant on renewables in Egypt as shown in Fig. 4. If the nation is to meet its 2030 targets for producing RE, an investment of upwards $70 billion as projected by International Energy Agency in 2023 [6–8]. However, Egypt is ahead in the solar sector. This offers an opportunity for a comprehensive appraisal of sustainable energy solutions appropriate for the specific needs of countries.

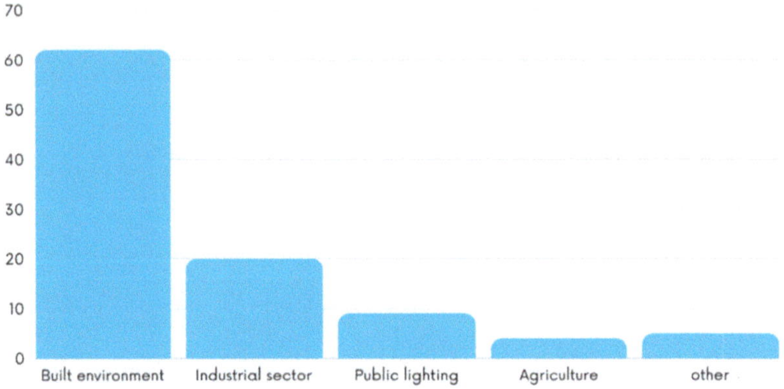

Fig. 3 Electricity consumption in Egypt [9]

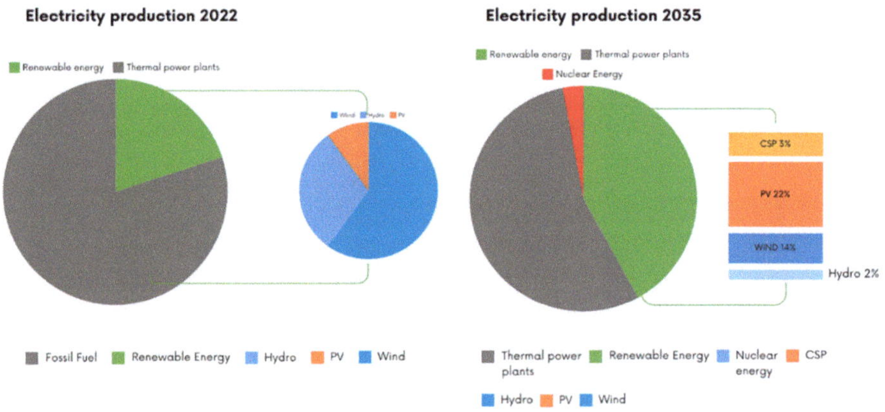

Fig. 4 Brief review on Egypt's renewable energy status and future vision [10]

2 Energy Consumption in Buildings in Egypt

An important factor in Egypt's overall energy landscape is the country's diverse building sector. The energy consumption of several building types—residential, retail, office, industrial, public and institutional and hospitality buildings—is thoroughly broken down in this section. Each group exhibits different consumption patterns, which correspond to their functions and operational needs. It provides insightful information about how much electricity and natural gas are used as well as how much of the total energy consumption is attributable to each type of structure. For Egypt's arid and semi-arid urban areas, it is crucial to comprehend these trends to design targeted energy efficiency techniques and sustainable solutions.

2.1 Residential Buildings

2.1.1 Total Energy Consumption

Residential buildings in Egypt consume approximately 56,380 GWh of energy annually, representing a significant portion of the country's total energy use. This high energy demand is attributed to increasing appliance ownership, cooling and heating requirements and the continuous growth of living spaces and standards [11].

2.1.2 Energy Consumption Breakdown

Electricity consumption in residential buildings reached 38,040 GWh, accounting for about 67.5% of the sector's total energy use. The main electricity end-uses include lighting, appliances and cooling systems. Natural gas consumption in residential buildings amounted to 18,340 GWh, representing approximately 32.5% of the sector's total energy use. It is primarily used for space heating and water heating [11].

2.2 Retail Buildings

2.2.1 Total Energy Consumption

Retail buildings in Egypt consume about 12,450 GWh of energy annually. This category includes malls, shops and office buildings, which have significant energy demands due to their cooling and lighting needs [11].

2.2.2 Energy Consumption Breakdown

Electricity consumption in retail buildings reached 9120 GWh, accounting for approximately 73.3% of the sector's total energy use. The primary electricity end-uses are cooling and lighting. Natural gas usage in retail buildings was approximately 3330 GWh, constituting about 26.7% of the sector's total energy consumption. It is mainly used for space heating and water heating [11].

2.3 *Mixed-Use Developments*

2.3.1 Total Energy Consumption

Mixed-use developments, which integrate residential, commercial and other activities within a single area consume 18,670 GWh of energy annually. The diverse nature of these complexes contributes to their varied energy consumption patterns [11].

2.3.2 Energy Consumption Breakdown

Electricity consumption in mixed-use developments reached 13,240 GWh, accounting for about 70.9% of the sector's total energy use. Electricity is used for lighting, cooling and operational purposes across various building types within these developments. Natural gas consumption in mixed-use developments amounted to 5430 GWh, contributing to approximately 29.1% of the sector's total energy use [11].

2.4 *Office Buildings*

2.4.1 Total Energy Consumption

Office buildings in Egypt consume an average of 15,890 GWh of energy per year. These spaces require substantial energy for lighting, cooling, ventilation and powering electronic equipment that supports business activities [11].

2.4.2 Energy Consumption Breakdown

Electricity consumption in office buildings reached 11,910 GWh, accounting for about 75% of the sector's total energy use. Electricity is the primary energy source in this sector used for lighting, cooling, ventilation and powering office equipment. Natural gas usage in office buildings was approximately 3980 GWh constituting about 25% of the sector's total energy consumption. It is mainly utilized for water heating and space heating [11].

2.5 Industrial Buildings

2.5.1 Total Energy Consumption

Industrial buildings in Egypt consumed approximately 65,000 GWh of energy in 2022, representing about 40% of the country's total energy consumption [12]. This sector's energy demand has been growing at an average annual rate of 3.5% over the past decade driven by economic expansion and increased industrial output.

2.5.2 Energy Consumption Breakdown

Electricity consumption in industrial buildings reached 42,250 GWh in 2022 accounting for 65% of the sector's total energy use [13]. The manufacturing subsector was the largest consumer followed by mining and quarrying activities. Natural gas consumption in industrial buildings amounted to 19,500 GWh in 2022, representing 30% of the sector's total energy use [14]. The chemical and petrochemical industries were the primary consumers within this category.

2.6 Public and Institutional Buildings

2.6.1 Total Energy Consumption

Public and institutional buildings in Egypt including government, educational and healthcare facilities consume approximately 24,500 GWh of energy in 2022, contributing to about 15% of the country's total building energy consumption [12]. The energy use intensity (EUI) of these buildings varies based on their function and operational hours.

2.6.2 Energy Consumption Breakdown

Electricity consumption in public and institutional buildings reached 19,600 GWh in 2022, accounting for 80% of the sector's total energy use [13]. Air conditioning represented up to 60% of this electricity consumption during peak summer months. Natural gas usage in public and institutional buildings is relatively low. It was approximately 3675 GWh in 2022 constituting 15% of the sector's total energy consumption [14]. It is primarily used for water heating and in some cases space heating during winter months in cooler regions of Egypt.

2.7 *Hospitality Buildings*

2.7.1 Total Energy Consumption

The hospitality sector in Egypt including hotels and resorts consumed about 13,000 GWh of energy in 2022, representing approximately 8% of the country's total building energy consumption [15]. However, these buildings often have higher energy use intensities compared to other commercial buildings due to their 24/7 operations and diverse energy requirements.

2.7.2 Energy Consumption Breakdown

Electricity consumption in hospitality buildings reached 9100 GWh in 2022, representing 70% of the sector's total energy use [13]. Major end-uses include space cooling, lighting and hotel-specific equipment. Luxury hotels typically have higher electricity consumption per unit area than budget accommodations. Natural gas consumption in hospitality buildings amounted to 3250 GWh in 2022, contributing to 25% of the sector's total energy use [14]. It is primarily used for water heating, cooking and laundry operations. Some larger hotels and resorts also use natural gas for space heating and pool heating systems.

The stacked bar chart shown in Fig. 5 depicts energy consumption patterns in Egypt's main building sectors, focusing on both electricity and natural gas usage. It enables a simple visual comparison of energy usage across various building types. The height of each bar shows each building type's annual energy usage in GWh (Gigawatt-hours), making it easy to determine which has the highest and lowest demand. Industrial buildings emerge as the most energy-intensive sector, far outpacing other categories. Residential buildings are the second-largest consumer demonstrating a strong need for both electricity and natural gas. Public and institutional institutions consume the third most energy overall. These three sectors combined account for a substantial portion of Egypt's overall building energy use, highlighting key areas for potential energy efficiency improvements and sustainable development initiatives.

3 Renewable Energy

RE is essential for a sustainable future due to its ability to reduce greenhouse gas emissions, decrease reliance on finite fossil fuels, and mitigate climate change. Unlike coal or oil, which rely on conventional energy sources, renewable resources such as solar, wind, and hydroelectric power produce electrical output with almost no carbon footprint [7]. The trend towards renewables has significant implications for air pollution and global warming, which are among the leading threats to human health and

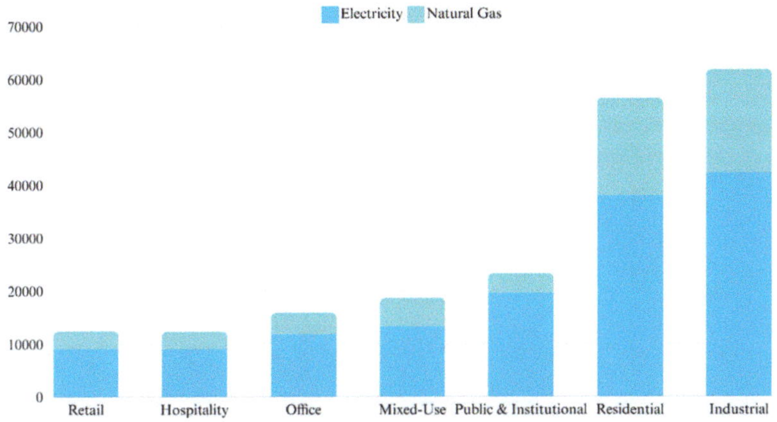

Fig. 5 A Stacked bar chart showing the energy consumption patterns across various building sectors in Egypt, highlighting both electricity and natural gas usage (by author(s), based on Egyptian Natural Gas Holding Company 2024)

ecosystems as the population increases globally along with ever-increasing energy demands [16]. In addition, RE enhances energy security and provides employment opportunities through jobs created by new technologies as well as infrastructure development. Therefore, investing in RE is a key step towards ensuring environmental sustainability and long-term economic stability.

3.1 Types of Renewable Energy

There are five main types of RE:

3.1.1 Solar Energy

Solar energy which can be extracted from the sun's radiation and transformed into heat or electricity is known as solar energy. It is an abundant, renewable resource that is essential to sustainable energy systems. There are three main ways to harness solar energy:

Photovoltaics

Photovoltaic or PV cells incorporated on solar panels work by converting sunlight into electricity by the photovoltaic effect. These PV cells are usually formed by joining two layers of semiconducting material—for instance, silicon that has various electrical features. When the photons of sunlight hit these materials, a flow of electric

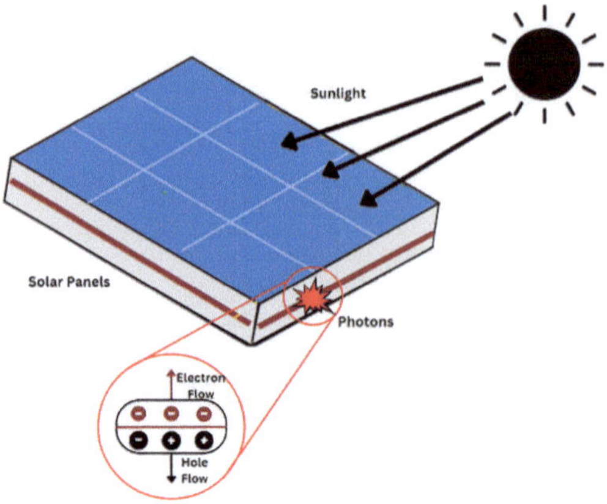

Fig. 6 How Do Solar Panels Generate Electricity [19]

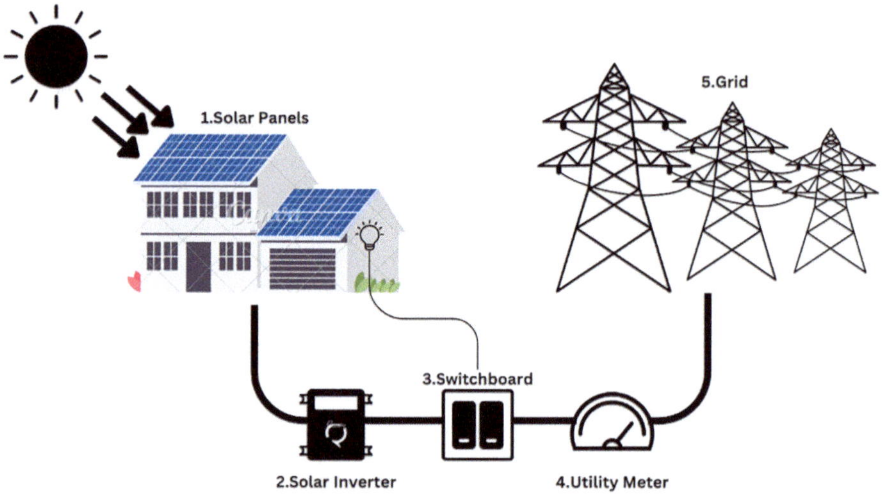

Fig. 7 How solar energy works [19]

current is produced by the movement of the electrons; enough electric current which can generate electricity, this is the idea behind solar energy [17]. This electricity that is generated is actually in a direct current (DC) and must be converted to an alternating current (AC) through what is known as an inverter. The electricity can be fed into the National Grid if the capacity is present or used by households and businesses that have installed solar [18] (Figs. 6, 7).

Solar PV system means a technology of direct conversion of sunlight into electricity with the help of semiconductors out of which the most widely used PV cell is silicon PV cell. These cells are classified into three primary types: It has three categories: monocrystalline, polycrystalline and thin film as depicted as shown in Fig. 8. While monocrystalline cells that can achieve efficiencies of between 15 and 22% are

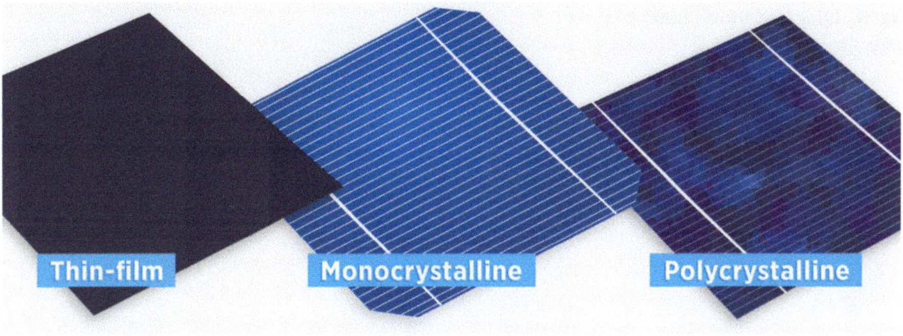

Fig. 8 PV cell types [21]

joined by the more expensive but small and versatile cadmium telluride and amorphous silicon cells that can achieve efficiencies between 10 and 12%. Latest advancements include bifacial modules, which work on both faces, enhancing efficiency perovskite solar cells that predict efficiency and costs to be high and applications being limited by stability and scalability [20].

Solar Thermal

Solar thermal power is one of the classes of RE, in which heat is produced using sunlight. Unlike photovoltaics that turn sunlight directly to electricity, solar thermal uses the sun to heat a working fluid, called a heat carrier and which is then used to produce electricity or useful heat for an application. This technique is a variation of solar energy but with improved thermal conversion [22].

The process begins with the capture of solar radiation by solar collectors, as shown in Fig. 9. These devices can take various forms, such as flat-plate or cylindrical-parabolic collectors, but they all share the same objective which is to capture the sun's energy and use it to heat a fluid circulating through them, such as water or thermal oil. This heat can be directly used to heat water for domestic and industrial uses or to generate electricity by using a conventional thermodynamic cycle. This is where other components like the solar panels and the thermosyphon come into play [22].

Unlike the expansion tank the thermosyphon has the important function of distributing the heat produced by the solar collector. This form of solar thermal technology makes use of natural circulation to transcribe the heat from the solar collectors to a water reservoir. Consequently, the thermosyphon permits this solar heat to be used to heat water without requiring any pumps or other mechanical means [22].

This Thermosyphon solar thermal hot water once it has released its heat to the storage tank can be utilized in various uses. Some of the uses of hot water may

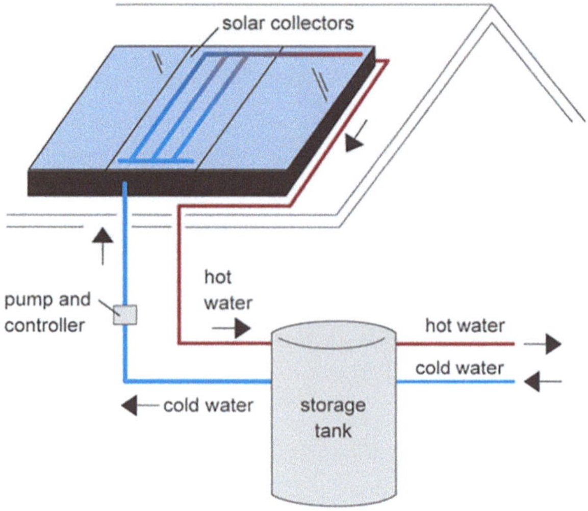

Fig. 9 Diagram shows how solar thermal works [23]

include powering domestic heating systems, producing domestic hot water, or may be used in industrial processes that call for heat [22].

Concentrated Solar Power

Concentrated solar power (CSP) is a technique of generating electricity through mirrors. Concentrated solar power technologies incorporate a mirror arrangement in that it focuses the sun's solar radiation on a receiver, which in turn produces heat. The heat is then very useful in producing the steam needed to turn a turbine for the generation of electrical power. Concentrated solar power plants can employ certain storage technologies to store the generated power during certain periods when there is little sunlight. What is more, concentrated solar power is a convenient form of RE since it can store energy [24].

There are four types of CSP technologies:

- **Parabolic Trough Systems**

Through this system, solar energy is concentrated by curved, trough-shaped reflectors, which are focused onto a receiver pipe. The pipe usually contains thermal oil, which is heated and then used in the thermal power block to generate electricity in a steam generator as shown in Fig. 10 [24].

- **Power Tower Systems**

These systems use mirrors called heliostats that track the sun and focus its energy onto a receiver at the top of a tower. A fluid (often, molten salts) is heated inside the receiver and is used to generate steam, which drives a turbine generator as shown in Fig. 11 [24].

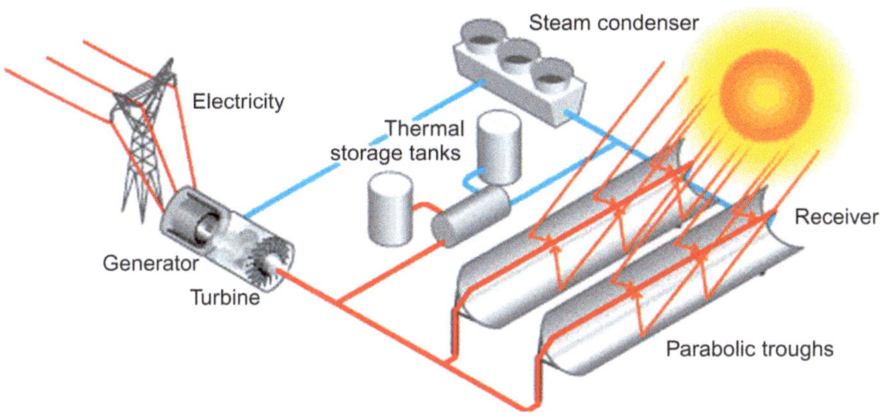

Fig. 10 Diagram shows how parabolic trough work [25]

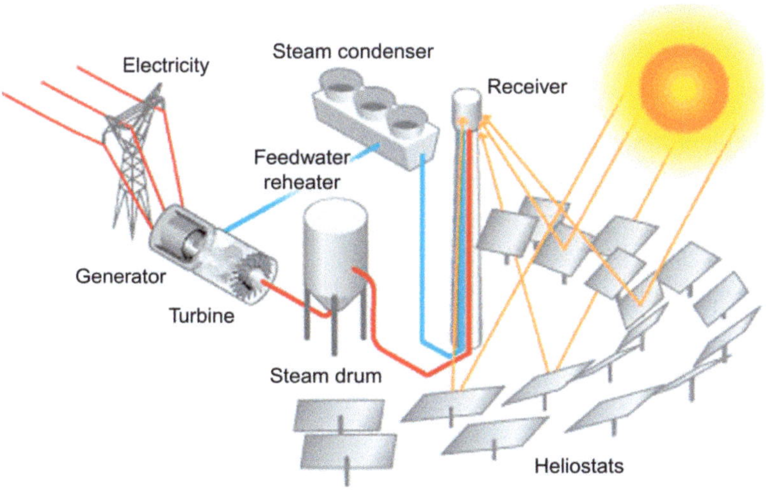

Fig. 11 Diagram shows how power tower work [25]

- **Linear Fresnel Systems**

Many collectors are set out in rows. The mirrors are laid flat on the ground and reflect the sun on the receiver pipe above as shown in Fig. 12. Like trough and tower systems, Fresnel can integrate storage in a power block or generate steam directly [24]

- **Parabolic Dish Systems**

A parabolic-shaped dish acts as a concentrator that reflects solar energy onto a receiver mounted on a structure with a tracking system that follows the sun as shown

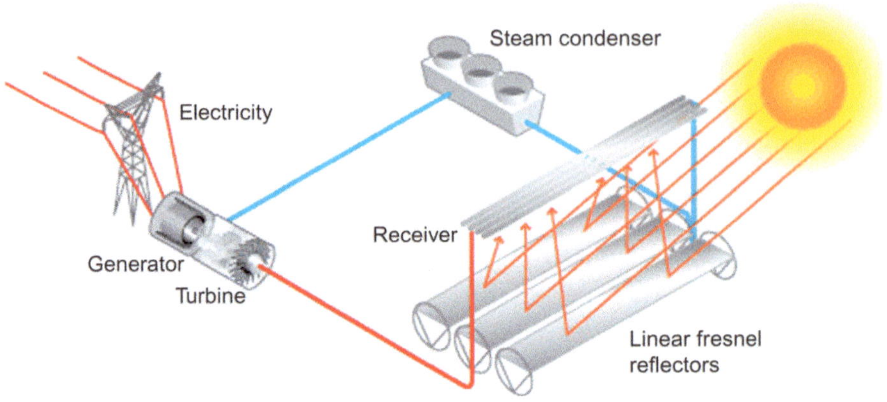

Fig. 12 Diagram shows how linear fresnel work [25]

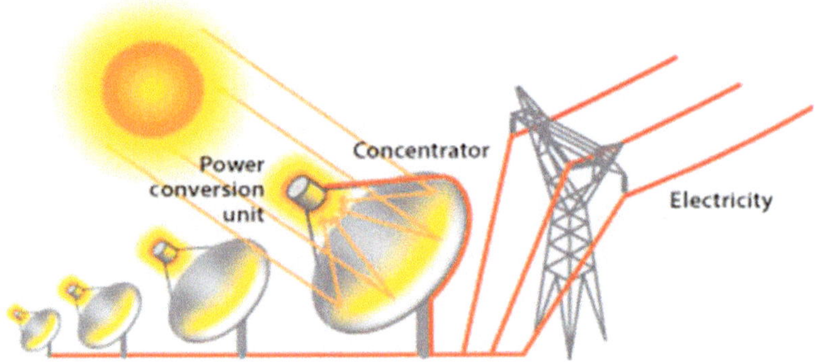

Fig. 13 Diagram shows how parabolic dish work [25]

in Fig. 13. The collected heat is then generated by a heat engine. The dish can attain very high temperatures, which makes the system potentially well-suited for use in solar reactors [24].

3.1.2 Wind Energy

Wind Energy is one of the RE used to produce electricity with wind turbines. These turbines convert kinetic energy of moving air into electrical energy by using the rotation of blade forced by the wind. A generator sits on top of the compact structure of the turbine's tower at the rear of the blades, as shown in Fig. 14. The electricity produced in wind turbines can be utilized directly to fulfill certain requirements like the illumination of streets or single farms, or the electricity can be channeled into

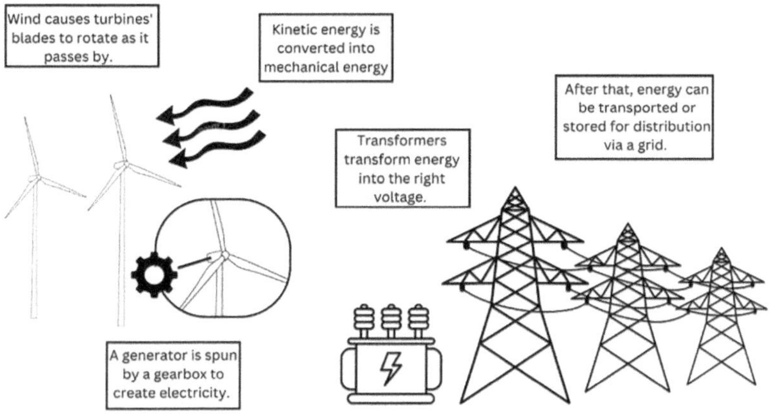

Fig. 14 Diagram shows how wind energy works (Office of GY4ES.ORG, 2023)

other transmission as well as distribution cables to power homes, businesses and learning institutions [26].

Modern wind energy technology primarily employs two types of wind turbines: structures of two types; onshore and offshore. While installing turbines on the onshore side is relatively cheaper than in the offshore side, the offshore turbines receive a more powerful and steady wind to generate more energy. Some of the advancement seen In the wind energy technologies includes the ability to install them on a floating platform, enabling them to be installed in deeper waters –these opens up new areas for wind power generation technology; and the vertical axis wind turbines (VAWTs), as explained in Fig. 15, which are well suited for small compact environments such as in urban centers as shown in Fig. 14 [27].

Fig. 15 Onshore wind (VAWTs) [28]

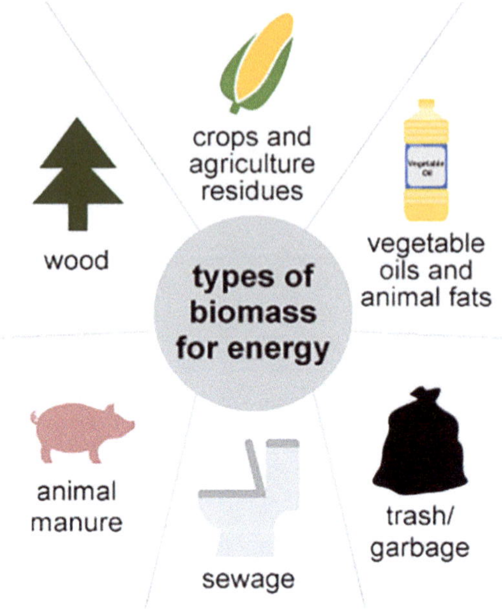

Fig. 16 Types of biomasses for energy (U.S. Energy Information Administration, n.d.)

3.1.3 Biomass

Biomass is natural, consisting of substances got from living organisms including animals and plants for example. Wood, bushes and rubbish are commonly used materials for biomass energy. They are called biomass feedstocks [29] (Fig. 15).

Energy generated from biomass is called bio energy, which includes a variety of biofuels, like ethanol, biodiesel, bio-oil and biogas. These energy sources are very attractive in that they can yield good economic returns, are renewable as well as having low emission rates of pollutants [30] (Fig. 16).

Biomass creates energy when organic materials burn, giving off heat. Power plants burn biomass on a big scale at high heat to make steam. This steam turns turbine blades to make electricity [30]. To turn biomass into bioenergy, we use different methods. We can burn it straight to make heat and power. We can also use bacteria to break it down without air to make biogas or use yeast to make bioethanol. New types of biofuels are coming soon, like ethanol from plant stems and fuel from algae. These might give us more energy for less work. We've also gotten better at turning biomass into gas and breaking it down with heat. This lets us use biomass energy in more ways and more efficiently [31] (Fig. 17).

3.1.4 Geothermal Energy

Geothermal power is a type of RE resource obtained from the Earth's core. This natural resource is derived from the heat produced when the planet was originally

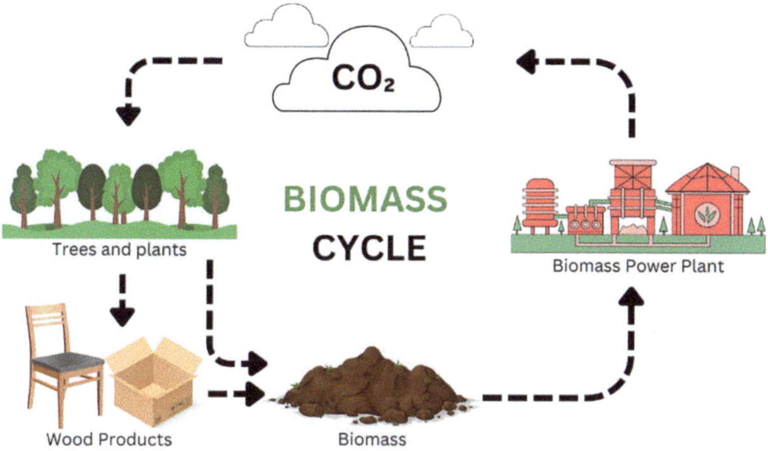

Fig. 17 Diagram shows the cycle of biomass energy (by author(s), based on GreenMatch.co.uk. 2024)

formed and from radioactive decay of materials. This thermal energy is trapped in rocks and fluids that lie at the center of our planet [32].

The underground existence of hot rocks, fluid and permeability make up natural geothermal systems. Small underground pathways, such as fractures, are used to carry fluids through the hot rocks. When generating electricity using geothermal energy, this fluid can be taken out as heat to wells on the surface of the earth where turbines used for producing electric power are turned by it as shown in Fig. 18 [33].

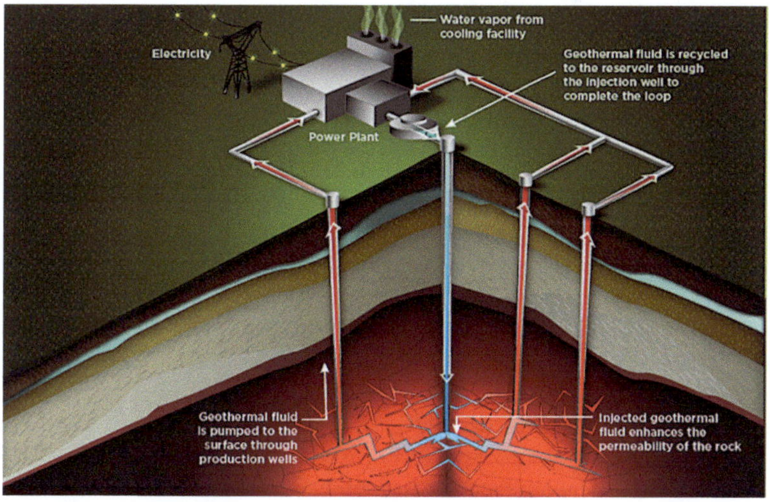

Fig. 18 Diagram shows how geothermal energy works [34]

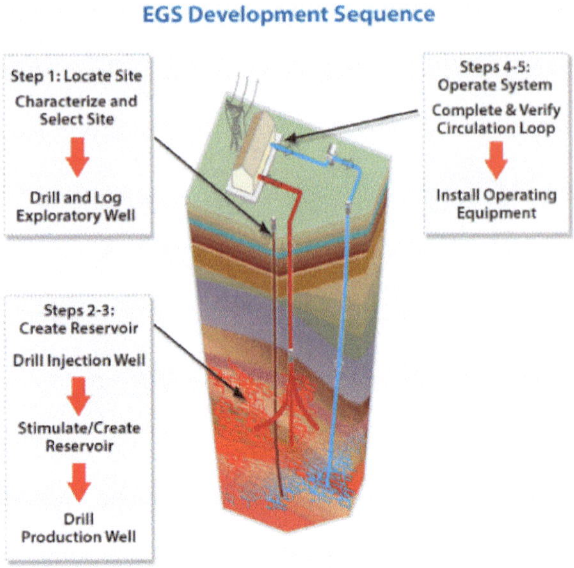

Fig. 19 Schematic diagram depicting an enhanced geothermal system (EGS) setup (U.S Department of Energy, n.d.)

People use geothermal energy to generate electricity and heat buildings. The most popular power plants are dry steam and flash steam types. These plants use steam taken straight from the ground or created from hot water to spin turbines. Binary cycle power plants are another important technology. They can work with cooler sources, which makes them good for places without very hot geothermal activity. A new development in this field is Enhanced Geothermal Systems (EGS). This method involves creating or preparing rock layers to let water flow through more. This has an impact on expanding where we can use geothermal power, as shown in Fig. 19 [35].

3.1.5 Hydropower

The process of hydropower allows electricity generation through various techniques including dam construction and water diversion. The flow of the river hence dictates how this source of energy will be produced. When water falls from a higher ground to a lower one, it transforms potential energy into kinetic energy converted into electricity through turbines and generators as shown in Fig. 20. The motion power in the water changes into electric energy in these gadgets. Thereafter, it is transmitted to the mains to illuminate houses, manufacture and keep trade running [36].

The technology used in hydropower is converting water into electricity through turbines. The most widely known use of this sort of power generation is usually seen in the huge-scale hydroelectric dams. However, on-going developments have paved way to smaller types such as run-of-river which has less impact on the natural environment. Another new feature is pumped storage hydropower where energy is

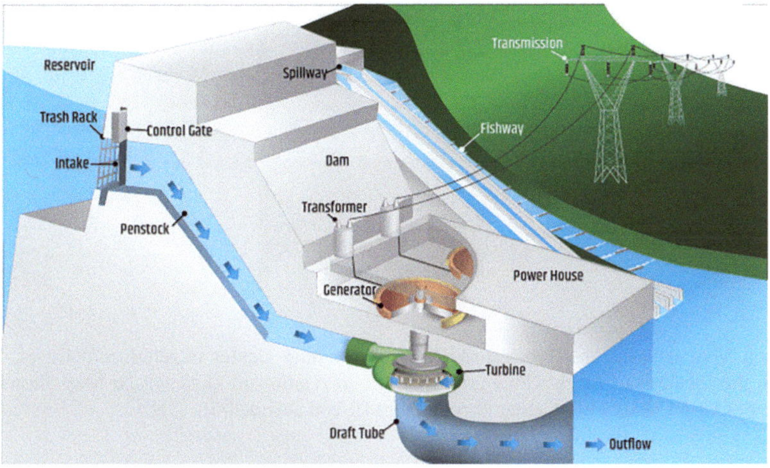

Fig. 20 Diagram shows how hydropower works (U.S Department of Energy, n.d.)

stored with the help of using water as a main product when it is pumped at a time when demand is low and when demand is high the energy which is stored is generated in the form of electricity. Others are coming up such as hydrokinetic turbines, which harvest power from currents and tidal motion without damming water [37].

3.1.6 Energy Storage Technology

However, the supply of this power is unpredictable and hence there is necessity to store it because the renewable sources of power are unpredictable. Lithium-ion batteries which have a high energy density and have reduced in price each year are the most populous in the market hence can be used for grid scale electricity storage and for home storage. Long time and discursivity cause the large flow batteries that use water-based electrolyte solutions stored in external tanks to be consonant for large-scale power applications. While CAES is still the most used, pumped hydro has risen over time and other technologies like solid-state batteries and thermal power can bring further improvements and sustainability [38].

3.1.7 Emerging Technology and Innovations

They are always improving renewables with some possible developments in the future. Now, perovskite solar cells are stated to hold the capability of outcompeting the silicon-based cells in terms of efficiency. Artificial photosynthesis, an attempt to mimic the natural process of using sunlight, for power like hydrogen is also currently being developed. In addition, the establishment of floating solar farms when

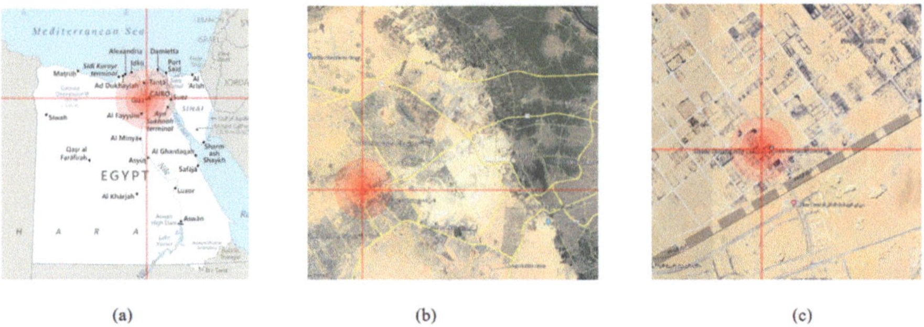

Fig. 21 **a** Map showing the location of Oasis Community Center in relation to Egypt. **b** Map showing the location of Oasis Community Center in relation to Giza gov. **c** Map showing the location of Oasis Community Center in relation to El Wahat road [40]

combined with hydropower offers another opportunity of increasing the advancement of solar energy though inoffensive land resource [39].

4 Case Study

This case study examines the Oasis Community Center in Giza, Egypt - a facility promoting sustainability, recreation and educational activities. Located on El Wahat Road. The center faces various environmental and operational challenges while trying to create a sustainable living experience through innovative design solutions and interactive learning stations.

4.1 Analysis

4.1.1 Location

Oasis Community Center, El Wahat Road, Giza Governorate, Egypt. Latitude: 29.8691, Longitude: 30.8297 (Fig. 21).

4.1.2 Existing Spatial Analysis

The Oasis Community Center is divided into several buildings, where the general use of the center varies from supporting sustainability, recreation, recreational activities, and simple educational activities. The building is divided into areas with different uses, as shown in Fig. 22.

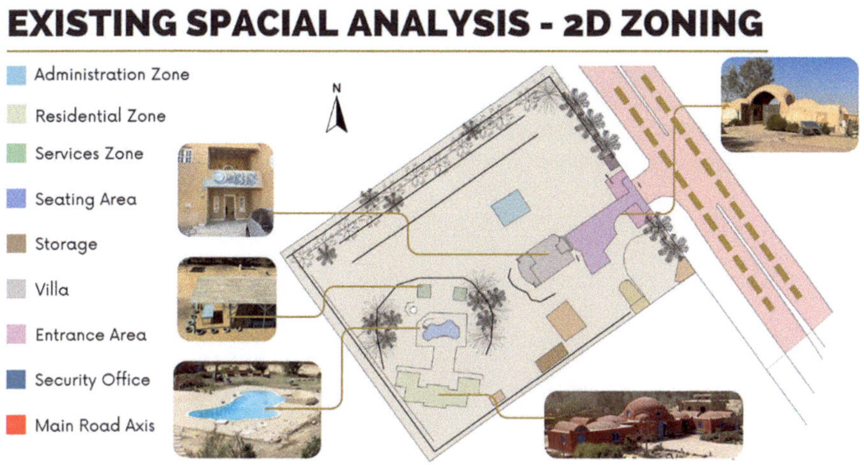

Fig. 22 Image collects layout analysis (by author(s) 2024)

Fig. 23 Image collects 3d layout analysis (by author(s) 2024)

The places of social interaction also vary and differ in their locations and forms, as shown in Fig. 23.

4.1.3 Climatic Analysis

The month with the highest sky cover is March (around 81%) while the month with the lowest sky cover is June (around 33%) (Figs. 24, 25). Throughout the year, the average sky cover is around 69% as shown in Fig. 26.

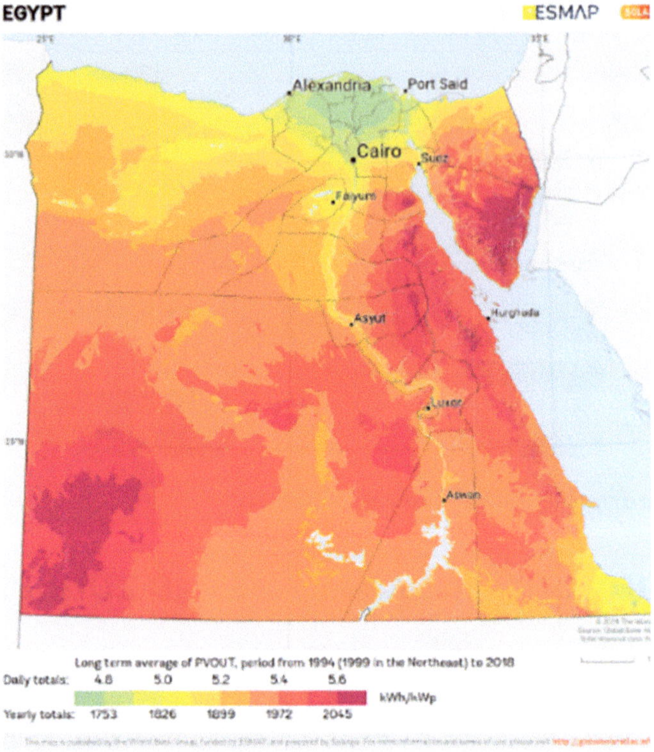

Fig. 24 Shows the distribution of solar radiation to Egypt between 1994 and 2010 [41]

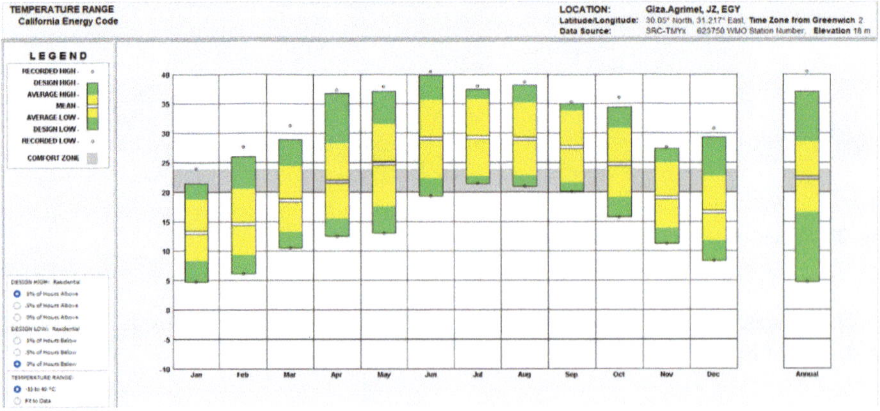

Fig. 25 Graph showing temperature range in Giza Egypt [42]

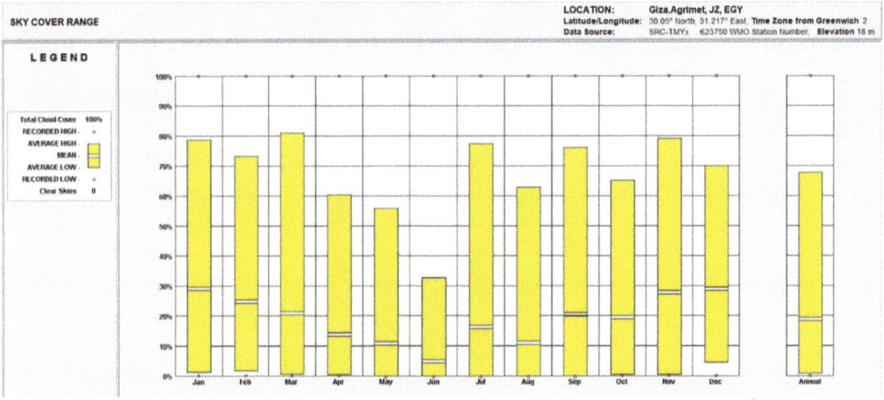

Fig. 26 Chart showing sky cover in Giza Egypt [42]

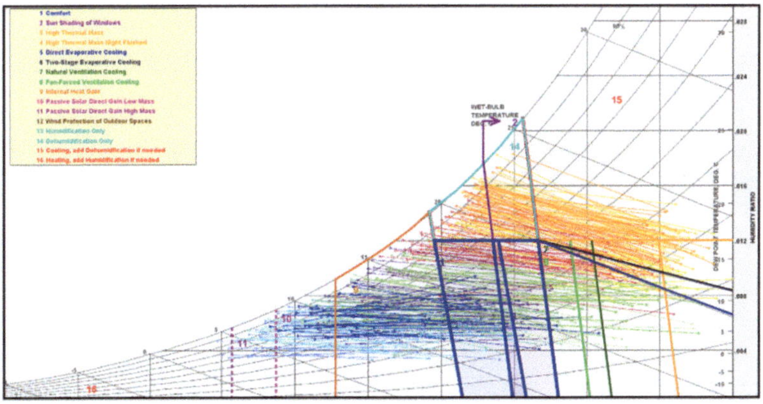

Fig. 27 Diagram showing thermal comfort zones and suggested modifications [42]

By navigating different times of the year, the greatest factor contributing to the level of discomfort is solar radiation and low humidity levels, as shown in Figs. 27, 28.

4.1.4 Site Analysis Conclusion

Water & Sun

In arid El Giza, some water conservation techniques are warranted such as xeriscaping and greywater systems. For effective use of energy, passive solar design should be employed and photovoltaic panels installed.

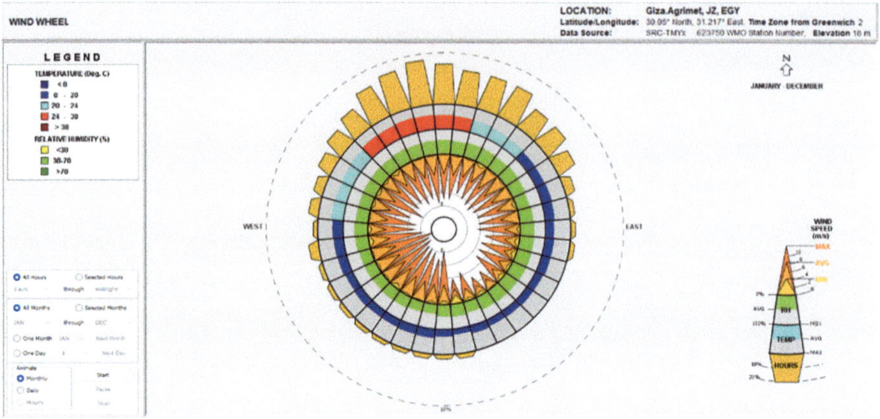

Fig. 28 A graph showing the direction and strength of the wind in Giza gov. (by author(s) using Climate Consultant 2024)

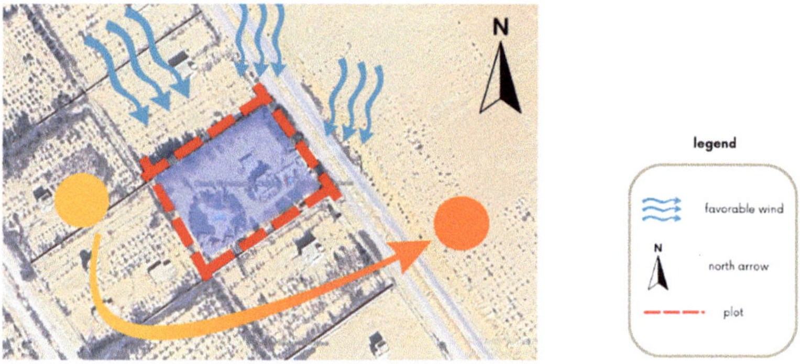

Fig. 29 Site analysis to the project's land. (by author(s) 2024)

Nearby Roads & Services

Nearness to existing roads should be reviewed for accessibility without being stuck in traffic jams. Adequate infrastructure from utility services like water, sewage and garbage collection must be available at that location. infrastructure is in place (Fig. 29).

4.2 *Challenges*

The project relies on non-RE sources: The main source of electricity is the grid, and even the solar panels available in the project transfer the resulting energy to the grid, which harms the environment.
Limited footprint: No heavy construction is allowed, with about 7% construction ratio for it being agricultural land.
Limited accessibility: The location is far from public transportation and there are limited transportation options.
Limited potable water: There are two sources of water. Recycled factory water which is unsuitable for edible crops and basin water stored in tanks that is potable, but it should be taken with a financial return
Limited staff: The place contains different zones but there is a limited number of staff available.

4.3 *Analyzing the Needs, Challenges and the Potentials of the Site*

4.3.1 The Project Needs

- **Potable water:** The project only relies on two water resources: recycled factory water that is not suitable for edible crops and basin water stored in tanks that's non-potable and unfit for drinking.
- **Shade:** Most of the site is exposed to the intense summer sunlight.
- **Power source:** Limited power source.
- **Green areas:** COVID 2020 caused significant tree loss.

4.3.2 The Project Challenges

- **Limited footprint:** No heavy construction is allowed (with a 7% construction ratio for it being agricultural land).
- **Limited accessibility:** The site is far from the main road and public transportation, and there are also limited transportation options.
- **Limited funding:** Solutions to be implemented without incurring high costs.
- **Limited staff:** Despite the fact that the project offers various activities, there is a limited number of staff available.

4.3.3 The Project Potentials

The project has the potential for various activities and technologies, such as sustainable agriculture practices, integrating special RE sources and integrating multipurpose spaces. This will foster job opportunities and will add new functions to the project.

4.3.4 The Project's Vision and Concept

The project concept focuses on creating a "Sustainable Living Experience" that provides an immersive, hands-on opportunity for visitors to engage with local culture, traditions and nature, while also gaining an understanding of sustainable living practices. This approach creates a well-rounded and impactful sustainable living experience for visitors, allowing them to slow down, connect with the local environment, and gain practical knowledge on sustainable living.

The experience seamlessly integrates **educational components** and **authentic experiences**, promoting the core values of slow tourism and fostering community engagement through the following elements:

Six modular repetitive interactive stations: each station focuses on an informative topic, with the aim of teaching visitors' valuable skills and knowledge.

- Saving the bees' station
- MPU station (existing)
- Local crafts station
- Traditional cooking station
- Aloe-vera products station
- Educational platform station

Four sustainable practices: The project incorporates four sustainable practices that benefit the project owners by reducing the negative impact on the environment and promoting self-sufficiency. These sustainable practices are also made available for visitors to engage with, allowing them to learn and participate in sustainable living.

- Water conservation
- Permaculture practices
- Vertical gardens
- Renewable energy

The project concept utilizes a modular design approach, sized at 9×12 m, allowing them to integrate seamlessly with the existing structures on-site in terms of scale and proportions. The modular units are constructed from recycled lightweight wood structures. Wood was selected as the primary building material due to its eco-friendly properties, cost-effectiveness, and natural aesthetic that blends seamlessly with the surrounding environment. A key feature of the design is the use of hinged connections, which provide adjustability, movability, and flexibility for future adaptations (Figs. 30, 31, 32).

The Six Interactive Stations

Saving the bees' station:

Bees play a vital role in both our food supply and ecosystems. Bees are the primary pollinators for plants, and they make up a third of the world's food crops. Bees sustain other insects, animals, and even us humans.

To attract bees and potentially produce honey, several things must be added to the site. First, planting a variety of flowering plants provides a continuous food source

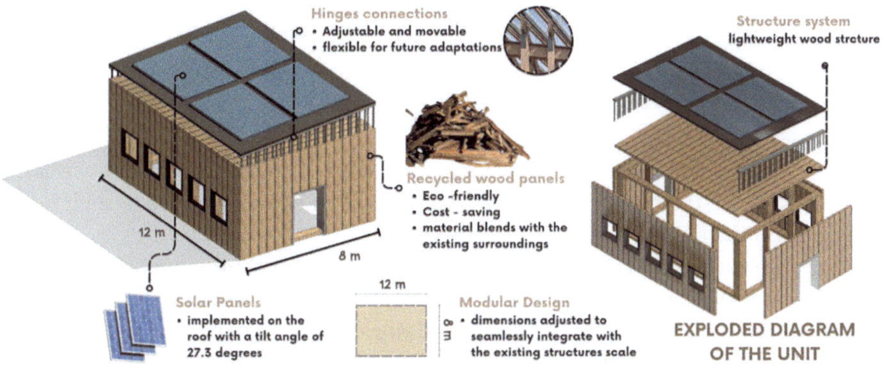

Fig. 30 Detailed diagram of the modular repetitive units (by author(s), 2024)

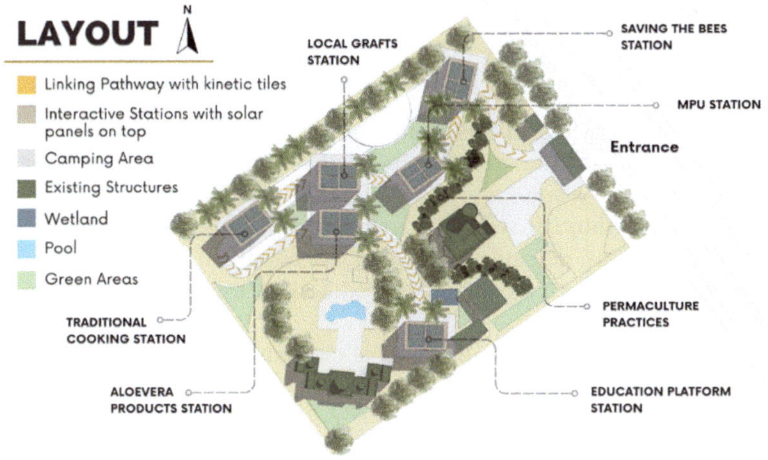

Fig. 31 Layout of the proposed design (by author(s) 2024)

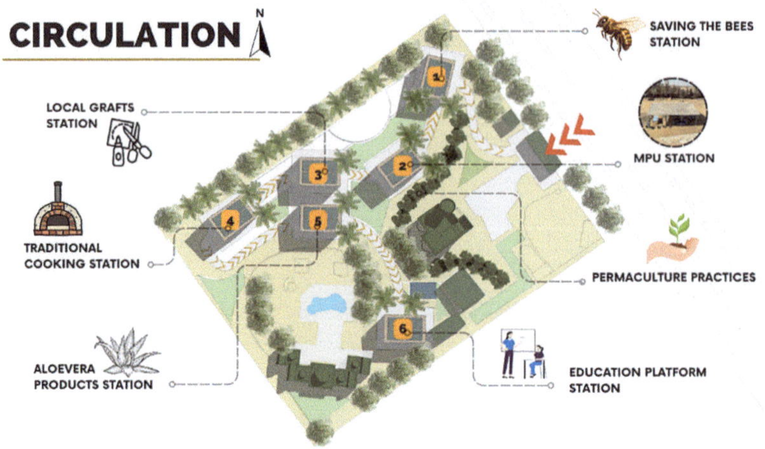

Fig. 32 Circulation of the proposed design (by author(s) 2024)

for bees. Additionally, establishing a water source helps quench their thirst. Creating bee hotels offers suitable nesting sites, while providing bee hives serve as their home for honey production.

Honey production entails overseeing a beehive and harvesting the honey created by bees. This process primarily involves Langstroth hives, which are composed of several stacked wooden boxes containing frames where bees construct their comb and store honey. Beekeepers wear protective gear, including suits and veils to shield themselves from stings. Additionally, a bee smoker is used to calm the bees during hive inspections, while a honey extractor machine facilitates the extraction of honey from the comb without damaging it (Fig. 33).

Local crafts station:

This station contains a weekly workshop series celebrating Egyptian crafts while promoting sustainability through the 3 Rs (Reduce, Reuse, Recycle).

- **The workshops existing in this station are:**
- **Handwoven baskets:** They mainly depend on palm fronds, reeds, or other natural materials available on site.
- **Upcycling Workshop:** It works by transforming discarded materials into new and functional items, reducing waste going to landfills.
- **Sand art:** It involves creating pictures and designs using colored sand from the surroundings.
- **Pottery workshop:** Potters will use local clays to make various functional and decorative items.

The station contains storage for the utilities, high tables for the pottery workshop and circular tables for better communication and storage units for the reeds and palm fronds for the handwoven workshop (Fig. 34).

Fig. 33 Circulation of the proposed design (by author(s), Using Prome AI tool 2024)

Fig. 34 Interior shot for the local crafts station (by author(s), Using Prome AI tool 2024)

Traditional Cooking Station

In this station, energy efficiency is a crucial consideration, especially in the context of sustainability and reducing carbon footprints. The integration of various energy-efficient cooking methods and solar-powered appliances can significantly enhance the overall efficiency of a kitchen.

The techniques applied in this station:

- **Energy-efficient cooking methods:** They are designed to minimize energy consumption while maximizing cooking effectiveness, such as pressure cookers.
- **Solar-Powered Appliances:** Solar-powered appliances harness RE from the sun, making them an excellent choice for sustainable cooking practices such as solar-powered refrigerators, solar cookers and solar ovens.

Fig. 35 **a** Solar Heater (by author 2024). **b** Heat pump water heater [43]

Fig. 36 Exterior shots for the traditional cooking station (by author(s), Using Prome AI tool 2024)

- **Efficient water heating:** Solar heaters harness solar energy to heat water, utilizing solar panels typically installed on the rooftop of the villa in the project, such as a solar heater and Heat pump water heater (Fig. 35).
- **Waste reduction and recycling:** Composting bins by reduce landfill waste and provide nutrient-rich soil for gardening, recycling bottles to conserve resources and reduce landfill waste, recycling cans to save energy and reduce air and water pollution of steel by using it again, recycling paper to conserve energy and decrease the demand for raw materials that can use it again in multipurpose things (Figs. 36, 37).

Aloe vera products station:

This station mainly depends on transforming aloe vera gel extracted from aloe vera cactus planted in the project.

The products produced from aloe vera:

Fig. 37 Interior shots for the traditional cooking station (by author(s), Using Prome AI tool 2024)

- **Aloe Vera Soap:** The aloe vera gel is added with essential oils and it costs approximately 80 Egyptian pounds (~$1.60) per unit and the recommended selling price is 100 Egyptian pounds (~$2.00) per unit.
- **Aloe Vera toner:** The aloe vera gel is mixed with witch hazel or rose water and it costs approximately 65 Egyptian pounds (~$1.30) per unit and the recommended selling price is 88 Egyptian pounds (~$1.75) per unit.
- **Aloe Vera After Sun:** The aloe vera gel is mixed with vitamin E oil, lavender essential oil, or coconut oil and it costs approximately 80 Egyptian pounds (~$1.60) per unit and the recommended selling price is 100 Egyptian pounds (~$2.60) per unit,
- **Aloe Vera hair mask:** The aloe vera gel is mixed with other natural ingredients like coconut oil or honey and it costs approximately 55 Egyptian pounds (~$1.09) per unit and the recommended selling price is 66 Egyptian pounds (~$1.30) per unit.
- **Aloe Vera gel:** It can be used in homemade skincare products and it costs approximately 15 Egyptian pounds (~$0.30) per unit and the recommended selling price is 18 Egyptian pounds (~$0.35) per unit (Fig. 38).

Educational Platform Station:

The station you are referring to appears to focus on educating individuals about three critical aspects of sustainable living: RE, water conservation and vertical gardening. Each topic plays a significant role in promoting environmental sustainability and addressing the challenges posed by climate change.

RE refers to energy derived from natural processes that are replenished at a faster rate than consumed. This includes sources such as solar, wind, hydroelectric, geothermal and biomass.

Vertical gardening is an innovative approach that allows individuals with limited space to grow plants vertically rather than horizontally. This method maximizes space usage and can improve air quality while providing fresh produce (Fig. 39).

Water conservation is essential due to the increasing scarcity of freshwater resources globally (Fig. 40).

Fig. 38 Interior shots for the aloe vera products station (by author(s), Using Prome AI tool 2024)

Fig. 39 Vertical gardening (by author(s), Using Prome AI tool 2024)

4.3.5 Integration of Renewable Energy

Solar Energy:

It is part of the sustainable design approach and one of the four sustainable practices in the project, solar photovoltaic panels have been installed on the rooftops of each modular wooden unit.

The panels are positioned at an angle optimized for capturing the peak sunlight hours in the local climate to maximize the amount of solar energy generated. This

Fig. 40 Interior shots for the educational platform station (by author(s), Using Prome AI tool 2024)

integrated RE system will provide a significant portion of the power required to operate the modular units, reducing the reliance on grid-supplied electricity and lowering the project's overall carbon footprint.

The solar panels seamlessly complement the natural aesthetic of the wooden structures, further enhancing the environmentally conscious design.

Kinetic Energy:

To further enhance the sustainable energy generation capabilities of the modular wooden units, a network of kinetic energy-harvesting floor tiles has been integrated along the pedestrian pathways.

As visitors stroll along these paths, their footsteps will activate the specialized tiles, converting the mechanical energy of their movement into electrical energy. This innovative system creates an interactive and sustainable walking experience, engaging users while generating supplementary power.

Each individual footstep upon a tile can produce up to 5 watts of electricity, an amount sufficient to power an LED light bulb for approximately 30 s. The integration of these kinetic tiles complements the rooftop solar panels, further reducing the reliance on grid-supplied electricity and minimizing the project's carbon footprint (Figs. 41, 42).

4.3.6 Permaculture and Vertical Gardening

Types of plants.

- Aloe Vera cactus
- Pollinator gardens

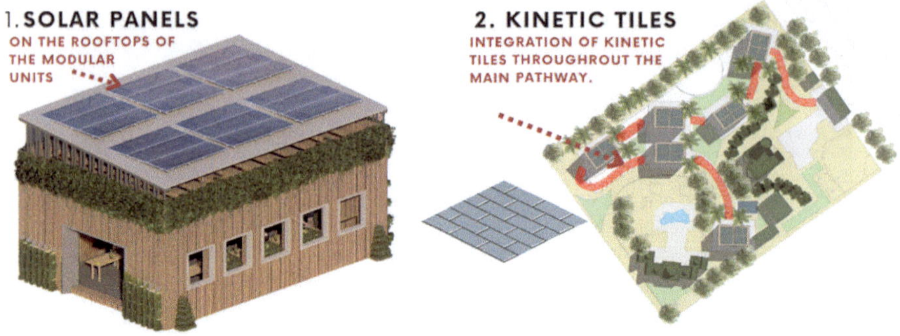

Fig. 41 Diagram of the RE integrated in the project (by author(s) 2024)

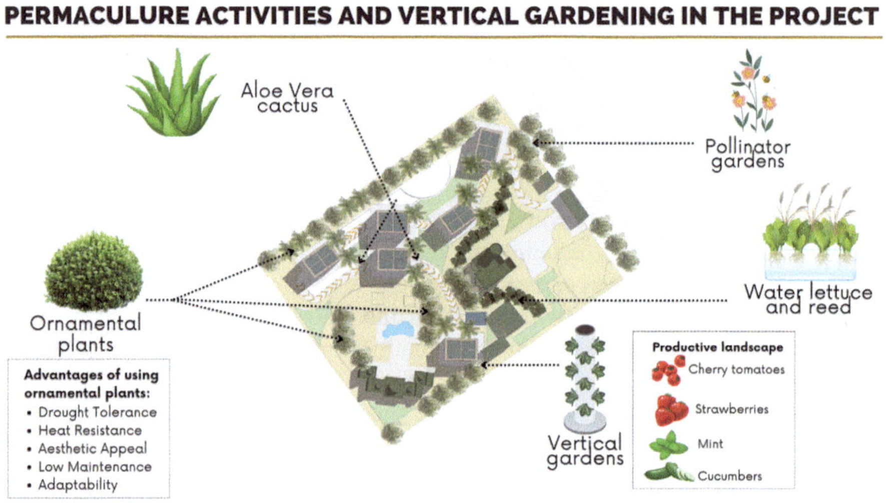

Fig. 42 Diagram of the permaculture activities and vertical gardening in the project (by author(s) 2024)

- Water lettuce and reed
- Ornamental plants
- Vertical gardens like Cherry tomatoes, Strawberries, Mint and Cucumbers.

Advantages of using ornamental plants

- Drought Tolerance
- Heat Resistance
- Aesthetic Appeal
- Low Maintenance
- Adaptability

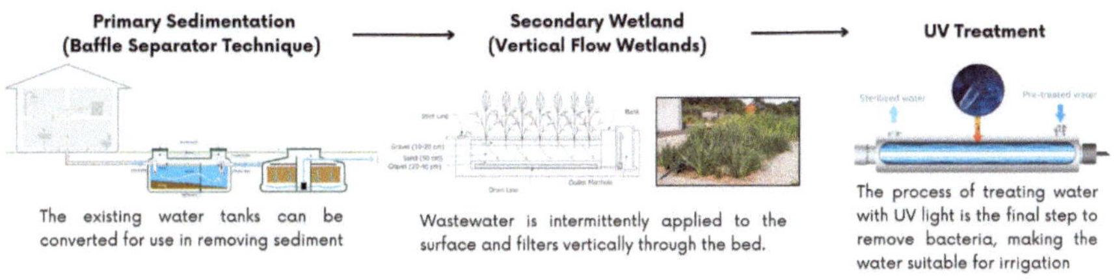

Fig. 43 Diagram of the Water treatment stages in the project (by author(s), Based on Premiertechaqua.com 2020 [44–46]

4.3.7 Water Treatment

To become suitable for irrigating human-consumable crops, treated water must pass through three stages:

Water treatment stages:

- **Primary Sedimentation (Baffle Separator Technique):**
- The existing water tanks can be converted for use in removing sediment
- **Secondary Wetland (Vertical Flow Wetlands):**
- Wastewater is intermittently applied to the surface and filters vertically through the bed.
- **Biological Treatment (UV Treatment):**
- The process of treating water with UV light is the final step to remove bacteria, making the water suitable for irrigation (Fig. 43).

4.3.8 Construction Calculations

The amount of greywater.

- Toiltes (75–115 L) of gray water per day * 4 persons* 3 toilets = 900 L
- Eco-friendly kitchen approximately 38 to 76 L

Total gray water: Toiltes + Eco friendly Kitchen = 1015 L per day

Wetlands area:

- Equation (A = Q × T × P/ H)
- A = Required area of the wetland system (in square meters)
- Q = Flow rate of gray water (in liters per day)
- T = Hydraulic retention time (in days)
- P = Porosity of the wetland media (dimensionless, typically between 0.3 and 0.5)
- H = Hydraulic loading rate (in meters per day)

Total Vertical Flow Wetlands area: 20M2 (Fig. 44).

Total Estimated Cost in 2024 According to Exchange Rate of 1 USD = 50 L.E.

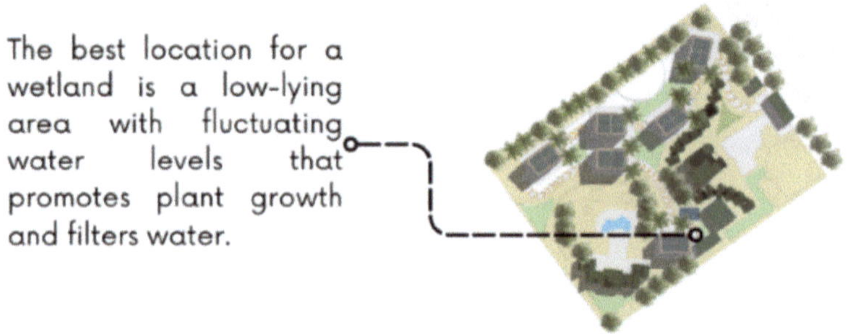

Fig. 44 The graph shows the best wetland location for the project (by author(s))

- **Pump for Tank Cost:** Approximately 200–300 EGP (~ $4.00–6.00).
  - **Base pump price:** 150–250 EGP (~$3.00–5.00).
    - Pump unit: 140–230 EGP (~$ 2.80–4.60).
    - Hoses and connectors: 10–20 EGP ($0.20–0.40).
  - **Installation and minor accessories:** 50 EGP ($ 1.00).
    - Labor cost: 30–40 EGP ($ 0.60–0.80).
    - Sealants and fittings: 10–20 EGP ($ 0.20–0.40).
- **Pump for Wetland Cost:** Approximately 150–250 EGP (~$3.00–5.00).
  - **Base pump price:** 120–200 EGP (~$2.40–4.00).
    - Pump unit: 110–180 EGP (~$2.20–3.60).
    - Pipes and fittings: 10–20 EGP ($0.20–0.40).
  - **Installation and fittings:** 30–50 EGP (~$0.60–1.00).
    - Labor cost: 20–30 EGP (~$0.40–0.60).
    - Mounting brackets and sealants: 10–20 EGP ($0.20–0.40).
- **Wetland Maintenance Cost:** Approximately 1200–1800 EGP (~$23.90–35.85).
  - **Labor cost:** 500–800 EGP (~$9.95–15.95).
    - Regular inspection: 200–300 EGP (~$4.00–6.00).
  - Cleaning and removal of debris: 150–250 EGP (~$3.00–5.00).
  - System adjustments and minor repairs: 150–250 EGP (~$3.00–5.00).
  - **Replacement of filters and materials:** 400–600 EGP (~$7.95–11.95).
    - New filtration media: 200–300 EGP (~$4.00–6.00).
    - Water treatment chemicals: 100–200 EGP (~$2.00–4.00).
    - Replacement of damaged components: 100–200 EGP (~$2.00–4.00).
  - **Miscellaneous supplies:** 300–400 EGP (~$6.00–7.95).

    - Safety gear (gloves, masks, etc.): 50–100 EGP (~$1.00–2.00).
    - Transportation of materials: 100–150 EGP (~$2.00–3.00).
    - Emergency repair materials: 150–200 EGP (~$3.00–2.00).

- **UV Treatment Cost:** Approximately 500–700 EGP. (~$9.95–13.95).
  - **UV lamp replacement:** 300–400 EGP (~$6.00–7.95).
    - New UV bulb: 250–350 EGP (~$5.00–6.95).
    - Electrical fittings and wiring: 50 EGP (~$1.00).
  - **Maintenance and cleaning:** 200–300 EGP (~$4.00–6.00).
    - Cleaning solutions: 50–100 EGP (~$1.00–2.00).
    - Labor cost for cleaning and servicing: 100–150 EGP (~$2.00–3.00).
    - Safety inspections: 50 EGP (~$1.00).
- **Solar Panel Maintenance:** Approximately 200–300 EGP. (~$4.00–6.00).
  - **Cleaning materials and labor:** 150–250 EGP (~$3.00–5.00).
    - Cleaning solution and tools: 50–100 EGP (~$1.00–2.00).
    - Labor cost for cleaning: 100–150 EGP (~$1.00–3.00).
  - **Minor repairs and inspections:** 50 EGP (~$1.00).
    - Inspection of connections and panels: 20–30 EGP (~$0.40–0.60).
    - Minor wire replacements or adjustments: 20–30 EGP (~$0.40–0.60).

5 Conclusion

This study dives into how renewable energy can be woven into Egypt's built environment, especially as cities expand and the demand for energy grows. It looks at how different types of buildings use energy and points out some of the key issues, like inefficiencies in the power grid and the financial hurdles of moving toward more sustainable solutions. Without stronger policies and tech advancements, making the switch to renewable energy is going to be a tough road ahead.

The case study of the Oasis Community Center in Giza shows how sustainable urban design can actually be put into practice. The center incorporates things like solar panels, kinetic energy tiles, and water-saving techniques, along with running educational programs to spread awareness about environmental issues. However, the project still faces some challenges, like limited access, water shortages, and its ongoing reliance on traditional energy sources. These problems really highlight the urgent need for better infrastructure and more funding.

This study ties into several of the UN's Sustainable Development Goals, especially those around clean energy, responsible consumption, and climate action. To really make an impact with similar projects, the research suggests stepping up renewable energy use, improving water management, expanding educational programs, and securing more funding. If these steps are taken, community-led initiatives could have a meaningful effect on Egypt's sustainability efforts, helping to tackle both environmental and economic challenges.

6 Relevant SDGs

The Oasis Community Center exemplifies a comprehensive approach to sustainability by tackling the UN Sustainable Development Goals. The project's innovative stations, sustainable practices and community-focused design have a significant impact on five key SDGs as shown in Fig. 45: Good Health and Wellbeing (SDG 3), Quality Education (SDG 4), Sustainable Cities and Communities (SDG 11), Responsible Consumption and Production (SDG 12) and Climate Action (SDG 13). Each project component has been carefully developed to contribute to these global goals while also bringing about long-term positive impact on a local scale.

6.1 SDG 3: Good Health and Wellbeing

The goal of guaranteeing healthy lifestyles and encouraging wellbeing for people of all ages is critical to sustainable development. The Oasis Community Center contributes to this goal by providing spaces and activities that promote both physical and mental health. The Aloe Vera Products station promotes natural wellbeing through environmentally friendly skincare and the Traditional Cooking station supports good eating practices. The project's emphasis on outdoor activities, connection with nature and stress-relieving surroundings benefits both visitors and community residents' mental and physical health.

Fig. 45 The five UN Sustainable Development Goals (SDGs) addressed by the Oasis Community Center project (by author(s) 2024)

6.2 SDG 4: Quality Education

Quality education is critical to achieving sustainable development. The project's Educational Platform station and six interactive learning zones show a significant dedication to this purpose. Visitors benefit from hands-on seminars, practical demonstrations and experiential learning opportunities to learn about sustainable living practices, traditional crafts and environmental protection. The project's educational approach goes beyond standard approaches, providing immersive experiences that make learning more engaging and successful.

6.3 SDG 11: Sustainable Cities and Communities

This goal aspires to make cities more inclusive, safe, resilient and sustainable. The Oasis Community Center exemplifies sustainable community development by including RE technologies, water conservation methods and green spaces in its design. The project's emphasis on local crafts and traditional techniques aims to maintain cultural heritage while encouraging sustainable urban growth. The center's accessibility and inclusive design make it a valued community asset that fosters collaboration while showcasing sustainable living concepts.

6.4 SDG 12: Responsible Consumption & Production

The project substantially supports the goal of establishing sustainable consumption and production patterns. The center encourages responsible resource use and waste reduction through its numerous stations, most notably the Traditional Cooking and Local Crafts stations. The implementation of water treatment systems, recycling programs and sustainable agriculture methods exemplifies practical responsible consumption strategies. Visitors learn about sustainable manufacturing processes through hands-on activities, which encourages them to embrace more sustainable living choices.

6.5 SDG 13: Climate Action

Taking immediate action to address climate change and its consequences is critical, and the Oasis Community Center makes a substantial contribution to this goal. The project's use of RE, including solar panels and kinetic tiles, provides feasible options for lowering carbon emissions. Vertical gardens and permaculture principles help to sequester carbon and promote sustainable agriculture. Educational activities on climate change and sustainable practices inspire visitors to take action in their own lives, resulting in a positive environmental ripple effect.

Through these interwoven activities, the Oasis Community Center demonstrates how local projects may help to achieve global sustainability goals while also providing practical benefits to their communities.

7 Recommendations

The following important recommendations are intended to improve the sustainability and impact of the Oasis Community Center and similar programs throughout Egypt. The program's strategies prioritize RE integration, water management, educational programs, accessibility, resource optimization, financial sustainability, community engagement and project documentation to address existing challenges and build on successes, resulting in a more sustainable future for the communities served:

- Adopting a hybrid RE system that combines solar PV with kinetic floor tiles is crucial for diversifying energy sources. Implementing energy storage systems will help manage intermittent electricity generation, while partnerships with RE suppliers can lower installation costs.
- Scaling up the three-stage water treatment system can significantly improve water quality. Smart water monitoring systems will enable real-time tracking and leak detection, while expanding greywater recycling will reduce reliance on potable water.
- Developing structured curriculum materials for each interactive station is essential for maximizing educational impact. Train-the-trainer programs will address staffing shortages and partnerships with local schools can facilitate regular educational visits.
- Creating transportation agreements for shuttle services will enhance access to community centers. Virtual tours and online educational content can reach a broader audience, while mobile educational stations can serve underserved schools and communities.
- Establishing detailed maintenance schedules for RE systems is vital for efficiency. Collaborating with local farms for composting and organic waste management will promote a circular economy and a volunteer program can engage the community in sustainability efforts.
- Community centers should develop revenue-generating activities around sustainable products, such as expanding the Aloe Vera line. Membership programs for regular visitors can provide steady income and corporate sponsorships can enhance financial support for educational programs.
- Regular community feedback sessions will ensure responsiveness to user needs. A mentorship program for local youth interested in sustainability will empower future environmentalists, while partnerships with local businesses can promote sustainable product development.

- Comprehensive monitoring systems for energy and water usage will facilitate ongoing efficiency assessments. Detailed case studies of successful implementations can serve as learning tools and best practice guides will offer insights for replicating successful initiatives.

Acknowledgements Special thanks to Oasis Community Center team: Eng. Rehab Aly Sharaf and Dr. Sally Hafez Bahgat for hosting the field visit and for providing an invaluable experience and feedback for the development of the project.

References

1. World Bank (2023) Urban Development (online) World Bank. https://www.worldbank.org/en/topic/urbandevelopment/overview
2. Egyptian Ministry of Electricity and Renewable Energy (2023) Energy Consumption Forecast Report, Cairo. http://www.moee.gov.eg/english_new/home.aspx. Accessed 11 August 2024
3. URBANET (2019) Urbanisation and urban development in Egypt I infographic (online). Urbanet. https://www.urbanet.info/urbanisation-and-urban-development-in-egypt/
4. UNDP (2024) Reimagining cooperation in a polarized world (online). https://hdr.undp.org/system/files/documents/global-report-document/hdr2023-24reporten.pdf.
5. IRENA (2024). International renewable energy agency (IRENA) (online) Irena.org. https://www.irena.org/. Accessed 10 August 2024
6. Egyptian Electric Utility and Consumer Protection Regulatory Agency (2023) Annual report on Egypt's electricity sector, Cairo. http://www.egyptera.org/en/. Accessed 11 August 2024
7. International Energy Agency (IEA) (2022) Renewables 2022: analysis and forecast for 2027. https://iea.blob.core.windows.net/assets/ada7af90-e280-46c4-a577-df2e4fb44254/Renewables2022.pdf. Accessed 20 August 2024
8. International Energy Agency (IEA) (2023) Egypt—countries & regions (online) https://www.iea.org/countries/egypt. Accessed 10 August 2024
9. El-Agami M, Hanafy G, Osman M (2021) Investigating the effect of high-rise buildings' mass geometry on energy efficiency within the climatic variation of Egypt. Sustainability 13(19):10529. https://doi.org/10.3390/su131910529
10. Moharram NA, Tarek A, Gaber M, Bayoumi S (2022) Brief review on Egypt's renewable energy current status and future vision. Ain Shams Eng J 13(6):101765
11. RCEEE (2023) AFEX 2023 (online) Rcreee.org. https://rcreee.org/afex-2023/. Accessed 28 March 2025
12. Ministry of Electricity and Renewable Energy (2023) Energy Efficiency in Government Buildings Project. Energy Consumption Patterns in Public and Institutional Buildings in Egypt, Cairo
13. Egyptian Electricity Holding Company (2023) Annual Report 2022. EEHC, Cairo
14. Egyptian Natural Gas Holding Company (2023) Natural gas consumption by sector: Annual Report 2022. EGAS, Cairo
15. Egyptian Hotel Association (2023) Energy consumption in the Egyptian hospitality sector: 2022 analysis. EHA, Cairo
16. Intergovernmental Panel on Climate Change (IPCC) (2023) Climate change 2021—the physical science basis: working group I contribution to the sixth assessment report of the intergovernmental panel on climate change. Cambridge University Press, Cambridge
17. Center for Energy Education (2017) How solar energy works—Center for Energy Education (online). https://center4ee.org/how-solar-energy-works/. Accessed 10 August 2024

18. Cardoso A (2020) About us—Center for Energy Education (online) Center4ee.org. https://center4ee.org/about-the-center-for-energy-education/. Accessed 28 March 2025
19. Tarana AC, Zoya F (2020) Design Modeling and Simulation of 150 KW SPV system (online). https://www.researchgate.net/publication/361396909_Design_Modeling_and_Simulation_of_150_KW_SPV_system
20. Green MA, Dunlop ED, Hohl-Ebinger J, Yoshita M, Kopidakis N, Bothe K, Hinken D, Rauer M, Hao X (2023) Solar cell efficiency tables (Version 60). Progress in photovoltaics: research and applications 30(7):687–701. https://doi.org/10.1002/pip.3595. Accessed 20 September 2024
21. Solar Energy Technologies Office (2022) Solar Photovoltaic Cell Basics (online). Energy.gov. https://www.energy.gov/eere/solar/solar-photovoltaic-cell-basics
22. REPSOL (2024) Solar thermal energy: What it is and its benefits. https://www.repsol.com/en/energy-and-the-future/future-of-the-world/solar-thermal-energy/index.cshtml. Accessed 20 August 2024
23. U.S. Energy Information Administration (2016) Solar explained—U.S. Energy Information Administration (EIA) (online). Eia.gov. https://www.eia.gov/energyexplained/solar/
24. Brunel (2021) What is concentrated solar power and how does CSP work? https://www.brunel.net/en/blog/renewable-energy/concentrated-solar-power. Accessed 20 August 2024.
25. U.S. Department of Energy (DOE) (2023) Department of Energy (online). Energy.gov. https://www.energy.gov/
26. Vartan S (2021) What is wind energy? Definition and how it works. https://www.treehugger.com/what-is-wind-energy-5097217. Accessed 10 August 2024
27. Manwell JF, McGowan JG, Rogers AL (2022) Wind energy explained: theory, design, and application, 3rd edn. Wiley, Chichester
28. Projet Eole (2020) Vertical axis wind turbine. https://commons.wikimedia.org/wiki/File:EolePat.jpg. Accessed 29 March 2025
29. Drummond D (2020) What is biomass? https://www.drax.com/sustainable-bioenergy/what-is-biomass/. Accessed 21 August 2024
30. REDDA (2022) What is biomass and how does it work. https://redda.com.br/en/what-is-biomass/. Accessed 10 August 2024
31. Demirbas A (2021) Biomass and the future of energy production, energy sources, part a: recovery, utilization, and environmental effects 43(15):2008–2021. Accessed 20 September 2024
32. U.S. Energy Information Administration (EIA) (2022) Geothermal explained. https://www.eia.gov/energyexplained/geothermal/. Accessed 21 August 2024
33. Office of Energy Efficiency & Renewable Energy (2019) Electricity Generation (online) Energy.gov. https://www.energy.gov/eere/geothermal/electricity-generation
34. Alekseenko S (2023) Deep geothermal heat overview. Thermopedia. https://doi.org/10.1615/thermopedia.010316
35. Dipippo R (2022) Geothermal power plants: principles, applications, case studies and environmental impact a book review. Accessed 14 September 2024
36. Water Science School (2019) Hydroelectric Power: How it Works | U.S. Geological Survey (online) www.usgs.gov. https://www.usgs.gov/special-topics/water-science-school/science/hydroelectric-power-how-it-works. Accessed 10 August 2024
37. Kumar A, Katoch SS (2022) Hydropower technology advancements: Current trends and future prospects. Renew Energy 146:1521–1535
38. Zakeri B, Syri S (2022) Electrical energy storage systems: A comparative life cycle cost analysis. Renew Sustain Energy Rev 144:111028
39. National Renewable Energy Laboratory (NREL) (2023) Emerging photovoltaic technologies. https://www.nrel.gov/pv/emerging-tech.html. Accessed 10 August 2024
40. CIA (n.d.) Details—the world factbook (online). https://www.cia.gov/the-world-factbook/countries/egypt/map.
41. Omran E-SE, Gaber IM, Elkashef TM (2022) Climate considerations in the planning and sustainability of Egyptian cities. Sustainable development goals series, pp 207–238. https://doi.org/10.1007/978-3-031-10676-7_13

42. Ucla.edu. (2024) Climate consultant (online). https://energy-design-tools.aud.ucla.edu/climate-consultant
43. Andy's Pipe Dream Plumbing (2024) Water heater regulations—Andy's pipe dream plumbing (online). https://andyspipedream.com/water-heater-regulations/. Accessed 29 March 2025
44. Aqua-nostra.de (2024) Constructed wetlands. aqua nostra eG (online). https://www.aqua-nostra.de/en/constructed-wetlands. Accessed 29 March 2025
45. Blue Steel Water | Water Treatment Systems Zimbabwe (2024) UV Water Sterilizers | Blue Steel Water | UV Water System (online). https://www.bluesteelwater.com/uv-water-sterilizers/. Accessed 29 March 2025
46. Sswm.info. (2015) Vertical Flow Constructed Wetland (VFCW) | SSWM - Find tools for sustainable sanitation and water management! (online). https://sswm.info/step-nawatech/module-1-nawatech-basics/appropriate-technologies-0/vertical-flow-constructed-wetland-%28vfcw%29

Role of Blockchain and Cyber-Physical Systems (CPSs) in Execution of Sustainable Smart Construction

Abdul Hannan Qureshi, Bilal Manzoor, Ahsan Waqar, and Tarek Zayed

Abstract The daily life, community dynamics, and attainment of necessities in arid and semi-arid cities face different challenges compared to pluvial or subtropical regions. However, by adopting technologies and smart operational infrastructure, the impact of such challenges can be mitigated, and a better lifestyle can be offered to such communities of arid or semi-arid regions. Likewise, integrating cyber-physical systems (CPSs) and blockchain has resulted in significant changes. This connection improves openness, efficiency, and confidence in construction processes by allowing live monitoring and decision-making. Blockchain provides a secure and unchangeable platform for documenting transactions, contracts, and asset management throughout the project life cycle, assuring accountability and traceability. CPSs combine computer, networking, and physical processes to provide a structure for automation that improves construction machines and operational workflows. This chapter explores the potential of blockchain technology and CPSs in the construction industry to achieve smart and sustainable operations. In addition, the chapter also includes a discussion on implementing CPSs for real-time monitoring, green energy, water harvesting, eco-management, and agronomics, considering the constraints of arid and semi-arid regions. This chapter will educate the readers about the significance of blockchain and CPSs in the construction sector and identify prospective areas that might benefit from integrated systems in such arid or semi-arid regions.

Keywords Industry 4.0 · Technology · Construction sector · Intelligent systems · DApps · Construction 4.0 · Building information modeling

---

A. H. Qureshi (✉) · B. Manzoor · T. Zayed
Department of Building and Real Estate, Faculty of Construction and Environment, The Hong Kong Polytechnic University, Kowloon, Hong Kong, China
e-mail: hannan.qureshi@polyu.edu.hk

A. Waqar
Department of Civil and Environmental Engineering, Universiti Teknologi PETRONAS, Perak Darul Ridzuan, Malaysia

Z. ElZein and A. Negm (eds.), *Recent Approaches of Sustainable Architecture in Arid and Semi-arid Cities*, Urban Sustainability,
https://doi.org/10.1007/978-981-95-1645-2_15

1 Introduction

Blockchain operates on a decentralized and secure software system. Blockchain enables decentralized data exchanges and storage that are resistant to tampering. This enhances the feasibility of utilizing blockchain technology in several financial sectors. Blockchains are digital ledgers that distribute and facilitate Internet transactions securely without relying on a central administrative authority. Blockchain technology functions as databases that allow numerous users to concurrently update a ledger chain, leading to various successive repetitions in a sequence [1]. On the other hand, a Cyber-Physical System (CPS), by incorporating communication technologies, is a combination of multidisciplinary physical systems, pervasive sensing, cyber components, and, most importantly, the Internet of Things (IoT) devices [2–4]. In simple words, CPS is a set of integrated computation, networking, and physical processes. The physical processes are managed and observed via sub-integrated network systems, allowing input adjustment as per user requirements. The subsystems function independently, with the capacity to respond to the external environment. Several small gadgets carry out physical processes via sensing, computing, and communicating (wired or wirelessly). Such physical devices can be characterized by physical information properties or sensing equipment, such as infrared sensors or Radio Frequency Identification (RFID). They can be coupled to a networking system to transport the acquired data to the computing subsystem. The creation of CPSs is a critical step in advancing operations in a variety of industries, including construction, leading to a sustainable livelihood. This progress is aided by technical improvements that are consistent with the CPS's goals of automation and intelligent interlinkage across cyber and physical domains. The true dream of sustainable cities and infrastructure seems to be more convenient and manageable under a CPS-based interlinked operational environment. Implementing CPSs in the construction sector necessitates the proper dissemination of information and communication among many stakeholders. This is because the construction business is highly fragmented, involving multiple key stakeholders [2, 5].

Industry 4.0 is a technologically advanced platform that automates conventional industrial manufacturing practices and processes. It may also be viewed as modifying current computerization, electrification, and mechanization processes [6, 7]. Nevertheless, the advent of blockchain technology has profoundly altered and transformed the operational frameworks of various industries. Therefore, Aste et al. [8] mentioned blockchain technology as a game changer in the construction industry under the umbrella of Industry 4.0 or the fourth industrial revolution. One of the big challenges in the construction industry is work transparency and lack of trust among stakeholders, which affects the working environment and relationships. These have resulted in inadequate collaboration among project stakeholders and difficulties with information exchange. However, the decentralized nature of blockchain technology has optimistically addressed the above-mentioned scenario, especially in the construction industry, by promoting trust and high transparency in the processes and operational systems. Moreover, researchers also believe that blockchain technology will play a significant part in future worldwide networks and in the development of sustainable lifestyles, with the motive of optimizing the information transfer between stakeholders [9].

In 2015, the UK government stated that blockchains might "reform our financial markets, supply chains, consumer and business-to-business services, and publicly held registers". Financial sector reports indicate that blockchain deployments in the banking market accelerate financial transactions and payments by expediting the confirmation process. Blockchain technology has the potential to reshape digital trust and remove middlemen, creating a new leadership paradigm that may challenge conventional governance structures [1]. In recent years, Industry 4.0 has garnered interest in the construction sector [10]. Extensive studies have been conducted on Industry 4.0, the IoT, and business analytics [11, 12]. Blockchain is a nascent technology extensively studied for features like decentralization, transparency, and publicly accessible transactions. However, elements like scalability and network infrastructure still necessitate additional research. Industry 4.0 applications in construction, also known as Construction 4.0, involve industrialized production like additive manufacturing and prefabrication, CPSs such as drones and IoT, and digitalization, including Building Information Modeling (BIM) and cloud computing [10]. The implementation of Industry 4.0 in the construction sector has been impeded by the fragmented structure of operations within the construction supply chain, where project delivery is divided into separate processes, lack of awareness, and organizational culture. Hence, it is advisable to utilize advanced Information and Communication Technologies (ICTs) more extensively to automate building jobs and minimize the resulting process fragmentation [7].

The construction industry has faced persistent challenges with inadequate productivity over many years. The industry faced a productivity gap of $1.6 trillion in 2017, as reported by the McKinsey Global Institute. Construction projects are characterized by their large scale, lengthy durations, and unique nature, leading to a high degree of complexity and unpredictability. These intrinsic characteristics lead prime stakeholders (e.g., client, contractor, consultant) to engage in opportunistic actions, including not fulfilling responsibilities, hiding crucial information, and needing more sincerity in negotiating/bargaining. Many actual accidents can serve as proof of opportunism. The American Society of Civil Engineers (ASCE) reports approximately $340 billion in corrupt practices and expenses in global construction annually. These opportunistic practices result in mistrust and hostile relationships. Researchers suggest that the dynamic between contractors and subcontractors is commonly cost-focused, transactional, and belligerent, marked by feelings of unfairness and doubt. Industry is constantly looking for ways to cultivate a culture based on trust, honesty, and shared values [13]. Researchers believe blockchain technology could speed up the implementation of digital construction by addressing existing obstacles related to trust, transparency, data traceability, intellectual property rights, and record-keeping in the industry. Hence, it is important to explore the possible applications of blockchain technology in the construction industry and assess its existing position within the sector.

Bitcoin is the pioneering implementation of blockchain technology, often known as Blockchain 1.0. Cryptocurrency is widely embraced as a means of payment for products and services. Satoshi Nakamoto first presented the idea in 2008 [14], and Bitcoin was not programmable at that point. Emphasizes applications that utilize public key encoding for peer-to-peer financial transactions within a dispersed setting. Blockchain 2.0 involves using digitized smart contracts created to revolutionize the financial sector. Blockchain 3.0 evolved to include decentralized applications (DApps) and smart asset transactions. Blockchain 3.0 seeks to enhance blockchain functionality regarding transaction speed, scalability, and implementation simplicity to meet the requirements of various industries. The latest iteration of blockchain technology, known as Blockchain 4.0, enhances its use by integrating with artificial intelligence (AI), enabling autonomous decision-making and action without human intervention. Blockchain 5.0 is currently in development [9]. However, the phrase "Blockchain 5.0" is not commonly acknowledged or well-defined in the blockchain or cryptocurrency circles. Blockchain technology's development is commonly divided into distinct generations, each characterized by notable breakthroughs or changes in capacity. Blockchain 1.0 denotes the initial blockchain idea pioneered by Bitcoin, emphasizing decentralized financial transactions. Blockchain 2.0 is linked to Ethereum's implementation of smart contracts, which broadened the scope of blockchain applications from basic transactions to more intricate, programmable functions. Blockchain 3.0 and beyond entail advancements like scalability solutions, interoperability among different blockchains, privacy enhancements, and the integration of blockchain technology into sectors beyond finance, such as supply chain management, healthcare, and governance. The names "Blockchain 4.0" and "Blockchain 5.0" are occasionally mentioned in conversations regarding the future of blockchain technology; however, they lack standardization, and their meanings may differ depending on the origin. Blockchain 4.0 can involve combining blockchain with technologies such as IoT, AI, and big data to develop advanced, autonomous, and efficient systems. "Blockchain 5.0" may indicate improvements that involve integrating blockchain technology into daily life and global economic systems, with exceptional scalability, interoperability, and user-friendliness [15, 16].

On the other hand, the usage of the Industry 4.0 theme in coordination with CPSs and other technologies is changing deserts into green lands. One of the current examples is the Kingdom of Saudi Arabia (KSA), with its various unleashed initiatives aiming to transform deserts and barren lands into thriving environments and green oases. Moreover, they have named this project "NEOM" and "Al-Dahnaa Project" with the slogan "The Sustainable City of the Future" [17]. Such areas are mostly semi-arid and arid regions with intense warm weather, high winds, and sporadic and low yearly rainfall. According to literature, around 41% of the Earth's land surface comprises arid zones [18]. Therefore, climate change is expected to heighten hazards and vulnerabilities in such dry regions. Researchers believe that one of the primary factors contributing to the presence of such a dry spell is the geographic location, i.e., distance from water bodies and the seldom occurrence of the adiabatic ascent of significant air masses. Nevertheless, such regions face anthropogenic and natural challenges related to agricultural production, which may directly affect food insecurity. Insufficient and unpredictable rainfall frequently reduces agricultural output and causes complete crop loss. The phenomenon of soil degradation is a significant component that threatens agricultural output, affecting food security, especially in arid and semi-arid countries. Land degradation and erosion are substantial issues in semi-arid and arid zones, characterized by strong winds, low soil quality, sandy soil, erratic rainfall, flat terrain, and inadequate vegetation cover. Land degradation is a significant cause of poverty in rural areas of developing countries [19]. However, with the emergence of Industry 4.0 and green construction concepts, various countries with semi-arid and arid regions have taken initiatives to achieve a sustainable environment and to fight climate change with the help of advanced and green technologies. A few such initiatives include the "Billion Tree Tsunami" program launched in Pakistan in 2015 [20] and "Golden Deserts of the Dubai and UAE transformed into green desert after record-breaking 75-year rainfall" due to efforts made by the United Arab Emirates administration advanced research and technologies [21].

This chapter delves into the implementation and utilization methods of blockchain and CPS in the construction sector, which contribute to creating sustainable cities. The full potential of automation in the context of Industry 4.0 and digitalization can only be realized by incorporating essential elements like blockchain and CPS systems. Both blockchain and CPS are emerging areas in the built environment, and the first article on blockchain technology in the domain of built environment was released in 2017 [1]. Furthermore, this chapter also examines the situation of arid and semi-arid dry regions in relation to CPS systems and the integration of blockchain technology, which ultimately contributes to the promotion of sustainable livelihoods and development. The primary goal of the CPS is to transition from traditional and inflexible methods to decentralized structures to implement Industry 4.0 and Industrial Internet visions. This chapter will elucidate the operational tactics of CPS and blockchains inside the Construction 4.0 framework for a better understanding of readers.

2 Smart Contract and Blockchain

The application of blockchain technology was initially seen in the popular digital currency system, Bitcoin, which is sometimes referred to as Blockchain 1.0. Nick Szabo, a cryptographer, proposed the concept of smart contracts, specifically designed for blockchain and incorporating advanced technological security measures [22], leading to the emergence of Blockchain 2.0. Blockchain 2.0 enables a customizable financial platform where smart contracts are the prominent application. A smart contract is a digitized protocol created, validated, and implemented as code without involving a third party. Blockchain technology and smart contracts enhance transaction flexibility, alter the continual execution process of traditional transactional platforms, and are suitable for complex transaction situations in practice. At present, smart contracts are arguably the most promising blockchain application.

Smart contracts enable a wide range of processes and applications to be executed on the blockchain. Blockchain technologies are widely used in various industries, including Vehicular Ad Hoc Networks (VANETs), Verifiable Encrypted Signature (VES), Virtual Wireless Networks (VWNs), Unmanned Aerial Vehicles (UAVs), Vehicular Fog Computing (VFC). Ethereum is a prominent public blockchain platform that is considered a representation of Blockchain 2.0. Ethereum-based smart contracts are commonly coded in Solidity, a specialized high-level programming language designed to execute smart contracts. Ethereum describes smart contracts as blockchain programs that can directly oversee digital assets. Smart contracts can be utilized for several purposes beyond digital asset transactions, including cross-border trading, information sharing, data storage, currency payments, and more. As a result, several smart contracts are created and executed on Ethereum, capturing the attention of scholars who study and scrutinize the data and transactions on the network [3, 23].

"DApps" are decentralized applications that utilize blockchain technology to allow users to communicate with the blockchain without intermediaries. End-users can interact with a variety of DApps on blockchain networks that are supported by smart contracts. Unlike traditional programs on centralized servers, DApps function in a decentralized infrastructure to maintain open, transparent, and unchangeable data. DApps have become a focal point of academic research due to this trait. Imagine a scenario in which the data and regulations of a game are no longer controlled by its developers. The game's decentralization is achieved by creating a DApp that stores data on a distributed ledger, enforces clear restrictions, and prevents data tampering. Therefore, every participant maintains complete authority over their in-game assets. Decentralized gaming in this context offers a fairer alternative to the traditional buy/sell mechanism. DApps are characterized by three prominent aspects. DApps are built on smart contracts, with each DApp possibly consisting of many agreements. DApps function through a decentralized consensus method. Finally, DApps frequently include a token incentive scheme. Developers have many choices for developing the front-end user interface of DApps, while smart contracts mainly control the back-end functionality. The close relationship between DApps and smart contracts implies that categorizing DApps can be successfully done by classifying smart contracts [3].

The construction industry has yet to widely embrace blockchain technology. Nevertheless, multiple endeavors have been made to utilize it to create company models. BIMChain is a proof of concept that combines BIM with blockchain technology using a customized plug-in designed for BIM software systems. Fox [24] has put forth empirical evidence in the use of smart contracts for construction businesses. Such agreements allow finance among the stakeholder partners, agreement streamlining, better copyright on project documents, and support for stakeholders in the alteration of terms. The smart contract will significantly help in the automation of construction processes because most are very complicated and involve the collective contribution of other players involved in decision-making. Smart contracts have enormous benefits in automation for construction operations. Processes are usually complex, involving several actors engaging in the decision-making process and a diverse range of connections [7, 25].

3 Blockchain in the Construction Industry

Blockchain technology significantly improves ICT, and the construction sector can benefit greatly from implementing it in its operations. It enormously improves the transparency, traceability, and reliability of real-time information. Blockchain technology and CPS transfer the operational characteristics, semantic, and geometric features of machines, sensors, actuators, and mobile devices frictionlessly from the physical world to the digital platform. This link encourages data flow and collaboration, particularly in prefabrication projects [26]. The possibility of applying blockchain technology in construction can shape traditional operations through secure access to critical project information. In academic works, evidence exists to prove the effectiveness of blockchain-augmented CPSs in construction. Evidence supports that these systems provide tremendous assistance in transforming information into electronic form, monitoring and controlling the physical movement of products and services, and tracking the allocation and consumption of resources. Such gadgets allow monitoring in real-time, so intelligent construction methods are now being implemented. Lu et al. [27] have forwarded a theoretical model for integrating blockchain into the existing CPSs. The system utilizes IoT sensors to gather data from intelligent construction assets. Knowledge gained from the construction site builds a strong case for decision-making and enables stockpile material on-site management and off-site logistic operations. According to the study results, it is evident that with the implementation of blockchain, improvement in resource prefabrication management will be possible, along with the reduction of information-related project costs. The work presented here is a blockchain-based product-service CPS using model-based design principles for enhanced sustainability of prefabricated buildings. This novel strategy seeks to showcase the capacity of blockchain technology to improve operational efficiency and advance sustainability in the construction process. By deploying systems based on CPSs, the goal is to achieve sustainable livelihood.

Research has revealed that hybridization significantly boosts the efficiency of all parties involved in creating a whole by simplifying procedures followed by offsite manufacturing, transportation, and onsite. Blockchain technology is recognized for reducing the spread of different messages and guarantees that stakeholders effectively handle the dynamic nature of construction operations. This is an architecture based on services created explicitly for evaluating positional data and integrating pre-existing components such as BIM, IoT sensors, and communication platforms (the system is supported by blockchain technology). Therefore, integrating blockchain and smart contract implementation has made prefabrication more efficient and transparent. A logistics CPS based on blockchain has been developed to improve the real-time information sharing between importers and exporters. This system incorporates the use of RFID and GPS technologies while dealing with social sustainability issues across the building supply chain. Smart contract technology is self-executing, reducing reliance on traditional compliance methods. Stakeholder agreements are encrypted and recorded into blockchain-led decentralized data systems. The proposed blockchain-based CPS concept decentralizes the environment of data visibility and management, which, in this regard, enhances stakeholder communication between customers, suppliers, contractors, and logistics businesses. Additionally, the performance of the system, namely the supply chain and stakeholder transaction intricacies, will be researched in the future in detail. This has boosted financial management by 50% and auditing of transaction records by approximately 66.9%. This is a blockchain-enabled design of the CPS that advocates for the decentralized way of data visibility and management across various construction players. Research studies of the workflow applied in the "decentralized wastewater sewerage systems" have lately focused on the potential benefits and challenges of implementing the blockchain technology supply stage of the project, which may be known as construction supply management (CSM). Further decentralization enhances the traceability of data integrity in the blockchain as a reliable platform for stakeholder control under CSM and process standardization [28, 29].

4 CPSs-Blockchain Applications and Sustainable Construction Practices

Xiao et al. [29] presented a novel approach to significantly improve systems integration into both the physical and cyber domains, termed Cyber-Physical Centered Design, or CCD. It will enhance data collection, information processing, and human–computer interaction by building on the classic User-Centered Design (UCD) methodology. UCD puts an emphasis on system analysis when designing a user-centric System Development Life Cycle to meet the needs and preferences of the users. UCD focuses on user needs and practical solution methods to improve system architecture and develop solutions that are in line with user needs and experience. UCD has five stages of design methodology, including the context of the user, requirements elicitation, system design, system implementation, and system evaluation. CCD focuses on unifying the physical and cyber worlds based on the relationship between human activity and system functionality. It does so by turning everyday processes into cybernetic solutions. This approach emphasizes the need for user interaction in both the cybernetic and physical worlds to be coupled to create more integrated and user-centric systems.

4.1 Funding Management

The construction industry should be intelligent enough to understand that blockchain technology could be useful in financial resource control and regulation in a construction project. Adequate financial resources are a requirement for any ambitious activity that is crucial for the development of the economy of any country. The lack of transparency in the financing of projects encourages corruption in the sector, leading to the loss of confidence among investors and project sponsors. With blockchain technology, crowdfunding is a reliable and secure method by which project developers can attain funds for their projects. Peer-to-peer fundraising removes the necessity of old ways of seeking support to initiate a business. Better transparency concerning crowdfunding websites through the use of blockchain reinforces trust among investors regarding how their money will be distributed. Blockchain technology allows careful tracking and validation of the expenditure of a project's finances because the money flow is tracked on a decentralized ledger. Each participant gets undeniable evidence of their financial activities because of the immutability of the transaction records of the blockchain. There may, however, be regulation over certain areas by government authorities to clear specific regulatory issues and then restrict the means of crowdfunding on blockchain. In order to safeguard the interests of investors and maintain the integrity of the crowdfunding system in the construction industry, it is essential to implement a range of preventive measures [9, 30].

4.2 Electronic Document Management

Blockchain can address transparency and trust concerns in building documentation. Effective document management is crucial in construction projects, particularly during the post-contract stage. This stage involves a significant revision of construction information that needs to be well documented to prevent construction stakeholders from misinterpreting outdated information that has been replaced during the revision process. The process involves modifications to the initial building plans, an abundance of consultants' directives, the submission of change orders and payment requests, and additional tasks. Multiple modifications of these documents could be perplexing for stakeholders without a structured document system that includes time-stamping. Additionally, most stakeholders have separate documentation systems for identical projects, which must be shared or integrated with other construction stakeholders. This led to them possessing non-identical knowledge. Construction stakeholders need help to promptly acquire precise data, information, and knowledge because of the absence of an appropriate data management solution in the construction business. The origins of mistrust stem from a lack of transparency in the construction paperwork [9, 31].

The centralized structure of electronic document management systems is a key reason for the need for more trust among stakeholders in the construction sector.

Business entities and organizations need full control over their project-related data. Scholars suggest that construction data should be considered assets owned by temporary project teams and by project clients, such as corporations, organizations, and governmental agencies. It is crucial to safeguard the client's exclusive rights to design documents, especially if the design consultant's contract ends in the design phase. A decentralized electronic document management system, supported by blockchain technology, reduces the potential financial and time-related losses for the client if the designer asserts ownership of the design papers after termination. Blockchain technology is suggested to offer a strong structure for managing information across all phases of a building's existence. This involves handling data related to facility management once the building has been occupied. The project client must obtain crucial project data, including as-built drawings, project specifications, and cost details, and store it in a decentralized system accessible to the facility management team to enhance the sustainability of maintenance data. Blockchain technology enables a high level of data transparency essential for the sustainable progress of building projects, ensuring all stakeholders can access trustworthy and unchangeable project information [32, 33].

4.3 Contract Management

Blockchain can address the trust problem in construction contracts, particularly concerning payment trust. Signing the construction contract is a crucial initial step in commencing construction. The significance of construction digitization is increasing with the advancement of Industry 4.0, necessitating a transformation in traditional construction contracting approaches. Automated contracts, also known as digital contracts, have proven effective in various sectors and have the potential to advance the digitization of the construction industry significantly. The term "smart contract" is used in the context of blockchain technology. Blockchain applications utilize rules and algorithms within smart contract applications to integrate digital legal procedures into construction contracts [30]. Smart contracts enable the execution of digital contracts on a distributed ledger by functioning as agents that perform operations prompted by events. Provided that the established contract parameters are met, the payments will be executed automatically, ensuring the security of payment credentials. This method of developing tamper-resistant smart contracts incorporates provisions that depend on decentralized consensus to enable the execution of digital agreements automatically. Payment disputes among primary contractors, subcontractors, clients, and material suppliers are frequently encountered in the construction sector. Among the various stakeholders involved in construction, subcontractors experience the most significant adverse effects. Smart contract implementation in the construction sector can mitigate payment problems among contractors, subcontractors, and clients by tackling non-payment and delayed payment concerns. Through automated procedures, smart contracts implemented on an unchangeable distributed ledger can ensure payment security for smaller contractors and promote increased trust among construction enterprises [9].

Smart contracts can transform the creation of appointment contract agreements between clients and project consultants, such as designers, cost engineers, and project managers, in addition to their current uses. The old model of legal contracting is likely to undergo substantial changes, moving from an emphasis on litigation to a proactive approach to preventing disputes. Smart contracts offer numerous options for optimizing the legal and contractual process in the construction industry. Because of privacy considerations, legal entities using contract systems may deem it unsuitable to utilize a public blockchain. A permissioned blockchain allows developers to provide participant access selectively and is considered more suitable for the building industry. This blockchain architecture safeguards contractual information from public disclosure and limits access exclusively to authorized entities, thereby ensuring the confidentiality of sensitive data [34].

The construction industry has numerous obstacles related to smart contracts, many requiring immediate resolutions. This is especially true when using Bitcoin as a means of payment, which needs to be better recognized in the construction industry. However, with blockchain technology being used in construction contracts, it can now stand alone from the cryptocurrencies in use. Adapting traditional contract terms to the technical demands of smart contracts is one of the key enablers for adopting smart contracts in the construction sector. This includes simplifying complex contractual requirements for smart contracts in financial and time-critical matters, such as payments, predetermined damages, monies held in security, and extensions of time. Even with smart contracts, keeping old standard contracts as an essential part of legally binding documents is recommended. Smart contracts are supposed to work as a form of intelligent way of enforcing legal obligations and not to be wholly substituted by conventional building contracts. The unalterable distributed ledger of a smart contract system may be crucial evidence in the event of building disputes and may, therefore, enhance efficiency in resolving such conflicts. Smart contracts will also enhance efficiency and time savings and reduce costs by minimizing the use of legal experts in resolving disputes in building projects. Confidence among parties entering into construction contracts is likely to be enhanced by the application of blockchain technology, thereby enhancing the negotiation of contracts [9, 34, 35].

4.4 Building Information Modeling (BIM)

BIM has found wide application in the construction industry and forms an integral part of digital construction. Trust and transparency remain persistent challenges in BIM collaboration among design stakeholders. Design changes are, however, often inevitable in construction due to the regulatory restrictions, market forces, and budgetary constraints imposed by clients. Existing BIM applications, which adopt a centralized platform to manage and record design change orders, have a few limitations. The adjustments typically involve altering and enhancing the existing data; no recorded evidence of the changes made in chronological order exists. The designer must collaborate meticulously and independently with other experts and contractors on adjustments, which may necessitate substantial resources. Lack of effective design coordination might result in misinterpretations and incongruous outcomes. The limitations of the current approaches have led to problems involving transparency and trust in BIM collaboration among the designers, hence hampering the widespread use of BIM in the building industry. Blockchain technology offers a solution for trust enhancement among BIM collaboration participants by dealing with existing challenges on design traceability. When coupled with its timestamp capability, the decentralized structure of blockchain technology will ensure the proper historical development of design modifications across a project. This advances better visibility and trust by all the participants in the record of design changes and provides more collaboration [9].

"BIMChain" is a new blockchain application for the construction sector, with the principal goal being to remove the current limitations in BIM concerning design ownership, tracking design changes, and final responsibility. BIMChain will require significant time to integrate into the industry, as it is currently in a very early development stage. In the future, the priority would be to adopt BIM, which incorporates supply chain management with blockchain technology, using RFID and the IoT. Integration of blockchain technology is bound to disrupt the current BIM function, which mainly focuses on design coordination. This blockchain implementation will bring a transformation in the collaborative practices that the participants of the BIM ecosystem have been following and will handle the concerns about trust properly [36, 37].

4.5 Supply Chain Management Operation

Building materials, workforce, services, and products are all components of the complex supply chain that is engaged in the construction industry. The construction industry still needs to catch up in adopting new techniques and technological solutions. Moreover, it continues to rely on manual document-centric processes. This involves handwritten documentation of construction logbooks, labor worksheets, manual file storage, and calculating material delivery orders. Large enterprises may utilize a centralized platform to document and oversee construction information. The current methods must be more open among project stakeholders and create risks to data security, especially when the central organization responsible for data custody accidentally deletes or misplaces information. Therefore, a decentralized network is necessary to oversee supply chain data and enhance the effectiveness of supply chain operations. Implementing blockchain-enabled applications is expected to significantly improve supply chain management by increasing the transparency and traceability of items. Blockchain technology accurately records and authenticates construction supply chain data on a decentralized ledger. This enables contractors to effectively manage costs, productivity, profitability, and overall supply chain performance with reduced chances of errors. Blockchain in supply chain management supports breaking down building tasks into smaller pieces and emphasizes a labor-focused strategy to enable direct purchasing of materials and goods from manufacturers. Clients can interact directly with suppliers and subcontractors via smart contracts, benefiting from enhanced trust facilitated by automation. The client's worries about their future interaction with the primary contractor are reduced because the primary contractor's involvement is crucial for project scheduling, monitoring, health and safety management, and overall supervision. Integrating blockchain technology into supply chain management is expected to result in significant time and cost savings for contractors and clients while also improving the security and accuracy of data [9, 31, 38].

5 CPSs Applications in Arid and Semi-Arid Regions

The advent of the fourth industrial revolution has brought forth many cutting-edge technologies and approaches, facilitating the attainment of sustainable solutions and lifestyles for communities. Similarly, adopting CPSs in semi-arid and arid regions can substantially impact the environment by addressing diverse environmental and survival-related obstacles.

5.1 Real-Time Monitoring

CPSs offer the capacity to monitor the real-time status of projects or infrastructure. These services can be used either during the execution stage of the project to monitor the progress of the construction or post-execution stage to monitor the condition of the structure or infrastructure in semi-arid and arid locations. One of the examples of post-execution monitoring is also known as "structural health monitoring". CPSs have the capability to monitor the state of critical systems and provide recommendations for proactive repair. Asset management solutions based on cyber-physical solutions can enhance the efficiency and durability of infrastructure in harsh environmental circumstances.

One of the examples of real-time monitoring is shown in Fig. 1, which illustrates a fundamental idea for real-time monitoring via various technologies and remote site data acquisition with the help of interconnected systems under the IoT umbrella being controlled and monitored by stakeholders from a distance or head office.

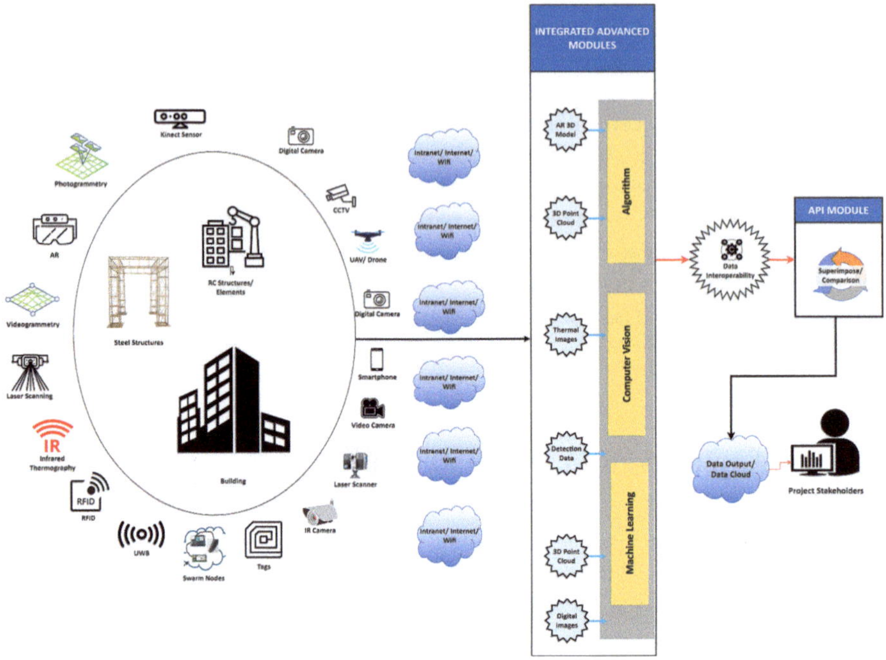

Fig. 1 Real-time monitoring of building infrastructures

5.2 *Green Energy*

By capitalizing on semi-arid and arid terrains, a range of green and sustainable energy sources, particularly wind and solar power, can be strategically implemented. The implementation of CPS-enabled smart grids, energy storage, and intelligent distribution systems, which adjust electrical load based on demand, can significantly improve the reliability of the system in rural villages or communities off-grid. Furthermore, this will aid in effectively regulating the uninterrupted transmission of energy through environmentally friendly and sustainable energy sources in regions with limited water resources.

Figure 2 illustrates a basic idea of a smart energy production and distribution system with the help of smart CPS systems. The whole idea works on supply and demand methodology; the sensors assess the demand of the community and, via an integrated system, feed the production side for the requirement. The smart green energy system controls production and distributes electricity as per the needs of the community.

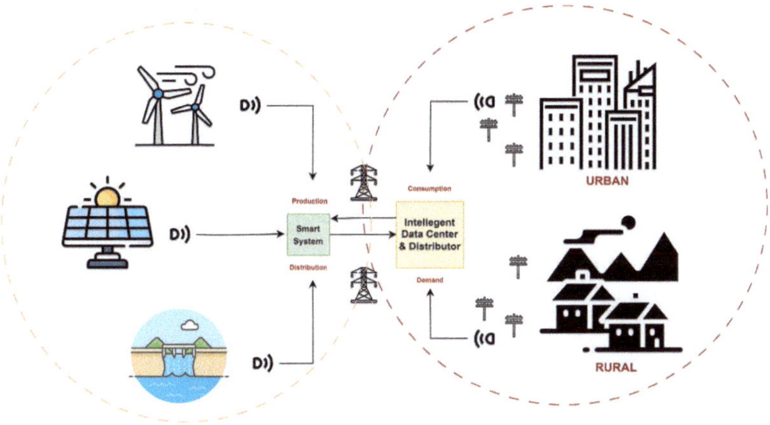

Fig. 2 Smart energy system

5.3 *Water Harvesting*

A planned and well-designed CPS-based water harvesting and distribution system has the potential to enhance water distribution efficiency and reduce avoidable water wastage. An array of sensors can be installed to identify instances of water leakage, and continuous monitoring can be carried out using a CPS-based system to assess the quality, availability, and levels of surface and groundwater in real-time. Alternatively, water treatment facilities and desalination plants that utilize CPS technology can provide continuous access to high-quality water in dry regions. Such technologies can be made part of decision support systems at the macro level to monitor and manage mega water infrastructure, such as water distribution networks linked with reservoirs and dams.

Figure 3 summarizes the theme of smart water management systems, pointing out water harvesting, supply and demand, leakage detection, and record management (i.e., inventory, utilization, storage, etc.) via long-range data sensing technologies at the head office.

5.4 *Eco Management*

Soil erosion and land degradation are significant concerns in semi-arid and arid zones, ultimately impacting vegetation. Deploying CPS-based systems enables the collection and analysis of data on soil erosion and land degradation, facilitating the research of prediction models. Similarly, these models can be integrated into an early disaster management system to forecast desertification, sandstorms, and drought. In addition, CPS-based systems can be adapted for climate and weather forecasting systems to provide early warnings, enhancing readiness for calamities.

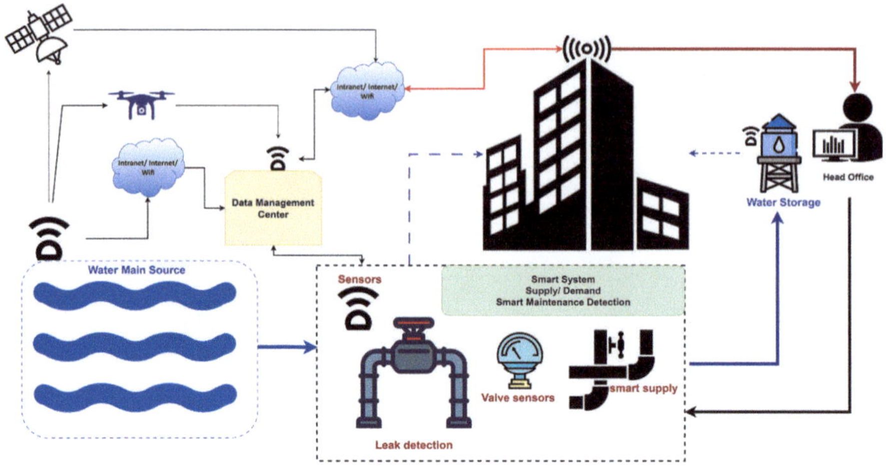

Fig. 3 Smart water management system

5.5 *Agronomics*

The scarcity of food supplies, limited vegetation, insufficient plantation, and low agricultural production pose substantial obstacles in semi-arid and dry regions. By utilizing smart devices facilitated by CPSs, it is possible to create a smart irrigation network that effectively manages water usage, minimizes waste, and optimizes the efficiency of water distribution according to specific needs. Furthermore, to further develop this concept, CPS-based sensors can be used to construct and sustain highly efficient agricultural centers. These sensors would enhance water utilization and regulate and supervise soil moisture levels based on weather conditions and vegetation needs. Agrarian operations will be improved, and productivity will be increased using technology and systems enabled by CPSs. This includes the capability to monitor and manage remotely.

Figure 4 gives a preview of a smart agronomic system, which is being monitored by smart sensors and modules for effective operations. The smart systems control the water supply, considering the conditions of arid and semi-arid regions. Moreover, operations such as insecticide application, fertilizing, and irrigation can be controlled remotely, making such operations cost-effective.

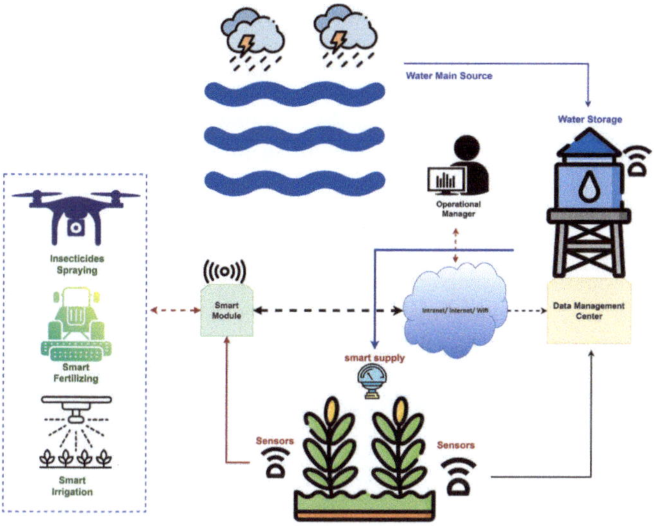

Fig. 4 Smart agronomic system monitored by smart sensors

6 Challenges of Blockchain Implementation in Sustainable Construction

6.1 Denial of Supporting Policies on Blockchain Adoption

The construction industry needs a robust regulatory framework encompassing legislation, norms, standards, and regulations for blockchain technology. Clear legislation is necessary to ensure the widespread implementation of blockchain technology. Without appropriate standards, construction stakeholders may lack clarity regarding the specific data that must be exchanged on a blockchain system throughout the construction and management stages. A regulatory framework is necessary to achieve service integration, address interoperability issues, maintain system stability, and encourage innovation. Decisions on the utilization and progress of blockchain technology may diverge due to insufficient infrastructure and policies. Stakeholders in the construction industry lack a comprehensive understanding of the impact of government legislation on their decision-making regarding the adoption of blockchain technology [39, 40].

6.2 Reluctance to Change Current Systems

Historically, information systems in the construction industry have consistently utilized centralized architectures. Introducing new systems often faces opposition, especially while these systems are still in the initial stages of development. Blockchain technology is a recent achievement in the building industry that has been affected by these hesitation phenomena. The long-term impact of blockchain technology on the construction sector and its integration into management and construction processes remains to be determined. Industry participants lacking comprehension of blockchain's economic potential may hesitate to take action. The construction industry lacks the expertise and experience to design and oversee blockchain technologies effectively. These reasons contribute to the construction industry's hesitance to adopt the new technology [13, 41].

6.3 Immutability Challenge of Smart Contracts

Smart contracts are crucial for blockchain applications since they eliminate human unpredictability and enhance specific business operations' efficiency, such as payment of construction progress and quality compliance checks. Smart contracts are written in computational programming languages and can be installed on a blockchain system to carry out predefined logic automatically. Smart contracts were not widely adopted until the emergence of blockchain technology, which allowed for the unchangeable nature of predetermined contracts. Once smart contracts are deployed on blockchain technology, they become immutable. Construction projects with high dynamic complexity may experience several unanticipated scenarios and events. Adjusting payment terms or addressing unexpected issues may be required throughout the construction process. One potential solution is to allow users to vote on various suggestions aimed at modifying the regulations of smart contracts. Exploring other methods to enhance the applicability of smart contracts in buildings is recommended [42, 43].

6.4 Lack of Supervision Expertise on Blockchain-Based Workflows

The construction industry is commonly believed to be slow to adopt digital technologies and is known for its conservative approach. Existing contractual ties primarily stem from adversarial circumstances, indicating a deficiency in a cooperative environment. Blockchains have the potential to transform the importance of trust and collaboration. It provides smaller enterprises with an opportunity to engage in negotiations with large corporations. An example of a progress payment based on blockchain technology. It could also remove intermediaries and impact business structures in inter-organizational collaboration. Prior research has yet to examine the potential impact of the blockchain system on inter-organizational cooperation. Most blockchain research in the construction sector has focused on addressing particular issues. These issues contain progress payments at the project level without considering the potential impact of blockchains on how construction stakeholders work together. Blockchain technology expands the network of possible collaborators by allowing strangers without social connections to interact within the system. Therefore, the impact of implementing blockchain on collaboration dynamics should be further investigated. Such a lack of presence could lead to certain people being inactive. Developers may find securing commitment from intermediate or upper management difficult. In addition, previous research on blockchain targeting the construction industry lacks practicality and, therefore, has not been effectively implemented despite the development of various workflow frameworks. Thus, there is a requirement and potential in this field for professionals and academics to conduct extensive research to uncover feasible blockchain-based systems and frameworks that are acceptable to the industry [13, 44, 45].

7 Conclusion

Utilizing CPSs and blockchain in the construction business represents a substantial advancement in tackling efficiency, trust, and transparency issues. Blockchain, a decentralized ledger technology, provides a secure framework for exchanging and storing data. It promotes participant confidence by guaranteeing that records and transactions are transparent and unchangeable. This is especially beneficial in a domain where conflicts and disputes may occasionally develop due to the intricate and extensive nature of projects. CPSs enhance operations by integrating physical processes with networking, computation, intelligence, and automation. The utilization of IoT devices and sensors enables real-time monitoring and management of construction processes, leading to improved decision-making, efficiency, and safety. By employing CPSs and blockchain implementation frameworks, it is possible to provide green construction solutions for arid and semi-arid locations. These solutions are not limited to the execution phase but extend to post-execution phases, thereby establishing an atmosphere that supports the development of sustainable cities. Moreover, from the data security point of view, blockchain-integrated CPS networks may reliably and confidentially document and exchange data, hence improving synchronization and collaboration throughout different stages of construction projects. Blockchain technology and CPSs synergistically enhance operations and drive innovation in the construction industry. Smart contracts provide the capability to automate payments and contractual duties. At the same time, blockchain-based BIM can offer a shared and verified source of accurate information for all individuals involved in a project. Despite the potential demonstrated by green technologies, the construction sector is still in its nascent phase of embracing them. Determining solutions for problems such as scalability, integration with existing systems, and the requirement for universal standards is crucial. The construction industry is expected to undergo a transformation with the adoption of blockchain and CPS technologies. Implementing green technologies along with blockchain and CPS-supported networks will enhance project delivery in arid and semi-arid zones by augmenting efficiency, transparency, and collaboration. With the ongoing digitalization of the industry, it is crucial to educate stakeholders about the benefits and potential of advanced technologies and frameworks. This will enable the construction industry to operate sustainably and establish a basis for future sustainable livelihoods.

8 Relevant Sustainable Development Goals (SDGs)

This chapter mainly addresses SDG 7 (Affordable and Clean Energy), SDG 8 (Decent Work and Economic Growth), SDG 9 (Industry, Innovation and Infrastructure), SDG 11 (Sustainable Cities and Communities), and SDG 12 (Responsible Consumption and Production). The summary and details related to each SDG are illustrated in Table 1.

Table 1 Relevant SDGs and summary

S No	SDG #	Leading Factors/Aspects	Summary
	7	Blockchain	• Peer-to-peer energy trading • Carbon credit tracking
		Smart systems	• Peak load optimization • Automated energy management • Energy consumption tracking
	8	Blockchain	• Contract compliance • Fair labor practices • Transparent payment systems
	9	Cyber-physical systems (CPSs)	• Automated quality control • Real-time monitoring • Predictive maintenance capabilities • Integrated sensor networks
		Blockchain	• Digital contract tracking
	11	Smart systems	• IoT-enabled management systems
		CPSs and Blockchain	• Decentralized facility management
	12	CPSs and Blockchain	• Supply chain transparency • Energy consumption monitoring • Lifecycle assessment • Waste reduction monitoring • Circular economy implementation

9 Recommendations

In light of the context of this chapter, the following are a few recommendations in these domains.

9.1 Smooth Technological Integration

To make the systems operate smoothly under CPSs and blockchains, there is a need to work on the integration capabilities of various technologies for better transfer of data and signals. There is a need for the development of universal protocols for blockchain-CPS integration systems and the implementation of unified IoT protocols for effective as well as efficient interoperability processes among different tools.

9.2 Digital Security Enhancement

The construction sector is moving towards the digitization of data and processes, which also enhances the need for digital data security. This demands the development of secure data storing and sharing protocols. To stop data from being stolen online, robust authentication systems and enhanced privacy preservation techniques need to be created.

9.3 Workforce Training

For effective and smooth implementation of digitalization and technologies to advance systems, there is also a need for training the workforce on such systems to remove their hesitations and fear of handling that pressure. The stakeholders should arrange prior training programs for the workforce for continuous professional development. Moreover, there should also be certification systems for new technologies and training programs to enhance digital skills.

9.4 Regulatory Framework

Implementation of blockchain-based systems is still a challenge in the construction industry. A central regulatory body should be established to monitor digital operations, which may help in the development of blockchain governance models and smart digital contract standards. Moreover, digital data protection guidelines should be proposed.

9.5 *Stakeholder Engagement*

For the success of any project, effective communication is one of the key success factors. Likewise, for effective implementation of technologies and the Industry 4.0 theme, a strong relationship between construction industry stakeholders is essential. There is a need to improve public–private sector partnerships. Most importantly, a strong academic-industry collaboration for cross-sector knowledge sharing and international cooperation will assist in the successful implementation of smart operations.

References

1. Akinradewo O et al (2023) A roadmap for present focus and future trends of blockchain technology in the built environment. Afr J Sci Technol Innov Dev 15(2):153–165. https://doi.org/10.1080/20421338.2022.2046249
2. Alzubi KM, Alaloul WS, Qureshi AH (2022) Applications of cyber-physical systems in construction projects. In: Cyber-Physical Systems in the Construction Sector. CRC Press, pp 153–171
3. Jiang Z et al (2022) Applying blockchain-based method to smart contract classification for CPS applications. Digit Commun Netw 8(6):964–975. https://doi.org/10.1016/j.dcan.2022.08.011
4. Qureshi AH, Alaloul WS, Alzubi KM (2022c) Internet of Things (IoT) for construction cyber-physical systems. In: Cyber-physical systems in the construction sector. CRC Press, pp 60–77
5. Qureshi AH, Alaloul WS, Alzubi, KM (2022b) Fundamentals of cyber-physical systems. In: Cyber-physical systems in the construction sector. CRC Press, pp 22–38
6. Alaloul WS, Saad S, Qureshi AH (2021) Construction Sector: IR 4.0 Applications. In: Handbook of smart materials, technologies, and devices. Springer, Cham, pp 1–50. https://doi.org/10.1007/978-3-030-58675-1_36-1.
7. Elghaish F et al (2021) Blockchain and the "Internet of Things" for the construction industry: research trends and opportunities". Autom Constr 132:103942. https://doi.org/10.1016/j.autcon.2021.103942
8. Aste T, Tasca P, Di Matteo T (2017) Blockchain technologies: the foreseeable impact on society and industry. Computer 50(9):18–28
9. Kiu MS, Chia FC, Wong PF (2022) Exploring the potentials of blockchain application in construction industry: a systematic review. Int J Constr Manag 22(15):2931–2940. https://doi.org/10.1080/15623599.2020.1833436
10. Alaloul WS et al (2021) Evolution of close-range detection and data acquisition technologies towards automation in construction progress monitoring. J Build Eng 43:102877. https://doi.org/10.1016/j.jobe.2021.102877
11. Qureshi AH et al (2022) Factors impacting the implementation process of automated construction progress monitoring. Ain Shams Eng J 13(6):101808. https://doi.org/10.1016/j.asej.2022.101808
12. Qureshi AH et al (2023) Characteristics-based framework of effective automated monitoring parameters in construction projects. Arab J Sci Eng 48(4):4731–4749. https://doi.org/10.1007/s13369-022-07172-y

13. Wu H et al (2022) Blockchain technology in the construction industry: current status, challenges, and future directions. J Constr Eng Manag. https://doi.org/10.1061/(asce)co.1943-7862.0002380
14. Satoshi Nakamoto (2008) Bitcoin: a peer-to-peer electronic cash system. https://bitcoin.org/en/bitcoin-paper. Accessed 3 May 2024
15. Banafa A (2022) Blockchain 4.0 | LinkedIn. https://www.linkedin.com/pulse/blockchain-40-ahmed-banafa/. Accessed 3 May 2024
16. Macy J (2019) Blockchain 5.0 is taking HR Tech to the Edge | by John Macy | Blockchain and the Distributed Workforce | Medium. https://medium.com/blockchain-and-the-distributed-workforce/blockchain-5-0-is-taking-hr-tech-to-the-edge-7cb360090203. Accessed 3 May 2024
17. Ali A (2023) How Saudi is turning desert into Greenland, Earth. https://vocal.media/earth/how-saudi-is-turning-desert-into-greenland. Accessed 5 August 2024
18. Begizew G (2021) Agricultural production system in arid and semi-arid regions. Int J Agric Sci Food Technol 7:234–244. https://doi.org/10.17352/2455-815x.000113
19. Mohamed E et al (2019) Land degradation. In: Encyclopedia of earth sciences series. Springer, pp 159–174. https://doi.org/10.1007/978-3-319-95516-2_9.
20. UNEP (2024) Pakistan's ten billion tree tsunami, UNEP. https://www.unep.org/news-and-stories/story/pakistans-ten-billion-tree-tsunami. Accessed 10 August 2024
21. EIN Presswire (2024) Golden Deserts of the dubai and UAE transformed into 'green desert' after record-breaking 75-year rainfall, EIN Presswire. https://www.wate.com/business/press-releases/ein-presswire/710219465/golden-deserts-of-the-dubai-and-uae-transformed-into-green-desert-after-record-breaking-75-year-rainfall
22. Szabo N (1997) Formalizing and securing relationships on public networks. First monday [Preprint]
23. Chen T et al (2019) Dataether: data exploration framework for ethereum. In: 2019 IEEE 39th International Conference on Distributed Computing Systems (ICDCS). IEEE, pp 1369–1380
24. Fox S (2016) Why construction needs smart contracts | NBS. https://www.thenbs.com/knowledge/why-construction-needs-smart-contracts. Accessed 3 May 2024
25. Lamb K (2018) Blockchain and smart contracts: what the AEC sector needs to know
26. Li Z et al (2021) Industrial blockchain: a state-of-the-art survey. Robot Comput-Integr Manuf 70:102124
27. Lu W et al (2021) Exploring smart construction objects as blockchain oracles in construction supply chain management. Autom Constr 129:103816
28. Kang K et al (2022) Blockchain opportunities for construction industry in Hong Kong: a case study of RISC and site diary. Constr Innov 23(2):443–466
29. Xiao J, Zhang W, Zhong RY (2023) Blockchain-enabled cyber-physical system for construction site management: a pilot implementation. Adv Eng Inform 57:102102. https://doi.org/10.1016/j.aei.2023.102102
30. San KM, Choy CF, Fung WP (2019) The potentials and impacts of blockchain technology in construction industry: a literature review. In: IOP conference series: materials science and engineering. IOP Publishing, p 12005
31. Wang J et al (2017) The outlook of blockchain technology for construction engineering management
32. Jiao Y et al (2013) A cloud approach to unified lifecycle data management in architecture, engineering, construction and facilities management: Integrating BIMs and SNS. Adv Eng Inform 27(2):173–188
33. Turk Ž, Klinc R (2017) Potentials of blockchain technology for construction management. Procedia Eng 196:638–645
34. Xu X et al (2016) The blockchain as a software connector. In: 2016 13th Working IEEE/IFIP conference on software architecture (WICSA). IEEE, pp 182–191
35. Nawari NO, Ravindran S (2019) Blockchain and building information modeling (BIM): review and applications in post-disaster recovery. Buildings 9(6):149
36. Akhavan P, Ravanshadnia M, Shahrayini A (2021) Blockchain technology in the construction industry: integrating BIM in project management and IOT in supply chain management. In: 2nd International conference on knowlege management, blockchain & economy

37. Hijazi AA et al (2022) Developing a BIM single source of truth prototype using blockchain technology. Buildings 13(1):91
38. Hughes D, Hughes D (2017) The impact of blockchain technology on the construction industry. animal Media [Preprint]
39. Li J, Greenwood D, Kassem M (2019) Blockchain in the built environment and construction industry: a systematic review, conceptual models and practical use cases. Autom Constr 102:288–307
40. Wang Q, Su M, Li R (2020) Is China the world's blockchain leader? Evidence, evolution and outlook of China's blockchain research. J Clean Prod 264:121742
41. Qureshi AH, Alaloul WS, Alzubi KM (2022a) Cyber-physical systems and digital twins (DT) in construction projects. In: Cyber-physical systems in the construction sector. CRC Press, pp 78–87
42. Mohsin F et al (2019) Ontology aided smart contract execution for unexpected situations. In: BlockSW/CKG@ ISWC
43. Zhong B et al (2020) Hyperledger fabric-based consortium blockchain for construction quality information management. Front Eng Manag 7(4):512–527
44. Lumineau F, Wang W, Schilke O (2021) Blockchain governance—a new way of organizing collaborations? Organ Sci 32(2):500–521
45. Martin L, Benson L (2021) Relationship quality in construction projects: a subcontractor perspective of principal contractor relationships. Int J Project Manage 39(6):633–645

Conclusions

Conclusions and Recommendations Towards Sustainable Architecture and Urban Practices in Arid and Semi-arid Cities

Zeina ElZein and Abdelazim Negm

Abstract This concluding chapter brings together the most important insights and lessons learned from the book's diverse chapters, offering a practical and thoughtful reflection on how sustainable architecture can shape the future of arid and semi-arid cities. Through the wide-ranging themes covered in this volume, from urban design and material innovation to water management, renewable energy, and digital technologies, the book highlights both the challenges these regions face, and the creative, locally grounded solutions already being explored. The opening section delves into the growing role of Artificial Intelligence in building smarter, more sustainable cities, illustrated by real-world experiences. It is followed by a focus on resilience in architecture and planning, emphasizing the value of community participation, adaptive reuse, and drawing wisdom from indigenous practices to meet modern climate and social challenges. Another key part of the book centers on sustainable building materials and the potential of a circular economy in arid regions, underlining the importance of using locally sourced, recycled, and bio-based resources supported by forward-looking policy. Water management receives particular attention, with chapters examining nature-based solutions, indigenous water systems, and integrated greywater reuse as essential tools for cities facing chronic water stress. Urban greening strategies, including haloculture and green roofs, are explored for their role in improving thermal comfort, air quality, and food security. The book also looks at renewable energy adoption within community spaces as a practical pathway to energy transition and highlights how blockchain, IoT, and cyber-physical systems could make construction smarter, safer, and more transparent. This chapter wraps up with concise, actionable recommendations, providing a clear roadmap for policy-makers, practitioners, and local communities to move toward resilient, sustainable urban environments tailored to the realities of arid and semi-arid regions.

Z. ElZein (✉)
Architecture Department, Faculty of Fine Arts, Helwan University, Cairo, Egypt
e-mail: Zeina.ElZein@f-arts.helwan.edu.eg; zeinabaher@gmail.com

A. Negm
Water and Water Structures Engineering Department, Faculty of Engineering, Zagazig University, Zagazig, Egypt
e-mail: amnegm@zu.edu.eg; amnegm85@gmail.com

Z. ElZein and A. Negm (eds.), *Recent Approaches of Sustainable Architecture in Arid and Semi-arid Cities*, Urban Sustainability,
https://doi.org/10.1007/978-981-95-1645-2_16

Keywords Sustainable architecture · Arid cities · Climate resilience · Resources management · Urban greening

1 Introduction

This book offers a comprehensive reference for integrating sustainable design principles into architecture and urban development in arid and semi-arid environments. It brings together diverse, practice-oriented perspectives on how architecture, planning, and policy can respond to some of the most pressing climate and resource challenges in dryland cities. Through case studies, theoretical insights, and applied research, the volume highlights a wide range of innovative solutions that contribute to more resilient, livable, and environmentally sound urban futures.

The book is organized around several key thematic areas:

- Introduction to Sustainable Architecture in Arid and Semi-arid Cities
- Resilience in Architecture and Planning
- Sustainable Building Materials and Circular Economy
- Sustainable Water Management in Urban Settings
- Urban Greening in Cities
- Energy Transition Toward Carbon Neutral and Green and Smart Cities

The chapters present regionally grounded insights from the Middle East, North Africa, and other dryland areas, illustrating how local knowledge, digital tools, community engagement, and policy frameworks can help shape more climate-adaptive urban environments.

The following section provides a synthesis of updated findings from the literature and case contributions on sustainable architecture and planning in arid and semi-arid contexts. It is followed by a set of key conclusions and actionable recommendations aimed at architects, urban planners, researchers, and decision-makers working toward sustainability and climate resilience in dryland cities.

2 Summary and Updates

This section includes key updates on each of the book themes as follows:

2.1 Introduction to Sustainable Architecture in Arid and Semi-Arid Cities

This section explores how artificial intelligence (AI) is beginning to shape new possibilities for sustainable urban development in arid and semi-arid regions. In places where high temperatures, limited water, and fast-growing populations create complex challenges, integrating AI into planning and design processes offers new ways to adapt. Drawing on examples from China, Malaysia, and Egypt, the chapter examines how AI technologies are helping cities manage energy, transport, water, and waste more efficiently while also responding more quickly to the impacts of climate change.

Each country presents a unique experience. China's centralized planning and strong investment in AI have led to rapid progress in building smart infrastructure—from energy-efficient transportation systems to climate-resilient cities. Malaysia has adopted a more collaborative model, involving various stakeholders in applying AI to improve urban services and disaster response. Egypt's situation is more complex, given its arid climate and growing urban stress. Here, efforts are being made to use AI for better water management and sustainable construction, with international support playing a key role [1, 2].

The chapter shows that while the use of AI varies depending on local conditions, it consistently helps cities become more adaptive and efficient. It also highlights that AI is not a stand-alone solution. Its benefits depend on local governance, funding, and the ability to bring experts from different sectors together. Challenges such as energy demands, data privacy, and lack of coordination between planners and tech developers remain serious concerns.

In the end, the chapter argues that building sustainable cities in drylands requires more than just smart tools. It calls for inclusive thinking, long-term planning, and policies that ensure technology helps everyone, not just those with access. It reminds us that sustainable architecture is not only about what we build, but how we build it, who we build it for, and how prepared we are to deal with future uncertainties.

2.2 Resilience in Architecture and Planning

This section brings together grounded reflections on what resilience means in the built environment, especially in regions where water scarcity, climate extremes, and socio-political pressures intersect. Through three different lenses, the chapters in this theme look at resilience not as a technical solution alone, but as a lived experience deeply tied to people, place, and time.

The first chapter, *Architecture and Resilience: A Comprehensive Overview of Purpose, Evolution and Methodologies*, revisits the core of architectural practice by asking a fundamental question: what does it mean to build in the face of adversity? Rather than framing architecture as the exclusive domain of trained professionals, it shows how communities, particularly those living in difficult conditions, build and adapt intuitively. Resilience, in this context, is not a new idea. It's a response to hardship, often born out of necessity, using what is available and known. The chapter reminds us that architecture is not only about form and function, but also about empathy, memory, and participation. It invites architects to pay attention—to observe, listen, and learn from local experiences and indigenous wisdom—and to view design as a collaborative, ongoing process [3].

In the second chapter, *Developing a Framework for the Design of a Resilient Social-Ecological System in the Arab Region*, the focus shifts to the broader relationship between ecosystems and human well-being. It makes a compelling case for integrating environmental awareness into planning and governance across the Arab region. The chapter argues that sustainable development requires more than technical fixes; it calls for political will, informed citizens, and public engagement. It acknowledges the ecological strain faced by many Arab communities and suggests that meaningful change begins with empowering people to care for the landscapes they live in. At its core, this chapter links resilience to awareness, responsibility, and the ability to respond collectively [4].

The third chapter, *Resilient Spaces Through Urban Agriculture Practices*, offers a hands-on look at how resilience can grow within the city. Rising human population have intensified the strain on agricultural systems to meet growing food demands [5]. Urban agriculture is presented not only as a way to boost food security, but also as a tool for community building, climate adaptation, and environmental healing. Particularly in arid countries like Egypt, these practices offer practical solutions to daily challenges. They make cities greener, provide fresh produce, and strengthen neighborhood ties. This chapter shows that when residents have the chance to shape their surroundings, they also gain the power to make their communities more adaptive and self-reliant.

Taken together, these chapters show that resilience in architecture and planning is not a fixed recipe. It is a mindset and a way of approaching uncertainty with creativity, humility, and care. Whether through design, policy, or everyday acts of maintenance and cultivation, resilience grows where people are given the tools and trust to shape their own futures.

2.3 Sustainable Building Materials and Circular Economy

This section explores how sustainable building materials and circular economy practices are shaping architecture in regions that face some of the harshest environmental conditions. Across the three chapters, the theme that emerges is clear: building responsibly in arid and semi-arid areas is not only possible but necessary—and doing so requires creativity, practical knowledge, and long-term thinking.

The first chapter, *Sustainable Environmental Architecture: Approaches for Sustainable Building Materials and Waste Management in the Era of Climate Change*, dives into how thoughtful choices in materials and waste handling can make a real difference. The widespread accumulation of plastic waste poses a major global issue, carrying serious consequences for both the environment and human health [6]. The chapter emphasizes the importance of using what's available locally and turning potential waste like plastics, into valuable building components. The idea is that even in places where resources are stretched, construction does not have to come at the environment's expense. When recycling technologies are accessible and supported by smart policies, they offer a real way to reduce environmental harm while still meeting the region's growing development needs.

In the second chapter, *Cost-effectiveness and Market Adoption of Sustainable Building Materials*, the focus turns to the real-world economics of sustainable construction. Through on-the-ground examples in places like Egypt and Saudi Arabia, the chapter shows that traditional materials and passive cooling techniques often outperform modern alternatives when it comes to long-term value. While initial investments may be higher, they usually pay off through reduced energy bills, less maintenance, and better indoor living conditions. For example, integrating bio-based and environmentally friendly composite materials into construction supply chains plays a crucial role in lowering embodied carbon emissions and decreasing long-term operational energy consumption [7]. This chapter brings to light the fact that building green is not just a luxury, it is a smart, forward-thinking choice, particularly in climates where energy consumption is a serious concern.

The third chapter, *Standards for the Use of Sustainable Green Finishing Materials and Their Impact on Internal Spaces*, takes a closer look at how interior finishes affect both people and the planet. It highlights the need for more awareness and education around sustainable materials, especially in countries like Egypt. The chapter also stresses the importance of updating architecture curricula to include environmental responsibility and encouraging a shift in how both professionals and the public view sustainable design. Changing attitudes, especially around cost, can play a big role in making green materials the norm rather than the exception.

Taken together, these chapters show that achieving sustainability in the building sector goes beyond technical know-how. It is about mindset, cultural awareness, and making choices that respect both the local context and the broader environment. These lessons are particularly urgent for arid and semi-arid cities, where resilience must be built into every layer of design and decision-making.

2.4 Sustainable Water Management in Urban Settings

Water shapes the life and rhythm of every city, and in regions where drought and scarcity are part of daily life, its role becomes even more vital. Over 50% of the world population live in urban areas, the fast pace of urbanization affects how well cities can withstand weather-related challenges [8, 9]. This section brings together three chapters that approach water not just as a resource, but as a thread that ties communities, traditions, and the built environment together. The takeaway is simple yet profound: sustainable water management is not only necessary, it is entirely possible, if we listen, learn, and act.

The first chapter, *Sustainable Water Management in Buildings: Innovative Strategies for an Ecological Future, The Case of Blida City*, paints a clear picture of how water stress is already impacting life in Algerian cities like Blida. In response, the chapter offers a series of practical, building-level actions, like fixing leaky taps and installing low-flow systems—that can make a real difference. It also pushes for broader strategies such as green roofs that cool buildings while capturing rainwater. The message is that effective change starts at the ground level, but it only lasts if everyone involved, from homeowners to developers to policy-makers, is fully on board.

In the second chapter, *Integrating Indigenous Knowledge for Sustainable Water Management in Buildings: Towards Liveable Communities*, we step back and take a long view of water wisdom. Here, ancient systems like qanats, carefully engineered underground canals, remind us of how communities have worked with water in sustainable ways for centuries. The chapter calls for more than just nostalgia. It argues for real collaboration between traditional and modern knowledge systems, showing that when the two work together, the results can be both practical and profound. Blending local insight with today's technology could help shape urban water systems that are both sustainable and socially rooted.

The third chapter, *Proposing Water Sensitive Urban Design Solutions in Residential Areas: Case Study of Dar Misr, Helwan, Egypt*, narrows in on a single community and asks: what if we made water a design priority? Using the Dar Misr compound as a testing ground, the chapter explores how Water Sensitive Urban Design (WSUD) can turn streets, courtyards, and rooftops into parts of the water cycle. With smart planting, permeable surfaces, and community-centered planning, this neighborhood model shows how cities might meet the challenge of water scarcity head-on, without sacrificing livability.

Taken together, these chapters remind us that smart water planning is never just about infrastructure. It is about stewardship, care, and creativity. In places where every drop counts, sustainable water design becomes not just good practice, it becomes an ethic, a responsibility, and a shared vision for how cities can thrive even under pressure.

2.5 Urban Greening in Cities

This section brings together two chapters that approach green infrastructure as more than just an environmental strategy, it's a lifeline for cities dealing with rising temperatures, shrinking green spaces, and growing populations. Whether through innovative agriculture or rooftop planting, both chapters present ways to reconnect nature with the urban landscape, particularly in regions where green space is scarce.

The first chapter, *Green Space Expansion Using Haloculture Technology to Combat Climate Change and Global Warming in the MENA Region*, introduces haloculture as a response to some of the region's toughest environmental challenges. In places where soil is too salty and fresh water is hard to come by, haloculture uses salt-tolerant plants and sustainable practices to create new green zones. The authors go beyond the technical explanation and show how education and public engagement can help spread these ideas. The approach is rooted in both science and community, offering a real-world path toward greener, more resilient cities.

In the second chapter, *The Impact of Green Roofs on Buildings' Thermal Performance in the Hot-Arid Climate*, the spotlight shifts to the rooftops of one of the world's most crowded cities, Cairo. Here, green roofs offer a much-needed breath of fresh air. They lower building temperatures, reduce energy use, and turn empty roofs into useful, even beautiful, spaces. But despite these clear benefits, the chapter explains why adoption has been slow. Limited awareness, high costs, and a lack of official guidelines make implementation difficult.

What ties both chapters together is the shared understanding that bringing green space back into cities is not just a trend, it's a necessary part of creating places where people can live better. In the context of arid and semi-arid cities, greening is not about adding luxury, it's about restoring balance. These chapters show how practical solutions, adapted to local realities, can help build healthier, more livable urban futures.

2.6 Energy Transition Toward Carbon Neutral and Green and Smart Cities

This section centers on how renewable energy can be more effectively integrated into Egypt's cities, particularly through community-led initiatives. More than half of the global energy resources are from non-renewable sources, causing climate change [10]. As urban populations continue to grow and the pressure on conventional energy systems intensifies, finding practical ways to transition toward cleaner alternatives becomes increasingly urgent.

The chapter, Application of Renewable Energy in the Built Environment: Towards Sustainable Development through Community Centers, offers a grounded example through the case study of the Oasis Community Center in Giza. The project showcases how environmentally responsible design can go beyond concept into action. From installing solar panels to using kinetic energy tiles and implementing water-saving technologies, the center blends sustainable infrastructure with community engagement. It also serves as a hub for raising environmental awareness among residents.

Despite its successes, the project still faces challenges. Water shortages, limited funding, and a lingering reliance on traditional energy sources reveal the broader issues facing sustainable development in Egypt. These difficulties point to a larger need for more robust infrastructure, policy reform, and financial investment.

The chapter also ties local efforts to global objectives by connecting its findings to the United Nations Sustainable Development Goals. It reinforces how clean energy, resource-conscious planning, and climate action are not only global aims but also community priorities. Local projects like the Oasis Center prove that when people are involved, the transition becomes not just a technical shift, but a social movement.

In the end, this chapter highlights a simple but important idea: renewable energy adoption isn't just about installing the right equipment. It's about creating environments where people are informed, empowered, and supported in making sustainable choices. With the right conditions, small community initiatives can inspire national change, offering a more resilient and inclusive path forward.

This final section takes a closer look at how digital technologies are beginning to reshape the construction industry, especially in regions facing intense environmental pressures. With growing interest in sustainability and smart design, tools like blockchain and cyber-physical systems (CPSs) are emerging as key players in the shift toward cleaner, more efficient, and transparent building practices.

In the chapter, *Role of Blockchain and Cyber-physical Systems (CPSs) in Execution of Sustainable Smart Construction*, the focus is on how these technologies can make a real difference in both the planning and execution of construction projects. Blockchain offers a way to record transactions and data in a secure, transparent, and tamper-proof manner, an important asset in an industry where collaboration and trust are often challenged by complexity. Meanwhile, CPSs, which connect physical systems with smart computing and sensors, help track and improve processes in real time, boosting efficiency and safety.

Together, these technologies provide an opportunity to rethink how we build, enabling more sustainable practices not just during construction but throughout a building's life cycle. Smart contracts and blockchain-integrated building models allow for smoother coordination and clearer communication among teams, improving how resources are managed and how quickly decisions can be made.

Yet, despite their promise, the road to full integration is far from smooth. Many construction firms still face challenges like system compatibility, limited digital infrastructure, and the need for consistent industry standards. What's clear from the chapter is that embracing these tools requires more than just technical upgrades, it requires awareness, training, and a commitment to long-term change.

Ultimately, this chapter highlights how innovation in construction isn't just about using advanced tools, it's about reimagining how the built environment can respond to today's challenges. In arid and semi-arid cities where sustainability isn't optional but essential, digital solutions like blockchain and CPSs offer a practical path forward, one where transparency, resilience, and collaboration form the foundation of future-ready cities.

3 Conclusions

The chapters presented in this book have offered valuable insights and practical conclusions to advance sustainable environmental architecture in arid and semi-arid cities. Through the lens of the book's core themes, each chapter contributes to a deeper understanding of how materials, water, energy, and urban greening can be thoughtfully integrated to address the pressing challenges posed by climate change, rapid urbanization, and environmental resource constraints. This section highlights the essential findings drawn from the chapters, reflecting on innovative approaches and best practices that are crucial for promoting resilience, sustainability, and livability in the unique and demanding climates of arid and semi-arid cities.

- Rapid urbanization, population growth, and climate change are intensifying environmental challenges in arid and semi-arid cities, placing increasing stress on urban water, energy, and green infrastructure systems while amplifying social and economic vulnerabilities.
- Artificial Intelligence has proven its potential as a valuable tool for promoting urban sustainability, supporting smarter planning, efficient resource management, and climate risk prediction. However, its integration still faces limitations related to high energy demand, data availability, ethical concerns, and gaps between policy and implementation.
- Resilience in architecture is not merely a technical or aesthetic concern but a reflection of human instinct to adapt and survive. This concept, embedded in indigenous practices and daily survival tactics, continues to inspire modern approaches for addressing social, environmental, and climatic stresses through context-aware design.
- Social-ecological frameworks are essential for sustainable development in the Arab region, where integrating environmental conservation with community well-being can provide practical, locally appropriate solutions to complex environmental problems. Empowering communities through awareness and policy participation is central to this process.
- Urban agriculture practices offer tangible resilience benefits by improving food security, enhancing local biodiversity, cooling urban environments, and fostering social connections. Regular monitoring, adaptation, and inclusive community engagement remain key for their long-term viability in arid regions.
- Utilizing locally sourced and sustainable building materials such as adobe, basalt, and limestone significantly reduces the carbon footprint, improves thermal performance, and enhances the durability of structures in harsh climates. Passive design elements like courtyards and wind catchers further boost energy efficiency.
- While the initial investment for sustainable materials is often higher, long-term financial benefits, including reduced operational costs, lower energy bills, and improved indoor air quality, outweigh these upfront expenses, making them a smart economic choice for arid and semi-arid regions.

The promotion of eco-friendly finishing materials in construction can meaningfully improve indoor environmental quality while reducing waste. However, public awareness, professional education, and policy incentives are necessary to overcome misconceptions around costs and availability.

Algeria's water scarcity and rising urban demands in cities like Blida call for immediate adoption of water conservation measures at the building scale, such as water-saving fixtures, leak detection, and rooftop greenery, supported by integrated urban water management strategies.

Indigenous water management techniques like qanats offer valuable, low-energy, and culturally grounded solutions for modern water scarcity challenges. When combined with contemporary technology, these methods can contribute to sustainable, resilient water systems in both rural and urban settings.

Water Sensitive Urban Design (WSUD) techniques such as permeable pavements, green roofs, and constructed wetlands, are highly effective in reducing urban heat, improving stormwater management, and mitigating water scarcity in rapidly expanding cities like Cairo.

Haloculture technology presents a practical and innovative solution for expanding green spaces in saline-affected, water-scarce environments across the MENA region, improving ecological health and social resilience while supporting food production and carbon sequestration.

Green roofs deliver multiple urban benefits in hot, dry climates, including lowering building temperatures, reducing energy demand, improving air quality, and creating restorative public spaces. However, challenges related to water supply, high costs, policy gaps, and public awareness remain significant barriers.

Integrating renewable energy systems into the built environment, particularly in community-focused spaces, enhances energy security, supports climate goals, and provides valuable public education opportunities around sustainability. Community centers like Oasis in Giza offer promising models.

Community-led renewable energy projects demonstrate that combining clean technologies with environmental education can strengthen environmental awareness, public engagement, and neighborhood-level resilience, while offering replicable models for broader urban areas.

Emerging technologies such as blockchain and cyber-physical systems (CPSs) present promising opportunities for improving transparency, operational efficiency, and data security in sustainable construction, supporting the transition to greener, smarter cities.

The use of smart contracts and blockchain-integrated Building Information Modeling (BIM) can improve coordination and decision-making in construction projects, reducing delays, disputes, and resource waste in arid, resource-constrained regions.

The transition toward digitally enabled, sustainable construction faces significant obstacles—including limited infrastructure, inconsistent standards, and lack of stakeholder awareness—highlighting the need for targeted education, policy reform, and institutional support.

Embedding climate-responsive design, indigenous knowledge, and sustainability concepts into architectural education and training programs is vital for preparing future practitioners to design responsibly in climate-vulnerable, resource-scarce contexts.

Overall, achieving sustainable, resilient, and livable cities in arid and semi-arid regions requires a balanced, inclusive, and context-sensitive approach that combines modern technologies, traditional wisdom, practical policy frameworks, and strong public engagement.

4 Recommendations

Build collaborative networks uniting AI developers, architects, planners, policy-makers, researchers, and community stakeholders to guide sustainable urban and climate-resilient initiatives in challenging climates.

Prioritize training for city officials, architects, and sustainability practitioners on AI's environmental, ethical, and social implications to support balanced and inclusive urban development.

Revive indigenous knowledge, biomimicry, and vernacular architecture to shape adaptive, resource-efficient solutions responsive to local environments and cultures.

Facilitate community-driven planning through participatory mapping, ethnographic studies, and grassroots awareness campaigns that ensure projects genuinely reflect local needs.

Embed social justice into design by addressing displacement, environmental inequality, and accessibility, ensuring resilience plans serve marginalized and vulnerable communities.

Integrate sustainability, climate adaptation, and social impact modules into architecture and planning curricula to cultivate future leaders of resilient, sustainable design.

Strengthen partnerships between NGOs, humanitarian agencies, universities, and the building industry to expand architecture's role in urban resilience and post-crisis reconstruction.

Mandate the use of locally sourced, low-carbon, bio-based, and recycled building materials through updated regulations and incentive schemes to advance the circular economy.

Update building codes for arid regions to enforce passive cooling strategies, water conservation systems, reflective materials, and energy-saving designs suited to harsh climates.

Introduce financial incentives, green bonds, and tax relief for developers adopting sustainable construction practices, renewable energy, and urban greening initiatives.

Localize and adapt green certification systems (e.g., LEED, BREEAM) to prioritize water efficiency, thermal comfort, and urban heat island mitigation in desert cities.
Expand urban green infrastructure through green roofs, vertical farming, permeable pavements, and urban agriculture projects to enhance biodiversity and public health.
Promote haloculture, saline agriculture, and salt-tolerant plantings to rehabilitate degraded soils, improve food security, and green arid urban spaces.
Integrate traditional water conservation systems like qanats with smart, AI-supported water management and regulatory incentives to maximize resource efficiency.
Support transdisciplinary collaborations uniting urban designers, environmental scientists, policymakers, and communities to co-create context-specific sustainability strategies.
Launch capacity-building initiatives to train local farmers, builders, and municipal staff in sustainable water management, renewable energy systems, and eco-friendly construction.
Encourage the adoption of greywater recycling, smart irrigation, and stormwater reuse in residential, commercial, and public infrastructure projects.
Implement large-scale public education campaigns highlighting the economic, environmental, and health benefits of sustainable materials, green infrastructure, and renewable energy.
Invest in decentralized water treatment plants, smart monitoring systems, and energy storage infrastructure to meet the demands of growing urban populations.
Adopt hybrid renewable energy solutions, combining solar, kinetic, and storage systems in public buildings, community hubs, and housing projects.
Develop innovative financing mechanisms such as revolving loan funds and micro-grants to support green construction and infrastructure upgrades in vulnerable areas.
Establish blockchain governance frameworks and standardized protocols for integrating CPS, IoT, and smart contract technologies in sustainable construction.
Implement strict data security and digital privacy protocols to safeguard infrastructure, construction data, and smart city services in increasingly digitized environments.
Create continuous monitoring, evaluation, and documentation frameworks for urban agriculture, renewable energy, haloculture, and WSUD initiatives, including field-based research.
Formalize municipal policies rewarding rooftop greening, halophyte planting, sustainable material use, and low-impact construction to embed sustainability into city planning frameworks.

Acknowledgements The editors want to thank all contributors to this book. Without their efforts, patience and several revisions until we receive the final revised version, it would not be possible to convert the dream into reality.

References

1. Goh KC, Kurniawan TA, Zainordin N, Arifah IDC, Abas MA, Masrom MAN, Mohamed S, Omar R, Khoo SL, Gui HC, Toh TC, Onn CW (2024) Expediting decarbonization in energy, waste, and water sector through digitalization in sustainable smart cities (SSC): case studies in Malaysia and China based on Industry 5.0 paradigm. Sustain Cities Soc 117:105969
2. Wang D, Peng D, Huang D (2025) Application and prospects of large AI models in virtual power plants. Electric Power Syst Res 241:111403. https://doi.org/10.1016/j.epsr.2024.111403
3. Zahran O, Micklewright J (2024) How the past meets the future: greening strategies with pocket parks in historic Cairo
4. ESCWA (2024) Arab Sustainable Development Report 2024. United Nations
5. Zuma M, Arthur G, Coopoosamy R, Naidoo K (2023) Incorporating cropping systems with eco-friendly strategies and solutions to mitigate the effects of climate change on crop production. J Agric Food Res 14:100722. https://doi.org/10.1016/j.jafr.2023.100722
6. Bidashimwa D, Hoke T, Huynh TB, Narkpitaks N, Priyonugroho K, Ha TT, Burns A, Weissman A (2023) Plastic pollution: how can the global health community fight the growing problem? BMJ Glob Health 8(Suppl 3):e012140. https://doi.org/10.1136/bmjgh-2023-012140
7. Hasan MR, Khan MA, Wuest T (2024) Towards Industry 5.0: A systematic literature review on sustainable and green composite materials supply chains. J Clean Prod 421:129001. https://doi.org/10.1016/j.jclepro.2024.129001
8. ElZein Z (2024) Sustainable built environments and the water, energy, and food nexus: toward an interdisciplinary approach to urban resilience. Chin J Urban Environ Stud 12(1):2450004-1–2450004-22
9. ElZein Z, Ahmed Milad N, Sherif Mohamed A, Kotb NM, Abdo N, Negm AM (2024a) Role of water, energy and food nexus in sustainable development of communities: case studies in the MENA region. In: Negm AM, ElZein Z (eds) Integration of core sustainable development goals in rural areas, earth and environmental sciences library. Springer, Cham, pp 15–66. https://doi.org/10.1007/978-3-031-60149-1_2
10. ElZein Z, Milad NA, Mohamed AS, Mahmoud N, Abdo N, Abdel-Ghafar HM (2024) Sustainable development on the basis of the WEF nexus in arid coastal areas for climate change mitigation: Case study of Rabia community in Matrouh, Egypt. Energy Nexus 14:100299. https://doi.org/10.1016/j.nexus.2024.100299

MIX
Papier aus verantwortungsvollen Quellen
Paper from responsible sources
FSC® C105338

If you have any concerns about our products,
you can contact us on
ProductSafety@springernature.com

In case Publisher is established outside the EU,
the EU authorized representative is:
Springer Nature Customer Service Center GmbH
Europaplatz 3, 69115 Heidelberg, Germany

Printed by Libri Plureos GmbH
in Hamburg, Germany